VOLUME ONE HUNDRED AND FIFTY FIVE

ADVANCES IN CANCER RESEARCH

Strategies to Mitigate the Toxicity of Cancer Therapeutics

VOLUME ONE HUNDRED AND FIFTY FIVE

ADVANCES IN CANCER RESEARCH

Strategies to Mitigate the Toxicity of Cancer Therapeutics

Edited by

DAVID A. GEWIRTZ

Department of Pharmacology and Toxicology,
Virginia Commonwealth University School of Medicine,
Richmond, VA, United States

PAUL B. FISHER

Department of Human and Molecular Genetics;
VCU Institute of Molecular Medicine;
VCU Massey Cancer Center,
Virginia Commonwealth University, School of Medicine,
Richmond, VA, United States

ELSEVIER

ACADEMIC PRESS
An imprint of Elsevier

Academic Press is an imprint of Elsevier
50 Hampshire Street, 5th Floor, Cambridge, MA 02139, United States
525 B Street, Suite 1650, San Diego, CA 92101, United States
The Boulevard, Langford Lane, Kidlington, Oxford OX5 1GB, United Kingdom
125 London Wall, London, EC2Y 5AS, United Kingdom

First edition 2022

ISBN: 978-0-323-90087-4
ISSN: 0065-230X

For information on all Academic Press publications
visit our website at https://www.elsevier.com/books-and-journals

Publisher: Zoe Kruze
Developmental Editor: Naiza Ermin Mendoza
Production Project Manager: James Selvam
Cover Designer: Miles Hitchen

Typeset by STRAIVE, India

Contents

Contributors

Hamid I. Akbarali
Department of Pharmacology and Toxicology, School of Medicine, Virginia Commonwealth University, Richmond, VA, United States

Erin Alesi
Department of Internal Medicine, Virginia Commonwealth University, Massey Cancer Center, Richmond, VA, United States

Rashmi Bhushan
Department of Pharmacology & Toxicology, National Institute of Pharmaceutical Education and Research, Hajipur, Bihar, India

Kim R.M. Blenman
Yale Cancer Center; Department of Internal Medicine, Section of Medical Oncology; Department of Computer Science, Yale University, New Haven, CT, United States

Stanley Cheatham
Department of Pharmacology and Toxicology, School of Medicine, Virginia Commonwealth University, Richmond, VA, United States

Chaoling Chen
Department of Pharmacology and Toxicology, Virginia Commonwealth University School of Medicine, Richmond, VA, United States

Sri Pragnya Cheruku
Department of Pharmacology, Manipal College of Pharmaceutical Sciences, Manipal Academy of Higher Education, Manipal, Karnataka, India

Laurie Cuttino
Sarah Cannon Cancer Institute at Henrico Doctor's Hospital, Richmond, VA, United States

David A. Gewirtz
Department of Pharmacology and Toxicology, Virginia Commonwealth University School of Medicine, Richmond, VA, United States

Mary Helen Hackney
Department of Internal Medicine, Virginia Commonwealth University, Massey Cancer Center, Richmond, VA, United States

Katherine Hunter
Division of Cardiology, Department of Internal Medicine, Virginia Commonwealth University, Pauley Heart Center, Richmond, VA, United States

Donald K. Jessup
Department of Pharmacology and Toxicology, School of Medicine, Virginia Commonwealth University, Richmond, VA, United States

Adriana M. Kahn
Yale Cancer Center; Department of Internal Medicine, Section of Medical Oncology, Yale University, New Haven, CT, United States

Nitesh Kumar
Department of Pharmacology & Toxicology, National Institute of Pharmaceutical Education and Research, Hajipur, Bihar, India

Priyanka Kumari
Department of Pharmacology & Toxicology, National Institute of Pharmaceutical Education and Research, Hajipur, Bihar, India

Ningjun Li
Department of Pharmacology and Toxicology, Virginia Commonwealth University School of Medicine, Richmond, VA, United States

Maryam B. Lustberg
Yale Cancer Center; Department of Internal Medicine, Section of Medical Oncology, Yale University, New Haven, CT, United States

Adolfo G. Mauro
Division of Cardiology, Department of Internal Medicine, Virginia Commonwealth University, Pauley Heart Center, Richmond, VA, United States

Karan H. Muchhala
Department of Pharmacology and Toxicology, School of Medicine, Virginia Commonwealth University, Richmond, VA, United States

Jennifer Myers
Department of Internal Medicine, Virginia Commonwealth University, Massey Cancer Center, Richmond, VA, United States

Vanishree Rao
Department of Pharmacology, Manipal College of Pharmaceutical Sciences, Manipal Academy of Higher Education, Manipal, Karnataka, India

V. Ravichandiran
Department of Pharmacology & Toxicology, National Institute of Pharmaceutical Education and Research, Hajipur, Bihar, India

Fadi N. Salloum
Division of Cardiology, Department of Internal Medicine, Virginia Commonwealth University, Pauley Heart Center, Richmond, VA, United States

Steve T. Sonis
Department of Surgery, Divisions of Oral Medicine and Dentistry, Brigham and Women's Hospital and the Dana-Farber Cancer Institute; Department of Oral Medicine, Infection and Immunity, Harvard School of Dental Medicine, Boston; Biomodels LLC., Waltham, MA, United States

Dengpiao Xie
Department of Pharmacology and Toxicology, Virginia Commonwealth University School of Medicine, Richmond, VA, United States

Victor Yazbeck
Department of Internal Medicine, Virginia Commonwealth University, Massey Cancer Center, Richmond, VA, United States

Preface

The concept that cancer might actually be a curable affliction is largely based on the original trials by Sidney Farber in the 1940s, when the first antitumor agent—aminopterin (or methotrexate)—was found to have activity in leukemia. Over the course of the succeeding 80 years, an expansive list of antitumor drugs have been developed (many if not most during the heydays of drug development through the 1970s) with the more recent advent of targeted therapies, antibody therapy, hormone antagonists, and immunotherapy. Radiation therapy has also long been a mainstay of cancer therapy and a highly reliable component of both adjuvant and neoadjuvant strategies.

The primary objective of the oncologist, of course, is to suppress disease progression and, ideally, to cure the disease. However, an additional challenge, which has often received inadequate consideration, is to balance the impact of therapy on the tumor with the not inconsequential toxicities resulting from both acute and chronic treatments. These toxicities disrupt the normal life of patients and cause frequent and severe reductions in their quality of life. These toxicities can and do lead to premature termination of treatment when the therapy is not tolerable, thereby resulting in a treatment that may prove inadequate.

A primary reason why there has not been greater success in overcoming therapy-associated toxicities to normal tissues is the lack of a basic understanding of the mechanistic foundation of the observed undesirable outcomes.

The chapters in this book first present an overview of the toxicities of chemotherapy and radiation (Yazbeck et al.), address common and frequent toxicities such as chemobrain resulting from a host of agents (Rao et al.), nephrotoxicity largely as a consequence of platinum based drugs (Chen et al.), GI toxicity from agents such as irinotecan in colorectal cancer (Akbarali et al.), and cardiac complications resulting from doxorubicin, as well as other agents (Mauro et al.), and finally an overview of strategies that have varied effectiveness in mitigating these toxicities (Kahn et al.). This by no means covers all of the different deleterious side effects of cancer therapeutics that patients may suffer such as chemotherapy-induced peripheral neuropathy. However, we trust that the areas that we have addressed will provide an overview of the challenges that still lie ahead in our struggles to defeat this disease while preserving patient safety.

David A. Gewirtz, PhD
Professor
Department of Pharmacology & Toxicology
Virginia Commonwealth University
School of Medicine
Richmond, VA 23298, USA

Paul B. Fisher, MPh, PhD, FNAI
Professor and Chair
Department of Human and Molecular Genetics
Director
VCU Institute of Molecular Medicine
Virginia Commonwealth University
School of Medicine
Richmond, VA 23298, USA

CHAPTER ONE

An overview of chemotoxicity and radiation toxicity in cancer therapy

Victor Yazbeck[a], Erin Alesi[a], Jennifer Myers[a], Mary Helen Hackney[a], Laurie Cuttino[c], and David A. Gewirtz[b,*]

[a]Department of Internal Medicine, Virginia Commonwealth University, Massey Cancer Center, Richmond, VA, United States
[b]Department of Pharmacology and Toxicology, Virginia Commonwealth University School of Medicine, Richmond, VA, United States
[c]Sarah Cannon Cancer Institute at Henrico Doctor's Hospital, Richmond, VA, United States
*Corresponding author: e-mail address: david.gewirtz@vcuhealth.org

Contents

Advances in Cancer Research, Volume 155
ISSN 0065-230X
https://doi.org/10.1016/bs.acr.2022.03.007

Abstract

While immunotherapy and targeted therapies represent major advances against different types of malignancies, the mainstay of cancer therapy continues to be radiation and surgery for localized disease, and chemotherapy for systemic disease, with the preponderance of chemotherapeutic agents (such as anthracyclines, alkylating agents, and antimetabolites) having been developed decades ago. Combination chemotherapy regimens have changed the natural history of once deadly diseases such as breast and prostate cancer and led to curative regimens in advanced hematological malignancies and testicular cancer. However, while oncologists maintain their focus on disease suppression, and where feasible, disease eradication, obstacles to achieving cure remain, such as tumor dormancy and ultimately disease recurrence, as well as both intrinsic and acquired resistance. In this review, complications of current cancer therapies toward major organs (heart, lung, kidney, gastro-intestinal, neuromuscular, brain, and skin) are emphasized, and efforts to mitigate these complications are described. This is particularly relevant for patients treated with curative intent, where adherence to treatment plan, and avoidance of interruptions in treatment schedule are essential for optimal outcome. Consequently, these patients are treated with an "aggressive" approach, with high tolerance for side effects. However, a deeper understanding of normal tissue toxicity resulting from the different cancer therapies remains an area of unmet medical need that will ultimately lead to improved therapeutic index for current and future therapies, planning for treatment adverse effects, and ultimately improvement in patient satisfaction, compliance and outcome.

1. Introduction

Although the incidence of cancer may not have declined significantly over the course of the last 50 years, and while the incidence for some forms of gastro-intestinal malignancies such as colorectal cancer in younger patients, as well as esophageal and pancreatic adenocarcinoma, has increased for reasons that we do not yet fully understand, the effectiveness of cancer therapeutics has had a profound impact on patient survival. For instance, in the case of testicular cancer, the advent of cisplatin has all but cured the disease in affected patients. The development of novel therapeutic strategies such as targeted agents (anti anaplastic lymphoma kinase, epidermal growth factor inhibitors), hormone therapy (anti estrogens and aromatase inhibitors in breast cancer, androgen deprivation therapy in prostate cancer), immune checkpoint inhibitors (anti program cell death protein 1 and cytotoxic T-lymphocyte-associated protein 4), and cell cycle checkpoint inhibitors (cyclin-dependent kinase 4/6 inhibitors) have clearly prolonged progression free survival and/or overall survival in diseases such as lung cancer, melanoma and breast cancer. Furthermore, progress in supportive care (such as new antiemetics and anti-infectives) have contributed significantly to improving patient outcome. However, it must be recognized that even current antineoplastic agents are not necessarily curative. Furthermore, the mainstays of cancer therapy (in addition to surgery, where feasible) continue to be radiation and chemotherapy (with the more recent contribution of immunotherapy), with the preponderance of chemotherapy drugs being those that were developed decades ago, including methotrexate, doxorubicin, paclitaxel, cyclophosphamide and 5-fluorouracil.

This is not to imply that the continuation of historically effective therapies is unproductive or undesirable; in fact, combination chemotherapy has led to curative regimens in some advanced hematological malignancies and testicular cancer. However, while oncologists maintain their focus on disease suppression, and where feasible, disease eradication, tumor dormancy, disease recurrence and both intrinsic and acquired resistance remain obstacles to cure.

In addition to these fundamental limitations to effective disease treatment, an often overlooked and perhaps underserved problem is that of patient toxicity from these therapies and quality of life issues over the extended time period during which the disease may be held in check. That is, patients often have to endure long-lasting and poorly tolerated "side effects" not only for the conventional historical drugs (and radiation), but for many of the newer treatments. Furthermore, for a significant number of patients, therapy will have to be terminated prematurely due to the severity/intensity of drug-related toxicities.

The fundamental limitation to the utility of anticancer drugs is their low selectivity (particularly in comparison to antibiotics) due largely to the unavoidable fact that tumor cells are derived from normal tissue. The relative lack of selectivity applies to those situations where toxicity to the tumor cells and normal tissue can be ascribed to similar mechanisms. However, what will become clear in subsequent sections of this article is that, despite many years of research, the basis for normal tissue toxicity (i.e., the low therapeutic index when considering the whole pantheon of toxicities of a particular agent) is often inadequately defined, and in some cases, entirely unknown. As a consequence, it becomes much more challenging to develop viable approaches for mitigation of these toxicities, often resulting in a "hit or miss" approach that has failed to provide tenable solutions.

Despite the advent of new and promising therapeutics in cancer, drugs that were discovered and developed in the last century (i.e., more than 20 years ago and, in some cases, 30, 40 or 50 years ago) continue to be central components of the anticancer armamentarium. Methotrexate, essentially the first cancer therapeutic, remains a mainstay for the treatment of acute lymphoblastic leukemia (in children), primary central nervous system lymphoma and osteogenic sarcoma. Cyclophosphamide, an alkylating agent used to treat lymphoma, multiple myeloma, leukemia, ovarian cancer, breast cancer, small cell lung cancer, neuroblastoma, and sarcoma, was approved for medical use in the United States in 1959. Cisplatin was discovered in 1965, and approved in 1978, and has been a godsend for the treatment (and cure) of testicular cancer in up to 90% of patients when used in combination therapy. Cisplatin is also used in the treatment of a wide range of other cancers, including lung, bladder, cervical, head and neck and ovarian cancers. Cisplatin and other similar platinum-based drugs are prescribed for an estimated 10–20% of all cancer patients and is actually one of the few drugs effective in head and neck cancer. Paclitaxel, a microtubule poison used to treat breast cancer, ovarian cancer, and lung cancer was discovered in 1962, antitumor activity confirmed in 1977, and first approved by the FDA in 1992. Doxorubicin and daunorubicin, drugs that act by interfering with the action of the enzyme, topoisomerase II, and are used in the treatment of breast cancer, bladder cancer, lymphoma and leukemia, were discovered in the 1960s.

5-Fluorouracil (5-FU) is an anti-metabolite that was discovered in 1957 by Heidelberger et al. and continues to be the primary chemotherapy backbone for the treatment of colorectal cancers today. It also has had a significant role in the treatment of anal cancers, esophageal adenocarcinoma, gastric cancer, pancreatic cancer, small bowel adenocarcinomas, as well as head and neck and breast cancers. 5-FU's anticancer effects are mediated via inhibition of thymidylate synthase as well as the incorporation of 5-FU metabolites into the RNA and DNA of tumor cells (Longley, Harkin, & Johnston, 2003). Capecitabine, the first oral chemotherapeutic agent, was discovered and approved for use as an oral fluoropyrimidine about 40 years after the discovery of infusional 5-FU. Capecitabine is converted into 5-FU within the liver and solid tumors through a pathway involving three enzymes, carboxylesterase, cytidine deaminase, and thymidine phosphorylase. These enzymes are predominantly located in liver and more highly concentrated in tumor cells as compared to normal cells (Miwa et al., 1998). Thus, the toxicities of capecitabine differ somewhat from infusional 5-FU which does not require any enzymatic breakdown to the active drug. Even capecitabine is now over 20 years old.

We mention these conventional and historical agents to emphasize that despite the remarkable advances in cancer therapeutics, many drugs that are from 30 to 60 years old still comprise the backbone of cancer therapeutics. Of course, the same can be said for many antibiotics. However, the efficacy of the antibiotics is largely not constrained by the severe and often prolonged toxicities associated with antitumor drugs. Here, we will discuss some of the drug toxicities that have challenged the scientific community for decades, and which do not currently have reliable mitigation strategies.

2. Cardiotoxicity

A number of cancer chemotherapeutic agents are associated with cardiotoxicity, which is often dose-limiting. Perhaps most prominent among these is doxorubicin, whose use is constrained by both acute and cumulative cardiotoxicity. Cardiotoxicity can be compounded by radiation exposure and comorbid conditions. Although doxorubicin (and its sister drug, daunorubicin) had been used in the clinic for years, a breakthrough came when the mechanism of action was identified as being the inhibition of topoisomerase II mediated DNA relegation, resulting in the induction of DNA strand breaks with the topoisomerase II enzyme linked to the break sites. However, despite decades of effort, effective strategies for preventing or mitigating this toxicity have not been productive. While the possibility of dissociating the drugs' antitumor effects from its cardiotoxicity via the suppression of free radical generation in the heart appeared to be promising strategy, this has not been a fruitful approach; the failure to achieve the desired outcome suggests that free radicals likely also contribute to antitumor drug action, and that further suppression of that free radical generation is likely to reduce antitumor drug efficacy. Recent work, as described in the chapter by Mendeles, has raised the possibility that senescence might, in part, be responsible for injury to the heart, which might open the possibility of utilizing senolytics as a mitigation strategy.

Dexrazoxane is an EDTA derivative that chelates iron and thus reduces the number of metal ions that are complexed with anthracyclines, thereby decreasing the formation of superoxide radicals, providing for its presumed cardioprotective properties, although precise mechanisms remain unknown. Dexrazoxane is the only agent that has received FDA and EMA approval as a cardioprotective agent against anthracyclines. It can be considered for patients that have already received more than $300\,mg/m^2$ of doxorubicin-based therapy (Hensley et al., 2009). However, caution is required in using dexrazoxane in settings where doxorubicin-based regimens have shown improved survival such as breast cancer and lymphomas.

5-Fluorouracil (5-FU) is the second most common chemotherapeutic agent associated with cardiotoxicity and is the third most common chemotherapy used to treat solid tumor malignancies throughout the world. In a recent review article, the range of clinically significant cardiotoxicity from 5-FU (including infusional 5FU, bolus 5-FU and capecitabine) ranged from 0% to 19.9% in various clinical trials (Sara et al., 2018). The symptoms range from EKG changes, palpitations, arrhythmias to angina, chest pain, and myocardial infarction. Although 5-FU has been a staple of many chemotherapy regimens as well as being used as a radiosensitizer in concurrent chemoradiation regimens, the mechanisms underlying the cardiotoxicity of 5-FU have not been fully elucidated. The primary theory is that 5-FU can induce vasospasm of the coronary arteries, which can cause myocardial injury. However, studies have not been able to consistently show coronary vasospasms in patients with symptoms by coronary angiography. In addition, despite 5-FU being widely distributed systemically, it is very uncommon to see multivessel coronary artery vasospasm. Consequently, it is thought that there must be additional mechanisms leading to cardiotoxicity, such as direct cellular damage, free radicals generated by 5-FU, and cardiotoxicity of a metabolite of 5-FU such as fluoroacetate. Given the difficulties in obtaining cardiac biopsies in patients on active chemotherapy treatment, this is still an area of investigation. There is some data to suggest that patients with a history of coronary artery disease or pre-existing cardiac disease may be at higher risk of cardiotoxicity. It is well established that the risk of cardiotoxicity is increased with infusional 5-FU (such as in FOLFOX regimens) as compared to bolus 5-FU, as 5-FU does have a short half-life of about 15–20 min (Sara et al., 2018).

Both cisplatin and cyclophosphamide therapies have also been associated with cardiotoxicity, although not to the extent that the utility of doxorubicin/daunorubicin have been restricted. It is unclear whether the cardiotoxicity of these agents has a similar or dissimilar foundation from that of doxorubicin since the three agents exhibit uniquely different mechanisms of action. Both cisplatin and cyclophosphamide bind to DNA and promote monofunctional and bifunctional DNA cross linking, with limited evidence for free radical generation.

Targeted agents, such as monoclonal antibodies, are designed to target a specific marker on the cancer cell surface, or intrinsic to the cancer cell, such as small molecule inhibitors, presumably providing for more selectivity and less toxicity than conventional chemotherapy. However, some of these agents have been associated with significant cardiotoxicity. Ibrutinib, the first in class Bruton tyrosine kinase inhibitor (BTKi) initially FDA-approved back in 2013 in mantle cell lymphomas and currently in several B-cell malignancies, is associated with increased risk of cardiovascular adverse events including arrythmias and hypertension (Salem et al., 2019). Next generation BTKis have shown less cardiotoxicity in randomized clinical studies when compared with ibrutinib (Byrd et al., 2021; Hillmen et al., 2020).

Recently approved immunotherapies that activate the patient's own immune system to eliminate cancerous cells have been associated with cardiotoxic side effects: While immune side effects from checkpoint inhibitors are common, cardiotoxic effects are uncommon, but are often serious with a relatively high mortality (Itzhaki Ben Zadok et al., 2019). In Chimeric antigen receptor T cell therapies, currently approved across several B cell malignancies, a recent report showed an association with increased risk of tachyarrhythmias, cardiomyopathy, pericardial and pleural disorders (Goldman et al., 2021). In order to better understand and mitigate the cardiotoxic side effects of different cancer therapies, the new discipline of cardio-oncology has recently been established. The chapter by Mauro et al. develops the theme of chemotherapy associated cardiotoxicity in greater detail.

3. Pulmonary toxicity

Pulmonary toxicity has been classically associated with bleomycin; however, the advent of tyrosine kinase inhibitors (TKIs) and immune checkpoint inhibitors (ICIs) has brought pulmonary toxicity into the modern era of antineoplastic therapy. Bleomycin, an antitumor antibiotic that inhibits DNA synthesis, was isolated from *Streptomyces verticillus* in 1966 and has been a mainstay of curative chemotherapeutic regimens in germ cell tumors and lymphoma. Its use has been greatly limited by the characteristic side effect of interstitial pneumonitis, with incidence rates ranging from 10% to 46% and mortality rates up to 24%. This pneumonitis typically evolves into pulmonary fibrosis, which carries with it significant long-term morbidity. The mechanisms of bleomycin-induced pulmonary fibrosis are not well-understood but likely involve processes beyond oxidative stress and inflammation, given the lack of benefit with antioxidant (*N*-acetylcysteine) and anti-inflammatory agents (Wang et al., 2018; Watson et al., 2018).

TKIs are newer anticancer agents associated with interstitial pneumonitis. Epidermal growth factor receptor (EGFR) TKIs (erlotinib, gefitinib, afatinib, and osimertinib) are one fairly common example. Incidence rates in this class of drugs is lower compared to bleomycin, at approximately 1–4%, but with high mortality rates up to approximately 30% (Suh et al., 2018). Again, the mechanism of lung injury is not well understood, but in the case of EGFR TKIs, the mechanism likely includes inhibition of EGFR on type II pneumocytes which leads to impaired alveolar repair mechanisms (Jia et al., 2020; Miettinen et al., 1997). Concomitant inflammatory and oxidative stressors, such as radiation, smoking, and underlying lung disease can potentiate lung injury in the setting of defective alveolar repair (Jia et al., 2020; Kudoh et al., 2008). Rates of Grade 2 or higher pneumonitis range from approximately 37–63% when EGFR TKIs are combined with thoracic radiotherapy (Jia et al., 2020). As with bleomycin, management of TKI-associated pulmonary toxicity is largely supportive with cessation of chemotherapy and respiratory support. High dose corticosteroid therapy is used in more severe cases, but support for its use is primarily observational (Camus et al., 2004).

Immune checkpoint inhibitors (ICIs) are one of the newest classes of anticancer therapy associated with pulmonary toxicity. These monoclonal antibodies target regulatory proteins on T cells, including pathways associated with cytotoxic T-lymphocyte associated protein-4 (CTLA-4; e.g., ipilimumab) and programmed death-1 (PD-1; e.g., nivolumab/pembrolizumab) or programmed death-ligand 1 (PD-L1; e.g., durvalumab). These interactions result in immune system stimulation against tumor cells; however, immune-related adverse events (irAE) can occur as a result of excessive stimulation of the immune response. ICI pneumonitis is the leading cause of fatal irAEs in patients receiving PD-1/PD-L1 therapy and occurs in approximately 2.6% of patients according to meta-analyses, but the true incidence may be as high as 13–19% in real-world settings. The incidence is highest when CTLA-4 and PD-1/PD-L1 inhibitors are combined. Radiation therapy has not been proven to increase the risk of ICI pneumonitis. Management consists of discontinuation of therapy and administration of immunosuppressive therapy, which is discussed in more detail in a subsequent section (Long & Suresh, 2020).

4. Renal toxicity

In addition to chemotherapy-induced peripheral neuropathy (Section 6), cardiotoxicity and ototoxicity (which will not be discussed here), Cisplatin is frequently associated with renal toxicity that can also be dose limiting. Acute kidney injury (AKI) and electrolyte wasting result from tubular damage, interstitial inflammation and vascular injury invoked by cisplatin. This effect is ostensibly mediated by toxic metabolites produced primarily from GGT-dependent metabolism within the tubulo-interstitial compartment of the kidney that results in DNA damage, cytoplasmic organelle dysfunction, apoptosis, oxidative stress, and inflammation. While AKI and hypomagnesemia are the most common manifestations of cisplatin nephrotoxicity, Fanconi-like syndrome, distal renal tubular acidosis, renal concentrating effect, and thrombotic microangiopathy are also potential manifestations. Cisplatin nephrotoxicity is dose-duration-frequency dependent. Other risk factors for cisplatin nephrotoxicity include older age, lower functional status, malnutrition, dehydration, medical comorbidities, pre-existing kidney disease, and concurrent use of nephrotoxic medications. Carboplatin, a newer drug in the platinum class, has been found to be less nephrotoxic, but may not be as clinically effective in certain settings. Mitigation efforts for cisplatin nephrotoxicity have focused on the varying nephrotoxic pathways mentioned above; however, very few have proven to provide renal protection from cisplatin without reducing its anti-tumor effects. The proven strategies are hydration, forced diuresis with mannitol or diuretics, and avoidance of nephrotoxic medications (Manohar & Leung, 2018), Table 1.

Table 1 Examples of supportive care medications commonly used in standard practice.

Gastro-intestinal toxicity	Diarrhea: *Treatment:* – Loperamide 4mg PO ×1, then 2mg PO after each loose stool (max dose: 16mg/day). Commonly used with fluorouracil, irinotecan (late onset), PI3Ki (early onset), ICI – Atropine/diphenoxylate 5mg (atropine) every 6h (max dose: 20mg/day): Early onset irinotecan-induced diarrhea – Steroids with budesonide 9mg QD (mild), or prednisone/methylprednisolone (1–2mg/kg/day) for more intense immune-mediated colitis, e.g., ICI and PI3Ki – BRAT diet: Banana, Rice, Apple, Toast

Continued

Table 1 Examples of supportive care medications commonly used in standard practice.—cont'd

	Mucositis: – Palifermin: 60 μg/kg daily for 3 days preceding the start of the conditioning regimen and 60 μg/kg daily for 3 days beginning on the day of stem-cell infusion[a] – Ice chips[b] Nausea and Vomiting: – Serotonin (5-HT3) antagonists: first generation (ondansetron 8 mg PO Q4-6H, granisetron (patch) 3.1 mg ×1)[c], second generation (palonosetron 0.25 mg IV ×1) – Corticosteroids: dexamethasone 2–4 mg daily – Neurokinin 1 (NK1) receptor antagonists: aprepitant 125 mg PO ×1 then 80 mg PO daily ×2 and fosaprepitant 150 mg IV ×1) – Antipsychotic: olanzapine 5–10 mg PO QD – Benzodiazepines: Lorazepam 0.5–2 mg PO QD – Dopamine antagonists: metoclopramide (10 mg PO Q6H), prochlorperazine (5–10 mg PO Q6H PRN), haloperidol (0.5–1 mg Q6H PRN) – Antihistamine: meclizine 12.5–25 mg Q6-12H for delayed N/V or vestibular component to patient's N/V – Relaxation techniques – Diet Xerostomia: *Prevention:* – Amifostine 200 mg/m^2/day given as slow IV push over 3 min, 15–30 min before each fraction of radiation therapy for prevention of acute and late xerostomia in patients undergoing fractionated radiation therapy alone for head and neck cancer

Table 1 Examples of supportive care medications commonly used in standard practice.—cont'd

Infections	*Prevention:* – High risk for febrile neutropenia: fluconazole 400 mg QD, posaconazole 300 mg QD, levofloxacin 500 mg QD – High risk for HSV/VZV reactivation: acyclovir 400 mg BID or valacyclovir 500 mg QD – High risk for PJP: sulfamethoxazole/trimethoprim DS 160/800 mg PO three times per week, pentamidine 300 mg IV/inhaled monthly – High risk for hepatitis B reactivation: entecavir 0.5 mg QD or lamivudine 100 mg PO QD
Neutropenia	*Primary or Secondary Prevention:* – Granulocyte colony stimulating factors (G-CSF): filgrastim 300–480 μg daily, pegfilgrastim 6 mg ×1, and biosimilars
Fluid retention	*Prevention:* – Dexamethasone 4–8 mg PO/IV BID for 3 days starting 1 day prior, the day of and the day after docetaxel, or loncastuximab administration *Treatment:* – Spironolactone 12.5–25 mg QD with loncastuximab
Hypersensitivity or infusion reaction	*Prevention:* – Diphenhydramine 25–50 mg PO, cetirizine 10 mg PO, famotidine 20 mg PO, and/or acetaminophen 650 mg PO *Treatment:* – Steroids (methylprednisolone 125 mg ×1), morphine 4 mg IV (rigors)
Antifolate side effects	*Prevention:* – Methotrexate: leucovorin (folinic acid) 25 mg PO Q6H beginning 24 h following methotrexate administration – Pemetrexed—oral folic acid 1 mg QD and cyanocobalamin 1000 μg Q3 weeks starting at least 7 days prior to first cycle *Treatment:* – Methotrexate: leucovorin 100 mg/m^2 IV Q3–6 h depending on methotrexate levels and/or glucarpidase 50 units/kg ×1 dose

Continued

Table 1 Examples of supportive care medications commonly used in standard practice.—cont'd

Nephrotoxicity	*Prevention:* – Cisplatin: prehydration with 500–1000 mL 0.9% NaCl, forced diuresis with mannitol – High-dose methotrexate (3.5–8 g/m^2): urine alkalinization with goal pH >7 and urine output >100 mL/h
Cardiotoxicity	*Prevention:* – Doxorubicin (>300 mg/m^2 lifetime accumulation): dexrazoxane at a ratio of 10:1 with the doxorubicin dose, given by slow IV push or short IV infusion, 15–30 min before doxorubicin
Bladder toxicity	*Prevention:* – Ifosfamide and high-dose cyclophosphamide: mesna IV or PO (dose-dependent) to prevent hemorrhagic cystitis
Arthralgia/Myalgia	*Treatment:* – Paclitaxel: prednisone or gabapentin (titrate dose based on response)
Chemotherapy Induced Peripheral Neuropathy	*Treatment:* – Vinca alkaloids or MMAE-based ADC: duloxetine 30–60 mg PO daily, pregabalin 75–300 mg BID

[a]Palifermin is only indicated in patient undergoing stem cell transplantation and should not be administered within 24 h of the initiation of the conditioning regimen.
[b]Specific to patients receiving Melphalan preparation.
[c]Available in oral (not in US), IV and SQ but most commonly prescribed formulation is the patch. It is only approved for prophylaxis not treatment.
ADC, antibody drug conjugate; ICI, Immune checkpoint; MMAE, monomethyl auristatin E; PI3K, Phosphatidylinositol-3-kinase.

5. Gastrointestinal toxicity

5.1 Chemotherapy-induced nausea and vomiting (CINV)

Given that chemotherapy acts by inhibiting rapidly dividing cells, the epithelial cells that line the gastrointestinal tract are frequently damaged, leading to common gastrointestinal side effects such as nausea, vomiting, diarrhea and mucositis. The emetogenic potential of a given regimen can vary from minimal (<10%), to high (>90%) (BC Cancer, 2020). Infusion time, dose and number of individual drugs are key contributors to the emetogenicity of a particular regimen. Furthermore, CINV can be classified into three major categories: anticipatory (prior to chemotherapy), acute (in the first 24 h), and delayed (after 24 h). First generation serotonin antagonists (ondansetron, granisetron) are used as acute preventive anti-CINV strategies in highly emetogenic regimens, while neurokinin1 receptor antagonists (aprepitant, fosaprepitant) are used as acute and delayed preventive anti-CINV agents. Other antinausea medications include steroids (dexamethasone), antipsychotics (olanzapine), antihistamines (dimenhydrinate), dopamine antagonists (prochlorperazine) and benzodiazepines (lorazepam) as mentioned in Table 1.

5.2 Severe delayed onset diarrhea

Loperamide is widely used to manage most common cancer therapy-associated diarrhea. Loperamide decreases intestinal transit time and slows intestinal motility. However, severe delayed onset diarrhea represents a substantive therapeutic challenge. This is the province of irinotecan, a topoisomerase I poison that is one of the few agents that is marginally effective in the treatment of colorectal cancer, through its conversion to SN38. This often severe toxicity that may contribute to cachexia is associated with the promotion of apoptosis in the GI tract. The drug loperamide may provide temporary refuge for patients experiencing this often debilitating toxicity, but the GI tract rapidly acquires tolerance to the loperamide, with few alternatives available. As indicated in the chapter by Akbarali et al., it appears likely, but is as yet unproven, the SN-38 may interfere with the protective autophagy that is necessary for GI tract integrity.

Idelalisib, the first-in-class phosphoinositide 3-kinase delta isoform inhibitor has been associated with frequent delayed diarrhea, thought to be immune mediated requiring steroids (Coutre et al., 2015). Similarly, immune check point inhibitors (ICIs) have been associated with frequent immune-mediated colitis, particularly when used in combination (Section 7).

5.3 Mucositis

As mentioned above, epithelial cells lining up the GI tract are very susceptible to chemotherapy, with mucositis frequently observed with high dose methotrexate and high dose chemotherapy prior to stem cell transplantation. Ice chips decrease the risk of mucositis associated with patients receiving melphalan preparations. Palifermin is the only FDA-approved agent for prophylaxis against severe mucositis among patients undergoing auto or allogeneic hematopoietic stem cell transplantation (Hensley et al., 2009). Palifermin is a recombinant truncated form of human keratinocyte growth factor (KGF), a member of the fibroblast growth factor family.

6. Neuromuscular toxicity

6.1 Chemotherapy-induced peripheral neuropathy (CIPN)

Chemotherapy-induced peripheral neuropathy (CIPN) is one of the most poorly understood toxicities associated with cancer chemotherapy. CIPN is associated with the taxanes and vinca alkaloids, both microtubule poisons but with different mechanisms of action, platinum-based regimens, proteasome inhibitors (e.g., bortezomib), and some of the newer antibody drug conjugates (e.g., brentuximab vedotin). Here again, drugs with entirely different mechanisms of action converge at the identical normal tissue endpoint. Multiple mechanisms have been proposed, but intuitively it appears that there should be a common intermediate or target. This "side effect" of these drugs is unpredictable, occurring to differing intensities in patients, resolving in some patients but with prolonged persistence in others, even after treatment has long been terminated and often making ordinary life extremely difficult. As of this writing, there is essentially no FDA-approved agent for the prevention or treatment of CIPN, and patients are encouraged to enroll in prospective natural history studies such as EPIPHANY (NCT03997981). Treatments for neuropathy focus on symptom reduction. Cymbalta has shown efficacy in decreasing symptoms from CIPN when compared to placebo (Smith et al., 2013).

Oxaliplatin, one of the components of the FOLFOX chemotherapy regimen, is widely used as frontline treatment for GI malignancies particularly colorectal cancers. Oxaliplatin is somewhat unique in that it has both an acute neurotoxicity during time of infusion or soon thereafter as well as a more late-onset neurotoxicity that often occurs later on in the course of chemotherapy treatments due to cumulative effects. The acute neurotoxicity occurs in the vast majority of patients and can range from very minor symptoms to much more severe symptoms. For example, a common side effect is cold sensitivity relating to touching or encountering something cold, as well as swallowing something cold causing discomfort and/or paresthesia of the hands/feet or mouth/esophagus, respectively. However, some patients also experience more severe symptoms including jaw pain/tightness or spasm limiting their ability to eat as well as hoarseness and voice changes. These acute toxicities, especially the voice changes and jaw side effects, can certainly lead to dose reductions of oxaliplatin chemotherapy in cases where the acute side effects are lasting longer or affecting quality of life.

In a review published in BMC Cancer in 2018, Gebremedhn et al. found that moderate to severe (grade3–4) acute oxaliplatin-induced neuropathy occurred more commonly in patients on higher doses of oxaliplatin of >85 mg/m^2 IV and that these effects were occurring within 24 h of infusion. Interestingly, 58% of patients who were receiving lower doses of oxaliplatin (at least 25 mg/m^2 IV) were still experiencing mild (grade 1–2) acute paresthesias in the fingers and toes. This review was also focused on the acute neurotoxicity that occurred even with cycle 1. Methods of improving the acute neurotoxicity effects include slowing down the rate of infusion, reducing the dose of oxaliplatin, or discontinuing it entirely (Gebremedhn, Shortland, & Mahns, 2018).

Oxaliplatin is also associated with a chronic peripheral neuropathy that can occur particularly with cumulative treatment. It has been shown that the risk of developing chronic oxaliplatin induced peripheral neuropathy is related to cumulative dose, infusion rate, starting dose of oxaliplatin, severity of acute neurotoxicity and other patient factors such as pre-existing conditions (e.g., Diabetes (Argyriou et al., 2013; Beijers, Mols, & Vreugdenhil, 2014)). Several studies have been published and are ongoing to try to determine how to prevent peripheral neuropathy from cumulative dosing of oxaliplatin, but as evidenced in a recent meta-analysis there have not been any proven prevention strategies thus far (Peng et al., 2022). The peripheral neuropathy sometimes improves with discontinuation of treatment, but is, for the most part, a long lasting and permanent side effect from chemotherapy that can affect quality of life for many years.

7. Cognitive dysfunction associated with chemotherapy: "Chemobrain or chemo fog"

Another relatively common toxicity associated with a relatively wide spectrum of agents, again, with highly diverse mechanisms of action, is chemobrain, where various aspects of cognitive function may be compromised either transiently or over a prolonged period of time. Doxorubicin, cyclophosphamide, 5-fluorouracil, and aromatase inhibitors have all been shown to contribute to this debilitating outcome that can severely constrain life choices for the patient. Since these agents may not have the capacity to actually cross the blood–brain barrier, this is a conundrum that makes it difficult to even hypothesize a mechanism of action or consequently, science-based approaches to prevent or mitigate this outcome. The chapter by Rao et al. further develops the topic of chemotherapy induced cognitive dysfunction.

8. Myalgias and arthralgias

Myalgias and arthralgias are common side effects of taxanes that can limit their wide use (BC Cancer, 2020). Acetaminophen or non-steroidal anti-inflammatory drugs have usually been prescribed for treatment. However, in the absence of relief, gabapentin or prednisone may provide a possible therapeutic benefit (Table 1).

9. Cutaneous toxicity

Essentially all types of anticancer therapy are associated with cutaneous toxicity. Cytotoxic agents are known to promote a myriad of cutaneous side effects given the high metabolic rate of skin. Side effects typically include reversible pigmentary changes of skin, nails, hair and mucous membranes most commonly associated with alkylating agents and antitumor antibiotics (Payne, James, & Weiss, 2006; Singal, Tunnessen, Wiley, & Hood, 1991). Other nail changes, which are commonly associated with taxane therapy, include onycholysis and paronychia (Payne et al., 2006; Singal et al., 1991). Radiation sensitization and recall, as well as photosensitivity are other fairly common side effects of cytotoxic chemotherapy. Non-specific exanthematous eruptions, that can occur with many different types of cytotoxic chemotherapy, are typically treated with topical corticosteroids (Alley, Green, & Schuchter, 2002). Systemic steroids are generally not indicated, but in the case of pemetrexed, premedication with corticosteroids has proven to prevent or attenuate the frequently observed papular rash associated with this drug (Thodtmann et al., 2003). Diffuse erythema can occur and may progress to exfoliative dermatitis in the setting of cisplatin or methotrexate use (Lee, Hook, & Long, 1994; Peters et al., 1999). Stevens–Johnson syndrome/toxic epidermal necrolysis (SJS/TEN) and drug reaction with eosinophilia and systemic symptoms (DRESS) are severe cutaneous drug reactions that also often accompany cytotoxic chemotherapy. While SJS/TEN is rare, mortality rates can be up to 40% due to extensive epidermal detachment secondary to keratinocyte necrosis (Rosen et al., 2014). DRESS has been associated with lenalidomide and is characterized by maculopapular erythema, fever, lymphadenopathy, multisystem organ failure and eosinophilia (Shaaban, Layne, & Guron, 2014).

Hand-foot syndrome is relatively common and has been linked with both cytotoxic agents (liposomal doxorubicin, 5-FU, capecitabine), as well as TKIs (sorafenib, sunitinib, lenvatinib). Clinical manifestations include dysesthesia, erythema or hyperpigmentation, cracking and desquamation of the palms and soles (Nikolaou, Syrigos, & Saif, 2016). Management includes dose interruption and/or dose-modification of chemotherapy and supportive measures, including high-potency topical corticosteroids, topical keratolytics (e.g., urea cream) and emollients, wound care and analgesics (Miller, Gorcey, & McLellan, 2014).

Acneiform rash is a very common dermatologic toxicity associated with EFGR inhibitors (especially monoclonal antibodies cetuximab and panitumumab) and mitogen-activated protein kinase (MEK) inhibitors (trametinib). Follicular erythematous papules and pustules without comedones develop in sebaceous-rich areas of skin (face, scalp, chest, upper back). Management consists of topical or systemic antibiotics (clindamycin gel, oral doxycycline) and topical corticosteroids (Lacouture & Sibaud, 2018). The presence of an acneiform eruption predicts improved response to treatment and survival with EGFR inhibitors (Stintzing et al., 2013).

Cutaneous toxicity is the most common and earliest irAE in ICI therapy. Manifestations include pruritis, inflammatory dermatoses (maculopapular, eczematous, lichenoid, and psoriasiform eruptions), bullous disease, depigmentation, and rarely SJS/TEN and DRESS. Management consists of high-potency topical corticosteroids without holding immunotherapy or high-dose systemic corticosteroids and interruption of immunotherapy for more severe cases (Geisler et al., 2020).

10. Secondary malignancies

Alkylating agents (cyclophosphamide), anthracyclines (doxorubicin), and purine analogs (fludarabine) are frequent components of curative regimens for patients with breast and hematological malignancies. However, these drugs have been associated with increased risk of therapy-related myeloid neoplasms (t-MN): myelodysplastic syndrome with alkylating agents and secondary acute myelogenous leukemia with topoisomerase II inhibitors (anthracyclines) that have a very poor prognosis (~8 months), with low chance of response to conventional therapies. T-MN is estimated to represent close to 15% of new MDS/AML cases, with an overall incidence 0.62/100,000, and cumulative incidence at 10 years of 0.5% in breast cancer and 10% in lymphomas (McNerney, Godley, & Le Beau, 2017).

11. Side effects of immune checkpoint inhibitors

Immune Check Point Inhibitors (ICIs), especially when used in combination (e.g., Ipilimumab and nivolumab), have been associated with increased risk of immune related adverse events due to overaction of the patient's own immune system. Patients with immune-related side effects may include a broad spectrum of affected organs (skin, endocrine, cardiac, GI, pulmonary, renal) are usually managed by withholding the drug, and administration of budesonide for mild cases and prednisone/methylprednisolone for more intense cases (Table 1). While initial cases respond to steroids, cases that are steroid-refractory constitute a therapeutic challenge with limited options such as infliximab or vedolizumab or tofactinib (Esfahani, Hudson, & Batist, 2020).

12. Radiation therapy

Because radiation is so broadly used in a variety of malignancies, this section develops radiation toxicity according to different tissue organ components. Advances in radiation therapy technology have resulted in the ability to deliver therapeutic doses to tumor targets with unprecedented accuracy and precision. This has resulted in both better tumor control and decreased short and long-term side effects. Still, radiation-related toxicity remains a significant source of concern for patients and frustration for providers who are sometimes relatively powerless to mitigate distress.

12.1 CNS

Radiation therapy for brain tumors can be delivered via a number of techniques depending on the tumor type. Treatment of primary brain tumors such as glioblastoma typically involves large volumes over several weeks. Although modern technology such as image-guided volumetric modulated arc therapy (VMAT) allows for highly conformal dose delivery, short-term side effects typically include significant fatigue, drowsiness, generalized weakness, and hair loss. Long-term side effects can include permanent hair loss, cataract formation, and chronic alternations in cognition. Metastatic tumors can be managed using stereotactic radiosurgery (SRS), which involves only 1–5 treatments using highly focused dose delivery. Radiosurgery is much better tolerated and does not usually cause significant short-term toxicity. The high doses used with SRS can result in late or subacute radionecrosis of normal brain, which is a very troublesome condition that sometimes requires surgical removal of the damaged tissue to resolve. Recent trials conducted to reduce neurotoxicity from radiotherapy have investigated the use of techniques which reduce dose to the hippocampus and on drugs, such as memantine and donazepil, which might improve cognition (Saad & Wang, 2015).

12.2 Head and neck

Improvements in technology (VMAT) have made the treatment of head and neck cancers much more tolerable, but this treatment remains one of the most toxic forms of radiation therapy. Short-term side effects include severe dermatitis, mucositis, taste changes, dry mouth, pain, difficulty swallowing, dehydration, and weight loss. These effects are usually managed with numbing mouth rinses and topical skin creams. Some physicians use a liquid form of the tricyclic antidepressant Doxepin as a mouthwash as some clinical trials have shown a modest improvement in oral discomfort with its use. The use of Benzydamine mouthwash has also been investigated but consistent, significant benefit was not observed. Opioids are often needed to manage pain. There has been some interest in the use of oral anti-oxidants such as superoxide dismutase, but there are no widely available preparations or guidelines for use. Many other approaches have been investigated to reduce mucositis, including amifostine, cryotherapy, glutamine, oral barrier agents, *aloe vera*, and honey, but a systematic review of the available data on these approaches concluded that none of these could be recommended for routine clinical use (Siddiqui & Movsas, 2017). Long-term toxicity from head and neck radiotherapy can include fibrosis, chronic lymphedema, permanent swallowing dysfunction, dry mouth, dental issues, and thyroid damage. This is particularly alarming given the increasing incident in HPV-related oropharyngeal cancers in younger patients who will have to live for decades with long-term adverse effects from treatment.

12.3 Breast

The most common short-term side effect associated with breast radiotherapy is dermatitis. This can mean anything from lightly pink skin to peeling moist desquamation. Dermatitis is managed with topical preparations, some of which have active ingredients such as calendula which are fairly effective at mitigating the irritation but cannot prevent it altogether. Small industry-sponsored trials have demonstrated that topical anti-oxidant serums are highly effective at reducing dermatitis and are available over the counter but are far too expensive for most patients to access. Another common short-term side effect is fatigue. Fatigue is very frequent in the breast cancer population, despite the fact that breast cancer treatment is typically far shorter than most other cancers, is usually a lower total dose, and is almost never delivered with concomitant chemotherapy. Some have postulated that cutaneous dermatitis leads to inflammation which in turn causes fatigue, but it is not clear why fatigue occurs so commonly in this population. Long-term toxicity includes breast edema, pain, shrinkage, and fibrosis. The risk of chronic lymphedema is increased if the nodal basins are treated. Rarely, pneumonitis of the underlying lung can occur. Late cardiac toxicity was a significant concern with older techniques but has been nearly eliminated with the use of modern, image-guided radiotherapy (Shah, Banda, Chandra, & Vicini, 2017).

12.4 Lung

Early stage lung cancers are typically treated with a short course of highly targeted treatment known as stereotactic ablative radiotherapy (SABR) or stereotactic body radiotherapy (SBRT). Like its cousin in the brain, stereotactic treatment in the body does not usually result in short-term side effects. Similarly, late effects with this technique are uncommon. Major advances in radiotherapy for advanced lung cancers (image guided VMAT) have dramatically improved the toxicity of this treatment. Consequently, normal tissue such as the esophagus and heart can be spared from receiving high doses. Still, short-term side effects typically include fatigue, cough, and esophagitis. Significant dermatitis is not usually observed with modern techniques. Esophagitis can lead to dehydration and weight loss. Long-term side effects include pneumonitis, esophageal stricture, and cardiotoxicity. Late effects of lung cancer treatment are now very important to consider as the treatment strategies continue to improve and overall survival increases. Patients treated with thoracic radiotherapy are at increased risk of cardiac events, particularly in the stetting of pre-existing cardiac disease (Banfill et al., 2021). The risk of symptomatic radiation pneumonitis has been reduced through the use of advanced techniques, although routine use of immune checkpoint inhibitors has been associated with recall pneumonitis (Fujiwara et al., 2021). Better radiotherapy technology allowing for dose escalation, image guidance to improve accuracy of delivery and allow for plan refinement with tumor response, and the addition of adjuvant immunotherapy have dramatically improved the prognosis for these patients. As such, treatment-related toxicity will continue to become more relevant.

12.5 Upper abdomen

Radiation therapy for upper gastrointestinal tumors such as esophageal, gastric, and pancreatic tumors is associated with nausea, vomiting, odynophagia, dysphagia, anorexia, abdominal pain, and fatigue. This can lead to significant issues with dehydration and weight loss. Some of these patients require feeding tubes during and immediately after treatment. Long-term side effects include esophageal or gastric ulcers, small bowel obstruction, kidney damage, or transient hepatitis. Pre-operative radiotherapy increases the rate of response to treatment but can also increase the risk of operative complications through microvascular changes.

12.6 Lower abdomen and pelvis

Radiation therapy is frequently used in the management of cancers of the rectum, prostate, cervix, endometrium, and anal canal. Acute toxicity includes fatigue, diarrhea, cystitis, and pelvic pain. Treatment of the anal canal results in significant perineal skin irritation which can be extremely painful and difficult to manage. Although topical preparations and warm baths can help, these patients are typically in significant discomfort or pain and frequently require narcotics. Long-term side effects can include small bowel obstruction, urinary dysfunction, erectile dysfunction in men, vaginal stenosis in women, and ovarian failure in premenopausal women. Long-term gastrointestinal toxicity is not only related to radiation dose delivered, but can also be increased in patients who use tobacco products or have pre-existing conditions such as diabetes, HIV, collagen vascular disease, or inflammatory bowel disease. Newer research suggests that the gut microbiome may play a significant role in late toxicity (Wang et al., 2021). Modern radiotherapy techniques have significantly reduced late toxicity through improved sparing of normal tissues. Other interventions such as dietary/lifestyle modifications, pharmacologic interventions, and supplement use continue to be investigated. At the present time, no interventions have demonstrated a consistent benefit to justify routine use (Nicholas et al., 2017).

13. Mitigating/circumventing chemotoxicity

Use of leucovorin rescue is a well-established modality to mitigate side effects to normal cells from prolonged exposure to high-dose methotrexate (BC Cancer, 2020). For DNA synthesis and cellular replication to take place, folic acid needs to be reduced to tetrahydrofolic acid, via dihydrofolate reductase. Given that leucovorin is a reduced form of folic acid, this allows leucovorin to be a potent antidote of folic acid antagonists such as methotrexate. Furthermore, Leucovorin is believed to have a preference for normal over malignant cells, thus allowing "rescue" of the normal cells from the toxicity of the folic acid antagonists (methotrexate, pralatrexate and pemetrexed).

Mesna (sodium 2-mercaptoethane sulphonate) is a uroprotective agent indicated as a prophylactic agent against ifosfamide and high dose cyclophosphamide bladder injury (BC Cancer, 2020). Each of these two drugs form acrolein, an oxazaphosphorine metabolite that has the potential to induce hemorrhagic cystitis (HC) that usually present with hematuria, frequency, dysuria, urgency and can ultimately lead to bladder fibrosis and obstructive renal failure. Mesna binds to acrolein in the bladder, leading to an inactive compound that is rapidly eliminated by the kidneys. Mesna in combination with diuresis decreases the incidence of HC.

Fluid retention has been observed in patients treated with docetaxel-containing regimens, and more recently, with the antibody-drug conjugate loncastuximab (BC Cancer, 2020). In order to mitigate this side effect, premedication with dexamethasone is given orally. Of note, severe hypersensitivity, including very rare fatal anaphylaxis, has been reported in patients treated with docetaxel who received dexamethasone premedication.

Monomethyl auristatin E (MMAE) is an antimitotic agent that inhibits cell division by blocking the polymerization of tubulin. ADCs (Antibody drug conjugates) consist of a drug, a cleavable linker, and a targeting antibody that interacts with the corresponding receptor on the tumor cell. The linker attached to the monoclonal antibody is stable in extracellular fluid, but once the conjugate enters the malignant cell, it is cleaved by cathepsin, thus providing for highly selective drug delivery to the malignant cells.

There are two ADC using the MMAE payload currently FDA-approved in lymphoid malignancies: Brentuximab Vedotin targeting CD30 approved in classic Hodgkin lymphoma, cutaneous and peripheral T cell lymphomas, and Polatuzumab Vedotin targeting CD79b approved in combination with bendamustine and rituximab in diffuse large b cell lymphoma (Al Juhaishi & Yazbeck, 2019). Despite their improved specificity, peripheral neuropathy remains a major side effect.

Pyrrolobenzodiazepine dimers (PBD) have anti-tumor as well as antibiotic properties. Synthetic PBD dimers, where two PBD monomers are linked through their aromatic A-ring phenolic C8-positions via a flexible propyldioxy tether, are highly efficient DNA minor groove cross-linking agents with potent cytotoxicity. Loncastuximab tesirine is the first-in-class ADC conjugated to PBD and targeting CD19 that is currently approved as a single agent in diffuse large B cell lymphoma (Willard, McKay, & Yazbeck, 2021). Unlike MMAE-based ADC, peripheral neuropathy is not usually reported with this type of ADC.

Proteolysis-targeting chimeras (PROTACs): Advances in the development of small-molecule degraders such as proteolysis-targeting chimeras (PROTACs), have allowed for the targeting of proteins previously considered as "undruggable." The proteolysis targeting chimera is a bivalent molecule comprised of a target-binding moiety, a linker and an E3 binder. Engagement of the proteolysis targeting chimera with its target protein activates the ubiquitin pathway resulting in degradation of the target protein by the proteasome system. The broad applicability of this approach is facilitated by the flexibility of individual E3 ligases to recognize different substrates, allowing their degradation by "hijacking" the cell's own ubiquitin–proteasome system. This approach is currently in early-phase clinical trials (e.g., NCT05006716).

14. Conclusion

Major advances in our understanding of the tumor, its microenvironment at the molecular level, lead to unprecedented therapeutic advances in precision medicine and modern radiotherapy. This, has translated into continued improvement in overall survival for several cancers, and steady decrease in cancer mortality over the last decade. Although we are curing patients at rates that would have seemed impossible only a few years ago, we are now faced with an enlarging population of patients who are troubled by chronic treatment-related toxicity. As clinicians, we are thrilled that we have the ability to help so many patients, but remain frustrated that more cannot be done to mitigate the suffering which still accompanies many forms of cancer therapy, particularly radiotherapy and chemotherapy. More research into how normal tissue repair can be assisted (or even prevented) is needed as we continue to move forward towards our goal of cancer-free world where every patient survives cancer while maintaining a normal quality of life.

References

Al Juhaishi, T., & Yazbeck, V. (2019). Choosing the right pharmacotherapy for non-Hodgkin's lymphoma: Does one size fit all? *Expert Opinion on Pharmacotherapy*, *20*(7), 773–775. https://doi.org/10.1080/14656566.2019.1582643.

Alley, E., Green, R., & Schuchter, L. (2002). Cutaneous toxicities of cancer therapy. *Current Opinion in Oncology*, *14*(2), 212–216. https://doi.org/10.1097/00001622-200203000-00012.

Argyriou, A. A., Cavaletti, G., Briani, C., Velasco, R., Bruna, J., Campagnolo, M., et al. (2013). Clinical pattern and associations of oxaliplatin acute neurotoxicity: A prospective study in 170 patients with colorectal cancer. *Cancer*, *119*(2), 438–444. https://doi.org/10.1002/cncr.27732.

Banfill, K., Giuliani, M., Aznar, M., Franks, K., McWilliam, A., Schmitt, M., et al. (2021). Cardiac toxicity of thoracic radiotherapy: Existing evidence and future directions. *Journal of Thoracic Oncology*, *16*(2), 216–227. https://doi.org/10.1016/j.jtho.2020.11.002.

BC Cancer. (2020). *Clinical pharmacy guide: Cancer drug treatment assessment and review* (5th Edition).

Beijers, A. J., Mols, F., & Vreugdenhil, G. (2014). A systematic review on chronic oxaliplatin-induced peripheral neuropathy and the relation with oxaliplatin administration. *Support Care Cancer*, *22*(7), 1999–2007. https://doi.org/10.1007/s00520-014-2242-z.

Byrd, J. C., Hillmen, P., Ghia, P., Kater, A. P., Chanan-Khan, A., Furman, R. R., et al. (2021). Acalabrutinib versus ibrutinib in previously treated chronic lymphocytic leukemia: Results of the first randomized phase III trial. *Journal of Clinical Oncology*, *39*(31), 3441–3452. https://doi.org/10.1200/JCO.21.01210.

Camus, P., Bonniaud, P., Fanton, A., Camus, C., Baudaun, N., & Foucher, P. (2004). Drug-induced and iatrogenic infiltrative lung disease. *Clinics in Chest Medicine*, *25*(3), 479–519. https://doi.org/10.1016/j.ccm.2004.05.006.

Coutre, S. E., Barrientos, J. C., Brown, J. R., de Vos, S., Furman, R. R., Keating, M. J., et al. (2015). Management of adverse events associated with idelalisib treatment: Expert panel opinion. *Leukemia & Lymphoma*, *56*(10), 2779–2786. https://doi.org/10.3109/10428194.2015.1022770.

Esfahani, K., Hudson, M., & Batist, G. (2020). Tofacitinib for refractory immune-related colitis from PD-1 therapy. *The New England Journal of Medicine*, *382*(24), 2374–2375. https://doi.org/10.1056/NEJMc2002527.

Fujiwara, M., Doi, H., Igeta, M., Suzuki, H., Kitajima, K., Tanooka, M., et al. (2021). Radiation pneumonitis after volumetric modulated arc therapy for non-small cell lung cancer. *Anticancer Research*, *41*(11), 5793–5802. https://doi.org/10.21873/anticanres.15396.

Gebremedhn, E. G., Shortland, P. J., & Mahns, D. A. (2018). The incidence of acute oxaliplatin-induced neuropathy and its impact on treatment in the first cycle: A systematic review. *BMC Cancer*, *18*(1), 410. https://doi.org/10.1186/s12885-018-4185-0.

Geisler, A. N., Phillips, G. S., Barrios, D. M., Wu, J., Leung, D. Y. M., Moy, A. P., et al. (2020). Immune checkpoint inhibitor-related dermatologic adverse events. *Journal of the American Academy of Dermatology*, *83*(5), 1255–1268. https://doi.org/10.1016/j.jaad.2020.03.132.

Goldman, A., Maor, E., Bomze, D., Liu, J. E., Herrmann, J., Fein, J., et al. (2021). Adverse cardiovascular and pulmonary events associated with chimeric antigen receptor T-cell therapy. *Journal of the American College of Cardiology*, *78*(18), 1800–1813. https://doi.org/10.1016/j.jacc.2021.08.044.

Hensley, M. L., Hagerty, K. L., Kewalramani, T., Green, D. M., Meropol, N. J., Wasserman, T. H., et al. (2009). American Society of Clinical Oncology 2008 clinical practice guideline update: Use of chemotherapy and radiation therapy protectants. *Journal of Clinical Oncology*, *27*(1), 127–145. https://doi.org/10.1200/JCO.2008.17.2627.

Hillmen, P., Brown, J. R., Eichhorst, B. F., Lamanna, N., O'Brien, S. M., Qiu, L., et al. (2020). ALPINE: Zanubrutinib versus ibrutinib in relapsed/refractory chronic lymphocytic leukemia/small lymphocytic lymphoma. *Future Oncology*, *16*(10), 517–523. https://doi.org/10.2217/fon-2019-0844.

Itzhaki Ben Zadok, O., Ben-Avraham, B., Nohria, A., Orvin, K., Nassar, M., Iakobishvili, Z., et al. (2019). Immune-checkpoint inhibitor-induced fulminant myocarditis and cardiogenic shock. *JACC CardioOncology*, *1*(1), 141–144. https://doi.org/10.1016/j.jaccao.2019.07.004.

Jia, W., Guo, H., Jing, W., Jing, X., Li, J., Wang, M., et al. (2020). An especially high rate of radiation pneumonitis observed in patients treated with thoracic radiotherapy and simultaneous osimertinib. *Radiotherapy and Oncology*, *152*, 96–100. https://doi.org/10.1016/j.radonc.2020.07.051.

Kudoh, S., Kato, H., Nishiwaki, Y., Fukuoka, M., Nakata, K., Ichinose, Y., et al. (2008). Interstitial lung disease in Japanese patients with lung cancer: A cohort and nested case-control study. *American Journal of Respiratory and Critical Care Medicine*, *177*(12), 1348–1357. https://doi.org/10.1164/rccm.200710-1501OC.

Lacouture, M., & Sibaud, V. (2018). Toxic side effects of targeted therapies and immunotherapies affecting the skin, Oral mucosa, hair, and nails. *American Journal of Clinical Dermatology*, *19*(Suppl 1), 31–39. https://doi.org/10.1007/s40257-018-0384-3.

Lee, T. C., Hook, C. C., & Long, H. J. (1994). Severe exfoliative dermatitis associated with hand ischemia during cisplatin therapy. *Mayo Clinic Proceedings*, *69*(1), 80–82. https://doi.org/10.1016/s0025-6196(12)61616-4.

Long, K., & Suresh, K. (2020). Pulmonary toxicity of systemic lung cancer therapy. *Respirology*, *25*(Suppl 2), 72–79. https://doi.org/10.1111/resp.13915.

Longley, D. B., Harkin, D. P., & Johnston, P. G. (2003). 5-fluorouracil: Mechanisms of action and clinical strategies. *Nature Reviews. Cancer*, *3*(5), 330–338. https://doi.org/10.1038/nrc1074.

Manohar, S., & Leung, N. (2018). Cisplatin nephrotoxicity: A review of the literature. *Journal of Nephrology*, *31*(1), 15–25. https://doi.org/10.1007/s40620-017-0392-z.

McNerney, M. E., Godley, L. A., & Le Beau, M. M. (2017). Therapy-related myeloid neoplasms: When genetics and environment collide. *Nature Reviews. Cancer*, *17*(9), 513–527. https://doi.org/10.1038/nrc.2017.60.

Miettinen, P. J., Warburton, D., Bu, D., Zhao, J. S., Berger, J. E., Minoo, P., et al. (1997). Impaired lung branching morphogenesis in the absence of functional EGF receptor. *Developmental Biology*, *186*(2), 224–236. https://doi.org/10.1006/dbio.1997.8593.

Miller, K. K., Gorcey, L., & McLellan, B. N. (2014). Chemotherapy-induced hand-foot syndrome and nail changes: A review of clinical presentation, etiology, pathogenesis, and management. *Journal of the American Academy of Dermatology*, *71*(4), 787–794. https://doi.org/10.1016/j.jaad.2014.03.019.

Miwa, M., Ura, M., Nishida, M., Sawada, N., Ishikawa, T., Mori, K., et al. (1998). Design of a novel oral fluoropyrimidine carbamate, capecitabine, which generates 5-fluorouracil selectively in tumours by enzymes concentrated in human liver and cancer tissue. *European Journal of Cancer*, *34*(8), 1274–1281. https://doi.org/10.1016/s0959-8049(98)00058-6.

Nicholas, S., Chen, L., Choflet, A., Fader, A., Guss, Z., Hazell, S., et al. (2017). Pelvic radiation and Normal tissue toxicity. *Seminars in Radiation Oncology*, *27*(4), 358–369. https://doi.org/10.1016/j.semradonc.2017.04.010.

Nikolaou, V., Syrigos, K., & Saif, M. W. (2016). Incidence and implications of chemotherapy related hand-foot syndrome. *Expert Opinion on Drug Safety*, *15*(12), 1625–1633. https://doi.org/10.1080/14740338.2016.1238067.

Payne, A. S., James, W. D., & Weiss, R. B. (2006). Dermatologic toxicity of chemotherapeutic agents. *Seminars in Oncology*, *33*(1), 86–97. https://doi.org/10.1053/j.seminoncol.2005.11.004.

Peng, S., Ying, A. F., Chan, N. J. H., Sundar, R., Soon, Y. Y., & Bandla, A. (2022). Prevention of oxaliplatin-induced peripheral neuropathy: A systematic review and meta-analysis. *Frontiers in Oncology*, *12*. https://doi.org/10.3389/fonc.2022.731223, 731223.

Peters, T., Theile-Ochel, S., Chemnitz, J., Sohngen, D., Hunzelmann, N., & Scharffetter-Kochanek, K. (1999). Exfoliative dermatitis after long-term methotrexate treatment of severe psoriasis. *Acta Dermato-Venereologica*, *79*(5), 391–392. https://doi.org/10.1080/000155599750010382.

Rosen, A. C., Balagula, Y., Raisch, D. W., Garg, V., Nardone, B., Larsen, N., et al. (2014). Life-threatening dermatologic adverse events in oncology. *Anti-Cancer Drugs*, *25*(2), 225–234. https://doi.org/10.1097/CAD.0000000000000032.

Saad, S., & Wang, T. J. (2015). Neurocognitive deficits after radiation therapy for brain malignancies. *American Journal of Clinical Oncology*, *38*(6), 634–640. https://doi.org/10.1097/COC.0000000000000158.

Salem, J. E., Manouchehri, A., Bretagne, M., Lebrun-Vignes, B., Groarke, J. D., Johnson, D. B., et al. (2019). Cardiovascular toxicities associated with ibrutinib. *Journal of the American College of Cardiology*, *74*(13), 1667–1678. https://doi.org/10.1016/j.jacc.2019.07.056.

Sara, J. D., Kaur, J., Khodadadi, R., Rehman, M., Lobo, R., Chakrabarti, S., et al. (2018). 5-fluorouracil and cardiotoxicity: A review. *Therapeutic Advances in Medical Oncology*, *10*. https://doi.org/10.1177/1758835918780140.

Shaaban, H., Layne, T., & Guron, G. (2014). A case of DRESS (drug reaction with eosinophilia and systemic symptoms) with acute interstitial nephritis secondary to lenalidomide. *Journal of Oncology Pharmacy Practice*, *20*(4), 302–304. https://doi.org/10.1177/1078155213502569.

Shah, C., Banda, B., Chandra, R., & Vicini, F. (2017). Minimizing toxicity in breast irradiation. *Expert Review of Anticancer Therapy*, *17*(3), 187–189. https://doi.org/10.1080/14737140.2017.1285231.

Siddiqui, F., & Movsas, B. (2017). Management of Radiation Toxicity in head and neck cancers. *Seminars in Radiation Oncology*, *27*(4), 340–349. https://doi.org/10.1016/j.semradonc.2017.04.008.

Singal, R., Tunnessen, W. W., Jr., Wiley, J. M., & Hood, A. F. (1991). Discrete pigmentation after chemotherapy. *Pediatric Dermatology*, *8*(3), 231–235. https://doi.org/10.1111/j.1525-1470.1991.tb00867.x.

Smith, E. M., Pang, H., Cirrincione, C., Fleishman, S., Paskett, E. D., Ahles, T., et al. (2013). Effect of duloxetine on pain, function, and quality of life among patients with chemotherapy-induced painful peripheral neuropathy: A randomized clinical trial. *JAMA*, *309*(13), 1359–1367. https://doi.org/10.1001/jama.2013.2813.

Stintzing, S., Kapaun, C., Laubender, R. P., Jung, A., Neumann, J., Modest, D. P., et al. (2013). Prognostic value of cetuximab-related skin toxicity in metastatic colorectal cancer patients and its correlation with parameters of the epidermal growth factor receptor signal transduction pathway: Results from a randomized trial of the GERMAN AIO CRC study group. *International Journal of Cancer, 132*(1), 236–245. https://doi.org/10.1002/ijc.27654.

Suh, C. H., Park, H. S., Kim, K. W., Pyo, J., Hatabu, H., & Nishino, M. (2018). Pneumonitis in advanced non-small-cell lung cancer patients treated with EGFR tyrosine kinase inhibitor: Meta-analysis of 153 cohorts with 15,713 patients: Meta-analysis of incidence and risk factors of EGFR-TKI pneumonitis in NSCLC. *Lung Cancer, 123*, 60–69. https://doi.org/10.1016/j.lungcan.2018.06.032.

Thodtmann, R., Sauter, T., Weinknecht, S., Weissbach, L., Blatter, J., Ohnmacht, U., et al. (2003). A phase II trial of pemetrexed in patients with metastatic renal cancer. *Investigational New Drugs, 21*(3), 353–358. https://doi.org/10.1023/a:1025480914273.

Wang, L., Wang, X., Zhang, G., Ma, Y., Zhang, Q., Li, Z., et al. (2021). The impact of pelvic radiotherapy on the gut microbiome and its role in radiation-induced diarrhoea: A systematic review. *Radiation Oncology, 16*(1), 187. https://doi.org/10.1186/s13014-021-01899-y.

Wang, K., Zhang, T., Lei, Y., Li, X., Jiang, J., Lan, J., et al. (2018). Identification of ANXA2 (annexin A2) as a specific bleomycin target to induce pulmonary fibrosis by impeding TFEB-mediated autophagic flux. *Autophagy, 14*(2), 269–282. https://doi.org/10.1080/15548627.2017.1409405.

Watson, R. A., De La Pena, H., Tsakok, M. T., Joseph, J., Stoneham, S., Shamash, J., et al. (2018). Development of a best-practice clinical guideline for the use of bleomycin in the treatment of germ cell tumours in the UK. *British Journal of Cancer, 119*(9), 1044–1051. https://doi.org/10.1038/s41416-018-0300-x.

Willard, P., McKay, J., & Yazbeck, V. (2021). Role of antibody-based therapy in indolent non-Hodgkin's lymphoma. *Leukemia Research Reports, 16*, 100275. https://doi.org/10.1016/j.lrr.2021.100275.

CHAPTER TWO

Chemobrain: A review on mechanistic insight, targets and treatments

Vanishree Rao[a], Rashmi Bhushan[b], Priyanka Kumari[b], Sri Pragnya Cheruku[a], V. Ravichandiran[b], and Nitesh Kumar[b,*]

[a]Department of Pharmacology, Manipal College of Pharmaceutical Sciences, Manipal Academy of Higher Education, Manipal, Karnataka, India
[b]Department of Pharmacology & Toxicology, National Institute of Pharmaceutical Education and Research, Hajipur, Bihar, India
*Corresponding author: e-mail address: niteshkumar43@gmail.com

Contents

Abstract

Chemo-brain refers to the thinking and memory problems that occur in cancer patients during and after chemotherapy. It is also known as cognitive dysfunction or chemo-fog.

Advances in Cancer Research, Volume 155
ISSN 0065-230X
https://doi.org/10.1016/bs.acr.2022.04.001

Risk factors include brain malignancies, either primary or metastatic, radiotherapy and chemotherapy, either systemic or brain targeted. There are various mechanisms by which chemo-brain occurs in patients post-chemotherapy, including inflammation of neurons, stress due to free radical generation, and alterations in normal neuronal cell process due to biochemical changes. While chemotherapy drugs that are non-brain targeted, usually fail to cross the blood-brain barrier (BBB), this is not the case for inflammatory cytokines that are released, which easily cross the BBB. These inflammatory neurotoxic agents may represent the primary mediators of chemobrain and include the pro-inflammatory cytokines such as interleukins 1 and 6 and tumor necrosis factor. The pronounced rise in oxidative stress due to continuous chemotherapy also leads to a reduction in neurogenesis and gliogenesis, loss of spine and dendritic cells, and a reduction in neurotransmitter release. Based on recent research, potential agents to prevent and treat chemo brain have been identified, which include Lithium, Fluoxetine, Metformin, Rolipram, Astaxanthin, and microglial inhibitors. However, more defined animal models for cognitive dysfunction are required to study in detail the mechanisms involved in chemo-brain; furthermore, well-defined clinical trials are required to identify drug targets and their therapeutic significance. With these focused approaches, the future for improved therapies is promising.

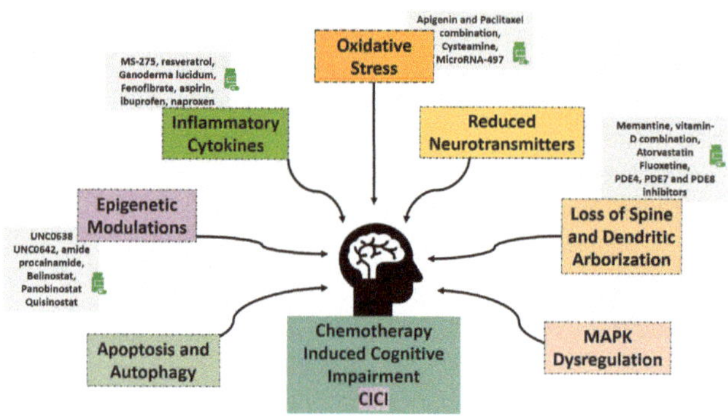

1. Introduction

Survival rates in cancer patients have drastically improved over the past decades as diagnostics and treatment modalities in this field rapidly advanced. However, increasing exposure to conventional and newer drugs/therapies is making patients more susceptible to developing chronic, irreversible adverse effects. This is largely a consequence of the systematic nature of cancer treatments that predominantly damage normal dividing cells (Nguyen & Ehrlich, 2020). When neuronal cells are affected, various neurological events corresponding to cognitive deficits can be observed (Milne, Nicklaus, Driscoll, Wang, & Zaharevitz, 1994). This effect of chemotherapy on the human brain was first published by Silberfarb and team in 1980. Subsequently, research in this area was further developed because rates of cognitive impairment in patients post-chemotherapy started increasing. This clinically significant cognitive decline seen in patients undergoing chemotherapy is termed ‘chemobrain’ or chemotherapy-related cognitive impairment (CRCI) (Ahles & Saykin, 2002).

In general, 70–75% of patients are estimated to develop cognitive decline, during and after prolonged chemotherapy (Fardell, Vardy, Johnston, & Winocur, 2011). 50% of the affected population have long-lasting cognitive effects, lasting for months and even years after the cessation of chemotherapy. This kind of persistent cognitive decline is not well tolerated because of the psychological stress induced in patients and the reduced quality of life, which ultimately affects daily living, education, work, and social interactions (Ho, 2015).

The exact cognitive functional domains that are responsible for causing the ‘chemobrain’ scenario are still not known; however, it is postulated that cognitive decline may be the result of functional and structural changes in the hippocampus and frontal lobes of the brain (Silberfarb, 1983). These domains control vital functions such as verbal skills, visuospatial memory, and concentration or attention. The degree of cognitive decline varies greatly among individuals; that is, high functioning individuals may experience minor setbacks whereas normal functioning aged patients may face serious issues and a great reduction in quality of life. While the transient decline is unlikely to cause significant changes in these individuals, the persistent decline will affect their quality of life. Transient declines are immediate and have the capacity to resolve spontaneously over a period of time (Ahles & Saykin, 2007).

Induction and rate of cognitive decline mainly depend on the type and duration of the drug regimen followed by the patient. A clear positive correlation was seen in early-stage breast cancer individuals between the duration of drug treatment and the rate of cognitive decline (Janelsins et al., 2011). Popular CMF (cyclophosphamide, methotrexate, and 5-fluorouracil) drug regimen showed a high incidence of cognitive decline when compared to individual drug treatments. Chemobrain situation was noted in the case of both individual and combination therapy, as it proved to be inevitable after chronic use. . Chemobrain was very evident in the case of methotrexate, a known neurotoxic drug that suppressed cell proliferation and affects the density of blood vessels in the hippocampus. When methotrexate was replaced by etoposide and adriamycin, significant cognitive decline was again noted stating that the chemobrain situation does not improve in combination treatment which was thought to show lesser cognitive effects in comparison with individual drugs (van Dam et al., 1998). Henceforth, a good understanding of chemotherapy-mediated cognitive decline can help in choosing the correct dosage and duration of treatment, aiding in targeted therapies and potentially minimizing negative effects on the quality of life. This chapter deals in detail with multifactorial mechanisms involved in the pathogenesis of chemobrain and current protocols or treatments in place to tackle this issue.

2. Impact of chemobrain in patients

The evolution of early cancer screening and detection procedures and improved chemotherapeutic interventions has led to a significant increase in worldwide cancer survivorship manifested by reduced recurrence rate, increased overall survival and, a gradual decline in the mortality rate. It is estimated that the total population of cancer survivors in the US will rise to 22.1 million in the next 10 years with 64% of survivors being more than 65 years of age. Despite efforts to treat cancer and extend patient survival time with chemotherapy, the long-term cognitive side effects of chemotherapy have become much more concerning to the survivors (Cranmer, 2019; Haines & Shapiro, 2019; ur Rashid, Rasool, Ali, Khan, & Martines, 2020). Termed "chemobrain" or "chemofog," the chemotherapy-related cognitive impairment (CRCI) is manifested by structural and functional alterations of the brain, impairing neurogenesis, thereby causing neurocognitive deficits. Initial demonstration of cognitive dysfunction was seen in breast cancer patients with 28% to 75% loss of cognitive function, months to years after being treated with chemotherapeutic agents. However, the results were considered insubstantial due to the small sample size or cross-sectional study designs. Further, longitudinal studies were conducted pre- and post-chemotherapy against healthy individuals and patients with breast cancer with no systemic chemotherapy. The experiments showed little variance in cognitive impairment or any recovery pattern in neuropsychological tests. Patients who received chemotherapy, however, manifested cognitive deficits with loss of learning ability, attention and memory, executive functioning, and processing speed. These insufficiencies can persist for a long or short duration thus affecting the quality of life (QoL) of cancer survivors portrayed by a change in perception of one's identity, their ability to interact socially, workplace adjustment, etc. The survivors exhibited a decrease in confidence, competence and, feeling of embarrassment. However, the deficits are more noticeable in the short term which recover with time in most patients (Henderson, Cross, & Baraniak, 2019; Moore, 2014).

It should be acknowledged that there are other potential factors such as age, pre- or post-menopausal status, psychological issues such as fatigue, depression, and anxiety that could play a significant role in escalating the risk of chemotherapy-induced cognitive decline (Sousa, Almeida, Bessa, & Pereira, 2020). Ahles et al. proved that factors such as age and pre-cognitive reserve in cancer survivors may contribute to post-cognitive debilities. Older patients showed decreased cognitive reserve in comparison to younger cancer survivors and showed worsened cognitive performance upon exposure to chemotherapy. It was ascertained that in the older population, the speed of processing and verbal competence was affected the most. However, further discontinuation of chemotherapy restored verbal ability (Ahles et al., 2010). Likewise, a study was conducted on chemotherapy-induced amenorrhea wherein a neuropsychological assessment of breast cancer survivors revealed the effect of menopause on CRCI. The study examined the effect of chemotherapy in pre- and post-menopausal breast cancer patients against healthy controls and proved that post-menopausal conditions failed to maintain visual and verbal cognitive function in cancer survivors. Additionally, the menopausal condition is also related to age which contributes to much lower cognitive performance (Conroy, McDonald, Ahles, West, & Saykin, 2013). Nevertheless, long-term follow-up evaluations of cognitive function suggest plausible partial or complete recovery from these deficits post-chemotherapy. However, which group of patients recuperates, and which subset of patients continue to endure through is still ambiguous. Contemplating the need to address these concerns, several randomized controlled trials have verified and recommended interventions. This includes mental stimulation by learning new skills like a musical instrument, cognitive rehabilitation delivered by web-based brain training programs, adaptation strategies such as journaling, exercise, and meditation therapy (Vardy, Bray, & Dhillon, 2017). Nonetheless, the effect of these interventions on CRCI is still unclear and needs more trials in the future that enables improved QoL in cancer survivors.

3. Symptoms and epidemiology

Post chemotherapy, it is quite common to see cancer patients complaining about problems with cognition or memory in particular.

Many longitudinal and cross-sectional studies that were performed on cancer patients reveal that working memory, short-term memory, speech or verbal fluency, and learning abilities are highly affected (Stewart, Bielajew, Collins, Parkinson, & Tomiak, 2006). Other cognition-related factors such as attention span, executive function, and visuospatial memory are less frequently affected as compared to working, short-term memory, and verbal efficiency. All the listed deficits are clinically subtle in nature and most cancer survivors demonstrate chemobrain signs on the lower side of the normal range. The signs related to chemobrain usually barely reach the pathological range (Horowitz, Suls, & Treviño, 2018; Jim et al., 2012). These mild signs usually remain undetected and consequently are generally underestimated by the attending clinicians, whereas major deficiencies resulting from strokes, traumatic brain injury, and Alzheimer's disease (AD) are detected by relying on tests that are exclusively designed for them. While estimating the percentage of individuals who suffer from chemotherapy-induced chemobrain, it was found to range from 17% to 76% (Ahles et al., 2002; Wefel & Schagen, 2012). This estimation is difficult to make because of the lack of objective definition and proper measurement of the chemobrain condition. When observed, the percentage of recovered patients who tested positive for indications of chemobrain is inversely proportionate to the duration after stopping the treatment. This suggests that slow recovery is happening over a period of time. In some subsets of cancer survivors, chemobrain symptoms are persistent for up to 10 years after the complete stoppage of treatment, indicating that chemobrain can also be a permanent condition in rare cases (Ahles et al., 2002).

In many of the structural studies conducted immediately after chemotherapy, an evident decrease in the density of grey matter in several regions related to the brain was observed, which included the temporal region, frontal cortex, cerebellar region, and right thalamus (McDonald, Conroy, Ahles, West, & Saykin, 2010; Zunini et al., 2013). Follow-up of these individuals showed partial recovery after a year. Decreased activation was noted in these regions during studies using functional magnetic resonance imaging (fMRI). The same regions showed elevated activation according to other studies and this is postulated to be a backup mechanism so that the patients can use more brain resources for a similar task (Zunini et al., 2013). All these pieces of evidence clearly support the conclusion that signs of chemobrain are not entirely psychological but have a biological basis too.

4. Mechanisms contributing to chemobrain

4.1 Oxidative stress

Oxidative stress is defined as an imbalance between the generation of reactive oxygen species (ROS) such as superoxide anion, hydrogen peroxide (H_2O_2), and hydroxyl radicals (OH) and the ability of the organism's antioxidative defence mechanisms to counteract their effects. The generation of ROS leads to disruption in the prooxidant to antioxidant equilibrium resulting in tissue damage (Chikara et al., 2018; Pisoschi & Pop, 2015). These reactive species are generated primarily from oxygen used during various metabolic processes and are found primarily in mitochondria, peroxisomes, and the endoplasmic reticulum (ER) (Chikara et al., 2018; Gorrini, Harris, & Mak, 2013). A rise in plasma Free superoxide radical levels due to chronic chemotherapy also oxidises Apolipoprotein A1 (ApoA1). Oxidated ApoA1 influences the production of TNF alpha which crosses the BBB to upregulate apoptosis and neural damage/death in the brain. Tumor formation and proliferation are also aided by oxidative stress, and have a role in all stages of carcinogenesis, either directly through DNA damage or indirectly through modulation of cell signaling. As a result, decreasing oxidative stress might contribute to chemo-prevention (Chikara et al., 2018).

4.1.1 Occurrence of ROS and its activity

ROS are chemical species including oxygen that have reactive characteristics. These include free radicals such as super oxide (O_2 •) and hydroxide (HO•), as well as non-radical compounds like hydrogen peroxide (H_2O_2). It is estimated that about 2% of the oxygen consumed by mitochondria is reduced to form superoxide; mitochondria are therefore considered to be a major source of ROS (Gorrini et al., 2013). Apart from mitochondria, peroxisomes facilitate the decomposition of H_2O_2 which is mediated by catalase and helps in the scavenging of ROS. Oxidation of fatty acids and flavin is also involved in the production of ROS. Also, the ER of the cell constitutes an oxidizing environment that favors disulfide bond formation and protein folding, thereby generating ROS through protein oxidation (Gorrini et al., 2013). At low concentrations, ROS act as signaling molecules that stimulate transcription factors to regulate cell growth, death, and gene expression (Pisoschi & Pop, 2015). ROS induce cell proliferation, survival, apoptosis, and cell differentiation, produce immunological responses, increase cell motility, and activate stress-responsive pathways. However, excessive free radical (ROS) production due to aerobic activities such as cellular respiration, phagocyte activation during microbial infections, intense physical activity, or on the impact of pollutants/toxins like cigarette smoke, alcohol, ionizing and UV radiation, pesticides, and ozone causes irreversible DNA damage such as single and double-strand breaks, base modifications, and DNA cross-links, which sets off chronic inflammatory pathways and cell death, thus promoting carcinogenesis if not corrected. Imbalance in the levels of ROS also contributes to the increased infiltration of TNF alpha which plays a major role in the formation of chemobrain. As a result, controlling cellular ROS is important for cellular homeostasis (Chikara et al., 2018).

Oxidative stress is the most common diagnostic factor leading to significant neurological changes. It has been shown that anti-neoplastic medicines create superoxide free radicals and also oxidize apolipoprotein A1 (Ren et al., 2019). TNF-α levels are raised by apolipoprotein A1, and cytokine alpha acts over its receptors on BBB before entering the brain and triggering neuronal death. Furthermore, it is important to note that oxidative stress is changed in many oncohematologic disorders and a huge proportion of chemotherapeutics work by generating oxidative stress in tumor tissue and even in brain tissue (Wahdan, El-Derany, Abdel-Maged, & Azab, 2020). Alterations in neuronal dendrites and spine density in brain tissue are morphological changes suggestive of oxidative stress damage following chemotherapy. In any event, the brain is vulnerable to ROS damage as it consumes a high amount of oxygen (20% of the body's total) and has a lower content of defensive antioxidant systems (Gaman, Uzoni, Popa-Wagner, Andrei, & Petcu, 2016). The brain contains high levels of iron and a large amount of polyunsaturated peroxidizable fatty acids, which act as pro-oxidants and may trigger autophagic cell death. The process of lipid peroxidation (LPO) causes the production of chemicals such as aldehydes (MDA), which has a direct effect on neuronal cell death. Chemotherapeutic effects on trophic factors and structural proteins of brain cells might potentially be a significant factor (Daiber et al., 2013). BDNF can promote the development of neural stem cells as well as the survival of neurons. Furthermore, increased BDNF levels are linked to increased long-term potentiation. Chemotherapeutic agents inhibit the synthesis of BDNF, which has negative consequences for hippocampus functions including memory and learning (Wang et al., 2021).

4.1.2 Antioxidants and their protective role

Of many antioxidant systems that can scavenge free radicals present in the body, glutathione (GSH) is a key intracellular antioxidant found throughout the brain at millimolar concentrations. GSH converts intracellular H_2O_2 to H_2O and O_2 after being oxidized to glutathione disulfide (GSSG) by glutathione peroxidase (GPx) (Joshi et al., 2010). Likewise, patients treated with Bortezomib, a novel proteosome inhibitor used in the treatment of refractory multiple myeloma, showed a high risk of developing sensory and reversible peripheral neuropathy through ROS production (Couch, Groteluschen, Stewart, & Mulkerin, 2003; Kannarkat, Lasher, & Schiff, 2007; Renouf & Gill, 2006). Adriamycin (ADR)/Doxorubicin is a chemotherapy drug that has a long history of being used to treat solid tumors. ADR is widely recognized for the excessive production of ROS in the body. Somnolence, often known as "chemobrain," is a frequent symptom in patients on long-term ADR therapy. Although it is commonly acknowledged that ADR creates significant levels of ROS in its mechanism of action against solid tumors, this mode of action has been linked to damage in numerous non-targeted organs, primarily the heart, limiting its use in cancer patients (Fig. 1) (Joshi et al., 2010).

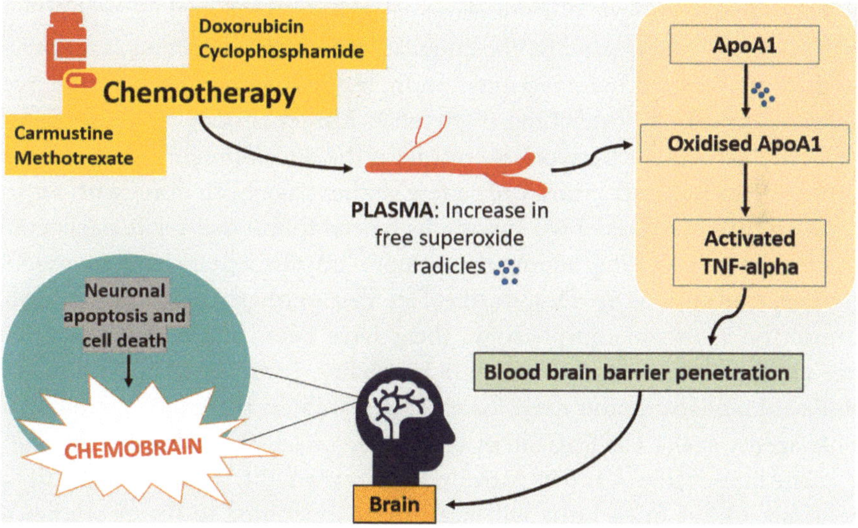

Fig. 1 Oxidative stress pathway. Increase in superoxide radicles oxidate ApoA1 which in turn activates TNF-alpha causing neuronal apoptosis and cell death. ApoA1: Apolipoprotein A-1, TNF-alpha: Tumor Necrosis Factor-alpha.

4.2 Absence of spinal and dendritic arborization

A majority of neuronal cells are polarized entities with a complicated structure that is necessary for their survival. Synaptic plasticity is necessary for learning, memory, and executive processes, and it is regulated by spines and dendrites (Kennedy, 2016). Spines and dendrites multiply early in life, then undergo regulated pruning during childhood and adolescence, before stabilizing at maturity. Nonetheless, in mature neurons, both features, particularly spines, remain active, enabling the flexibility necessary for reacting and adapting to new situations. Reduced spines and dendrites are caused by glutamate toxicity, decreased levels of amyloid oligomers, dysregulated calcium, the presynaptic release of neurotransmitters flow, cytoskeleton disruption, and disruption of the ubiquitin-proteasome system. Also, ageing causes progressive spine and dendrite losses. The cortex thins as a result of these losses, which might explain why the brain of cancer survivors has reduced grey matter integrity following chemotherapy treatment (Murillo Espitia, n.d.; Winocur et al., 2018). Several studies conducted in mouse models have found a decrease in dendritic and spinal complexity upon treatment with chemotherapeutic agents. Upon intravenous administration of cisplatin, the number of neuronal spines and dendritic branching in granule cells, as well as CA1 and CA3 pyramidal neurons in the hippocampus, were reduced. With fluorouracil, doxorubicin, and cyclophosphamide combination treatment, the cingulate cortex, an important component of the limbic system involved in emotion, learning, and memory, had fewer spines and dendritic branching (Nguyen & Ehrlich, 2020). Despite the fact that the microtubule network is critical for the development and stability of spines, dendrites, and axons, only a few studies have been done with regard to these consequences. Two studies have been found that relate paclitaxel's microtubule-stabilizing action to memory impairment in rodent models (Acharya et al., 2015). Despite the fact that neither study looked at the impact on neuronal morphology, there have been suggestions that these agents can be used to combat spine instability, particularly in Alzheimer's disease, highlighting the need for more research on the effects of microtubule agents in the CNS (Chiu et al., 2017).

The hippocampus and its surrounding areas are the focus of most current research. Other brain areas will need to be examined in future studies to determine if drug effects are widespread or unique to certain regions in the brain. In chemobrain animal models, for example, behavioral tests to evaluate cortical-based performance, such as attention and executive skills, have largely been absent. Only one study examined prefrontal brain damage induced by cisplatin using the 5-choice serial response time measure (Huo, Reyes, Heijnen, & Kavelaars, 2018). In mouse models of mental illnesses including schizophrenia and bipolar disorder, cortical-based tasks have been designed, updated, and used. Cognitive impairments in chemobrain should be studied using similar assays (Powell & Miyakawa, 2006).

The number of dendritic spines and the complexity of the dendritic tree have a strong relationship with neuronal connection and function. Dendritic spines are specialized protrusions seen in excitatory synapses in the brain. The shape and density of these small structures have been identified as key locations in synaptic plasticity, where they are required for memory and learning processes. Mushroom spines, thin spines, stubby spines, and cup-shaped spines have been postulated as the four kinds of dendritic spines based on their morphology, with mushroom spines being the most stable. The least stable spines are thin ones. Long-term potentiation (LTP) has been used to describe the possible relevance of dendritic spines in memory and learning (Kang et al., 2018). The spines with the most activity will remain preserved over time in this process (Kasai et al., 2010). The intricacy of the dendritic tree has also been shown to have a function in memory formation and consolidation. In ovariectomized mice, intravenous injections of cyclophosphamide or doxorubicin cause a decrease in the number of stubby dendritic spines in the dentate gyrus (DG) (Wang et al., 2018). In mice, intraperitoneal injections of doxorubicin reduce the density of dendritic spines and synaptic density in CA1 area neurons. In addition, treatment with cisplatin resulted in a decrease in the density of dendritic spines and dendritic branches in apical and basal dendrites in pyramidal neurons of the CA1 and CA3 areas of the hippocampus in both in vivo and in vitro models (Andres, Gong, Di, & Bota, 2014). Even five days following treatment, the synaptic damage caused by cisplatin was irreversible (Andres et al., 2014). In a juvenile mouse model, intrathecal methotrexate treatment significantly alters dendritic morphology and decreases the number of mushroom dendritic spines in the CA1, CA3, and dentate gyrus of the hippocampus (Kiffer et al., 2018). In another mouse model, intraperitoneal injection of docetaxel, doxorubicin, and cyclophosphamide (DAC) substantially eliminated dendritic spines of medial prefrontal cortex neurons, which is consistent with these findings (Ali et al., 2020). In CMF-treated mice, there was a substantial increase in the number of stubby spines and a decrease in the number of mushroom-shaped spines in the DG, as well as a significant rise in the density of dendritic spines in the CA1 and CA3 areas of the hippocampus. In CMF-treated mice, however, there were reductions in the number of branch points, dendritic endpoints, dendritic length, and dendritic complexity. Furthermore, intraperitoneal doses of fluorouracil resulted in a reduction in the density of mushroom dendritic spines in the DG, whereas the density of thin and stubby dendritic spines increased in an elderly mouse model. As a result, the apical dendrites of pyramidal neurons in the CA1 area had a lower density of mushroom spines but a higher density of stubby spines; meanwhile, the basal dendrites of pyramidal neurons in the CA3 region had a lower density of thin spines (Groves et al., 2017). Furthermore, fluorouracil has been shown to reduce dendritic complexity in DG neurons (Groves et al., 2017). These findings suggest that chemotherapy alters dendritic spine growth dynamics and that pyramidal neurons in the hippocampus have a propensity to reduce dendritic complexity. The decline in the number of mushroom dendritic spines and the increase in thin and stubby spines were two of the most noticeable changes. Because mushroom spines are the most stable and thin spines are the least stable, alterations in the density ratio of these spines might help explain chemotherapy-induced cognitive impairment (CICI)development (Fig. 2A) (Chidambaram et al., 2019).

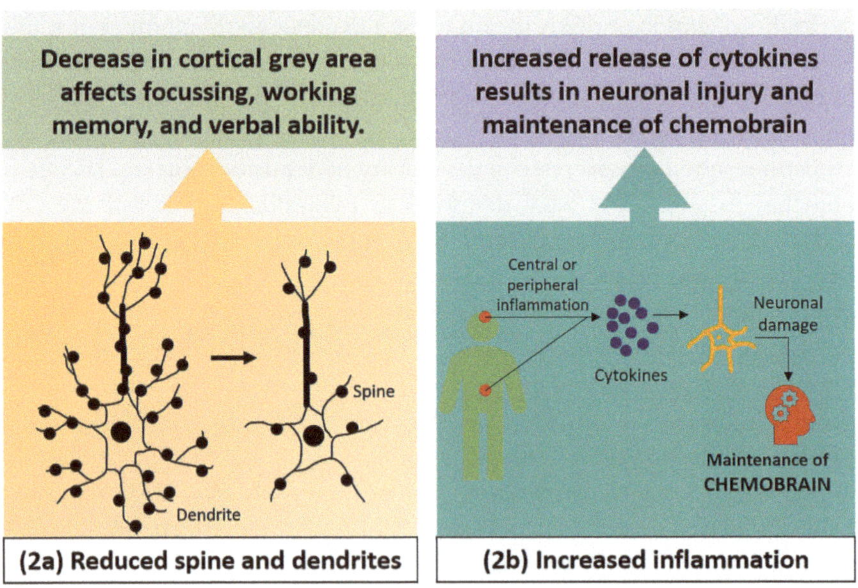

Fig. 2 Role of inflammatory cytokines and loss of dendritic arborization in causing chemobrain. (A) Loss of spines and dendritic structures due to chemotherapy affects focusing working memory and verbal ability. (B) Increased release of cytokines in the periphery or brain due to chemotherapy results in neuronal injury and maintenance of chemobrain.

4.3 Inflammatory cytokines

Chronic neuroinflammation is thought to be the cause of chronic cognitive dysfunctions in age-related and degenerative disorders, according to most experts. In the CNS, microglia, and astrocytes after activation, can directly generate cytokines. Peripherally produced cytokines, on the other hand, can enter the brain and cause local cytokine release. Cytokines can also weaken the blood-brain barrier, allowing additional cytokines and chemotherapeutic medicines to enter the body. The fact that total circulating cytokine contents can be regularly evaluated from blood or cerebrospinal fluid makes them useful as prospective biomarkers of translational importance (Gutmann, 2019; Wang et al., 2015).

Immunosenescence, or the progressive weakening of the immune system, is thought to be at the root of a persistent state of low-grade inflammation in older people. Increased levels of pro-inflammatory cytokines have also been linked to Alzheimer's disease. Increased contents of inflammatory cytokines are linked to worse cognitive function, as well as higher morbidity and death, under all circumstances. Surprisingly, increased peripheral cytokines were also found in cancer survivors who had received diverse chemotherapy regimens (El-Agamy, Abdel-Aziz, Wahdan, Esmat, & Azab, 2018).

The phenomenon of activated microglia releasing cytokines in the CNS has only been evaluated in a few studies in animal models. In the hippocampus, 5-fluorouracil with docetaxel, cyclophosphamide, and doxorubicin, increased the production of cytokines (IL-1β and IL-17) (Edwardson et al., 2017). Methotrexate stimulated microglia but had little effect on CNS cytokine levels (Seigers et al., 2010). On the other hand, chemotherapeutic agents like cisplatin, methotrexate, vincristine, and paclitaxel increased total peripheral cytokines that cause long-term discomfort (Grisold, Cavaletti, & Windebank, 2012). A high number of peripheral cytokines can also cross the blood-brain barrier (BBB), acting indirectly on the brain and activating microglia and astrocytes to produce more cytokines. However, the relationship between increased non-brain cytokines and neuro inflammation is not well understood (Nguyen & Ehrlich, 2020).

Demographical, physiological, psychological, pathological, and pharmaceutical variables have all been proposed as potential drivers of cognitive alterations in cancer patients. Notably, new research suggests that pro-inflammatory cytokines like interleukin (IL)-1β and IL-6 may play a role in chemotherapy-induced cognitive impairment (Cheung et al., 2015). The activation of cytokines in the central nervous system has been proposed as a mediator of "ill behavior" in individuals having high infection or malignancy, as well as the negative psychiatric consequences of interleukin and interferon therapy. It is well known that cytokines and other inflammatory indicators can cause a multitude of cancer-related symptoms such as tiredness, sadness, and stress, as well as cognitive abnormalities. In conclusion, cytokines are important players in the neuro-immuno-endocrine processes caused by cancer cells and cytotoxic treatment (Ren et al., 2019).

Chemotherapeutic drugs can cause brain damage indirectly through pro-inflammatory cytokine pathways since they are unable to pass the BBB due to their molecular size. Active transport allows cytokines to easily pass past the BBB and into the brain's circumventricular regions. The cytokines attach to endothelial receptors in the brain vasculature, causing the production of additional inflammatory mediators such as cell adhesion molecules, chemokines, nitric oxide (NOS), and prostaglandins, all of which obstruct the BBB's integrity. Cytokines cause local inflammation in the brain through oxidative mechanisms, particularly in the hippocampal region and other CNS areas with numerous receptors for cytokines. For instance, preclinical studies have shown that chemotherapy with doxorubicin may upregulate the amount of TNF in the circulation, which crosses the BBB and causes activation of nitric oxide synthase, resulting in an excess of reactive NOS and ROS that induce oxidative stress in the brain mitochondria. These responses might eventually manifest as clinical signs of cognitive impairment (El-Agamy, Abdel-Aziz, Ahmed, & Azab, 2019).

Both pre-clinical and clinical evaluations have confirmed the association of cytokine release to a variety of cancer-related symptoms and chemotherapy-induced cognitive impairment. (Taillibert, Voillery, & Bernard-Marty, 2007). Acute and chronic stresses have been found to raise IL-6 and TNF-α levels in the blood. Physical and psychological stresses encountered by patients' post-diagnosis, drug therapy, and chronic follow-up are all key variables that might cause cytokine levels to rise in the blood. Chemotherapy can also affect cytokine release by causing tumor cell death and collateral tissue damage. TNF-α, IL-6, MCP-1, IL-1β, and IL-10 levels are statistically significant in patients receiving anthracycline-containing chemotherapy for ovarian cancer, Hodgkin's disease, or breast cancer. In an animal model, adriamycin produces a peripheral rise in TNF-α, which is then identified in the brain (hippocampus and cerebral cortex), despite the fact that neither adriamycin nor its metabolites can easily pass the BBB (Cheung, Lim, Ho, & Chan, 2013). While cytokine release due to chemotherapy is not uniform among patients, they do appear to be linked to cognitive impairment. Multiple clinical investigations have shown that a normal dosage of drug therapy intensifies the cytokines like TNF-α, IL-8, IL-6, IL-10, and MCP-1 in individuals with cancer, with such changes being more pronounced in individuals with compromised cognitive function. While there is substantial evidence that alterations in the levels of cytokines circulating in the periphery are linked to cognitive decline in patients undergoing chemotherapy, the mechanisms of the same are still unknown (Vichaya et al., 2015).

A majority of systemically given chemotherapeutic drugs, on the other hand, do not penetrate the BBB. Rather than peripheral inflammatory cells, microglia produce the majority of cytokines in the brain, with a minor amount of cytokine release from astrocytes, oligodendrocytes, and neurons (Banks, Farr, & Morley, 2002). Chemotherapy does not appear to be the exclusive cause of cytokine alteration in the central nervous system. It has been postulated that communication between peripheral and central cytokines may have a greater influence in chemobrain than cytokines generated in CNS. According to this concept, cytokines in the periphery cause neuronal and supporting cells to produce central cytokines, which then affect neural plasticity (Licinio, Kling, & Hauser, 1998). According to recent studies, there is a significant bidirectional link between peripherally generated cytokines and cytokines in the brain, which occurs via (1) active trafficking over the BBB, (2) Passive crossing of the BBB at circumventricular organs through leaky areas, (3) peripheral cytokines stimulate central cytokine release via the brain's local inflammatory network (Myers, 2010). Cytokines also function as mediators that connect the neuroendocrine and immunological systems, with the ability to affect brain activity, neurotransmitter metabolism, neuronal and glial cell function, and neural repair and regeneration. Furthermore, an increase in cytokine levels may cause excitotoxic glutamate receptor-mediated damage, as well as changes in monoaminergic systems, GABA, acetylcholine, and nerve growth factors (BDNF- Brain-derived neurotrophic factor), all of which are linked to cognitive and neurological processes.

Chemotherapeutic damage may also result in an excessive generation of ROS and RNS species, resulting in increased oxidative stress via the nitric oxide (NO) pathway (Lee et al., 2004). Activated glial cells generate and release cytokines, which are linked to an increase in NO synthase induction (Harry & Kraft, 2008). The most common cause of DNA damage in neuronal cells is due to high levels of ROS and oxidative stress due to activation of the NO pathway (Reis, de Albuquerque, Gutierrez, Silva, & de Castro Faria Neto, 2017). This damage is linked to cognitive impairment and has been discovered in breast cancer patients mediated by the release of peripheral lymphocytes following treatment (Reis et al., 2017). As a result, additional characterization of cytokine expression following chemotherapy might provide new insights into the development of chemobrain, perhaps leading to the development of effective preventive or treatment techniques. In conclusion, there is substantial evidence that alterations in peripheral cytokines are linked to the development of chemobrain in the context of many, if not all, frequently used cancer chemotherapeutic treatments. The changes are generally heterogeneous, with modest increases in numerous cytokines occurring at the same time (Fig. 2B). (Meyers, Albitar, & Estey, 2005).

4.4 Neurotransmitters

Chemical mediators are released by sensory neurons' soma and are engaged in chemical transmission. The traditional transmitters glutamine, gamma-amino butyric acid (GABA), and ATP are among them, neuropeptides, galanin (GAL) gaseous neuro messengers (NO) other neuroactive molecules (BDNF) (Walker & Tesco, 2013).

It is observed that following nerve damage, the expression of these chemical mediators changes (up or down-regulation), which might explain neural plasticity. The effects of nicotine on different forms of memory (e.g., shorter long-term, prospective, working, olfactory, spatial, etc.) and other elements of cognitive performance are well documented. Acute nicotine improves cognitive function, and, specific brain areas have been linked to these benefits. Recent research has looked at how nicotine affects neuronal activity, synaptic plasticity, and gene expression in the prefrontal cortex and other cortical circuits, and how these changes could affect cognitive functioning (Kaplan et al., 2016).

The pathophysiological mechanisms that cause cancer-related cognitive impairment (CRCI) are constantly being unraveled. The idea that norepinephrine (NE), a neurotransmitter in the brain that is also produced peripherally through the adrenal glands and sympathetic nervous system, is a major component causing CRCI is briefly explored. If this hypothesis is true, it will not only give fundamental information on the neurochemical basis of CRCI, but it will also propose a number of neuropharmacological therapies using already available noradrenergic medications (Fitzgerald, 2021).

In most neurological diseases, neurotransmitter dysregulation (typically a decrease in availability) is found. A reduction in acetylcholine is common in Alzheimer's disease, which explains why acetylcholinesterase inhibitors make up three of the four FDA-approved medicines for treating the disease. It has been observed that as people age, they lose about 5–10% of their dopaminergic neurons every decade. An excess of glutamate causes calcium overload, which initiates the secondary damage phase of tramautic brain injury (TBI). The bulk of neurological medicines, for example, work by regulating neurotransmitters (Sheldrick et al., 2008).

Studies linking catechol-O-methyltransferase (COMT) polymorphisms to varying chances of developing chemobrain in cancer survivors provide more evidence for neurotransmitter involvement. COMT controls the metabolism of dopamine, epinephrine, and norepinephrine. The Val allele is linked to increased COMT enzymatic activity and, as a result, decreased cortical dopamine availability for the COMT Val158Met polymorphism (rs4680). As a result, cancer survivors who have at least one Val allele are more likely to develop chemobrain, probably because they have a lower dopamine reserve. In breast cancer patients, COMT variation, rs165599 G/G, similarly raises the incidence of chemobrain (Sheldrick et al., 2008; Small et al., 2011).

The amount of research related to alterations in neurotransmitter release due to chemobrain is limited. The glutamate absorption inside the cells in both the brain and the dentate gyrus was slower in mice when given a single injection of doxorubicin (Thomas et al., 2017). In the same way, carboplatin injections reduced dopamine release in the striatum (Kaplan et al., 2016). Following carboplatin injection, serotonin release in the raphe nucleus was similarly decreased (Kaplan et al., 2016). Although lower glutamate transporter expression has been linked to reduced glutamate transporter expression, the underlying processes are yet unknown and impaired exocytosis was implicated (Kaplan et al., 2016). Furthermore, enhanced acetylcholinesterase level was detected in the hippocampal region of doxorubicin-treated rats. After doxorubicin therapy, there was a decrease in choline level, which is the precursor of acetylcholine, showing that chemobrain is caused by a reduction in cholinergic activity (Jarmolowicz et al., 2019; Thomas et al., 2017).

Doxorubicin can induce dysregulation of neurotransmitter synthesis and release in the brain, according to several animal studies (Keeney et al., 2018). Acetylcholine (ACH) is a neurotransmitter in the cholinergic nervous system that promotes long-term potentiation (LTP) and hence supports brain processes (Hafner, 2009). Phosphatidylcholine (PtdCho) is dissolved by phospholipase D (PLD) during acetylcholine composition, releasing choline, which is then acetylated by choline acetyltransferase (ChAT) to produce acetylcholine. Following doxorubicin treatment, the levels of PLD, ChAT activity, and choline-containing compounds in the hippocampus area of mice decreased substantially, indicating that ACH synthesis had been exhausted. Furthermore, doxorubicin-induced oxidative stress enhanced the activity of acetylcholinesterase (AChE) mediated by reactive oxygen species (ROS) (Evens & Eschiti, 2009).

Choline-containing material changes are linked to membrane turnover (synthesis and degradation of phospholipids) and have been linked to myelin damage as a result of chemotherapy. Increased TNF-α may reduce PLD activity, preventing PtdCho production. TNF-α is also linked to lower phosphatidic acid levels, implying a link between phospholipase and TNF-α expression. When PLD is inhibited, the generation of cytokines, such as TNF-α, is reduced. Phosphatidic acid, a PLD pathway intermediate, increases $Ca2^{+}$ mobilization and has growth factor-like action, which aids in the reduction of doxorubicin-induced mitochondrial dysfunction in mice. PLD's enzymatic activity is essential for cell survival, and structural damage to PLD or decreased PLD activity can trigger apoptosis (Arafa & Atteia, 2020).

Doxorubicin can change glutamate levels in the synaptic gap in addition to influencing acetylcholine metabolism. Glutamate clearance was decreased by doxorubicin as evidenced by a decrease in glutamate absorption in the frontal cortex of mice. It was proposed that reduced glutamate clearance leads to decreased expression of glial transport proteins or increased glutamate synthesis from neurogliocytes, particularly in astrocytes, in this setting (Thomas et al., 2017). As previously stated, TNF-α induced astrocyte activation results in significant glutamate release (Nikolic et al., 2018). Glutamate can diffuse to and interact with NMDA acceptors in the synapse when glutamate concentrations are high. Increased calcium-dependent excitability and reduction of BDNF composition are caused by the activation of extrasynaptic NMDA receptors, resulting in synaptic plasticity loss and increased neuronal death. This might explain why doxorubicin reduced BDNF and its receptor tropomyosin receptor kinase B expression (TrkB) (Ongnok, Chattipakorn, & Chattipakorn, 2020).

Furthermore, doxorubicin injection significantly decreased the levels of two monoamines linked to cognitive function: serotonin (5-HT) and dopamine (DA) (Du et al., 2021). Through 5-HT1A receptor-mediated inhibitory regulation, 5-HTergic neurons play a major role in controlling hippocampus synaptic plasticity. 5-HT depletion impairs hippocampus-dependent declarative memory and results in poor performance in a new

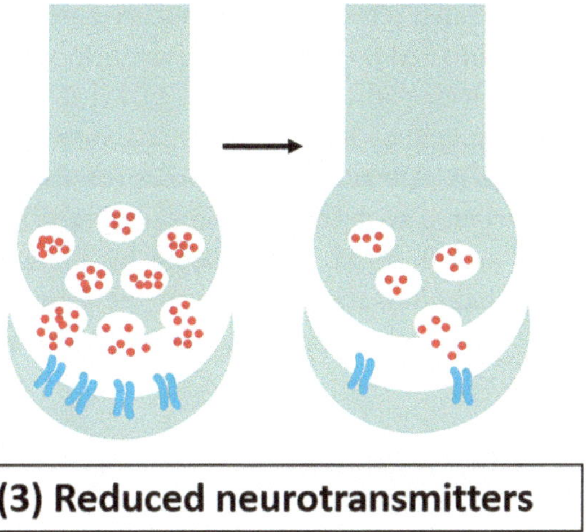

Fig. 3 Reduced neurotransmitters. Reduction in neurotransmitter release at synaptic junctions due to chronic chemotherapy impairs cognition.

object identification challenge. By stimulating the D1/D5 receptor during encoding, doxorubicin facilitates the acquisition of long-term memory in the hippocampus (Roggenhofer et al., 2010). Changes in neurotransmitter levels cause anomalies in nerve signal transduction, resulting in cognitive impairment (Fig. 3) (Cardoso et al., 2020; Du et al., 2021).

4.5 Apoptosis and autophagy

Prolonged chemotherapy with drugs like doxorubicin was found to damage progenitor lysosomes of neurons, impair pathways related to neuronal degradation, increase autophagy and hinder the elimination of p62 protein that acts as an autophagic marker (Tangpong et al., 2007). When these neurons were studied under an electron microscope, vacuoles were seen to be accumulated along with mitochondrial autophagosomes and lipid drops (Dong & Czaja, 2011). This phenomenon of degradation disorder heavily impacts the functioning of neurons and results in severe cognitive impairment in patients (Du et al., 2021; Manchon et al., 2016).

The chemotherapy-induced apoptosis process is called death receptor-mediated apoptosis because it depends on the exogenous apoptotic process in primary cortical neurons. Drugs like doxorubicin mediate the direct apoptotic process that is via an exogenous apoptotic pathway (Pistritto, Trisciuoglio, Ceci, Garufi, & D'Orazi, 2016). Fas-Fas ligand (FasL) connections are increased in the presence of doxorubicin, and this leads to the assembly of Fas-associated protein. The fas-associated protein interacts with the death domain (FADD) and this interaction signals the initiation of the exogenous apoptotic pathway (Du et al., 2021). The endogenous apoptotic pathways are initiated by high cellular stress, increased developmental signaling, cellular DNA damage, and lessening of survival factors. This pathway is in close relation with the mechanism of oxidative stress in mitochondrial apparatus and it is mainly regulated by Bcl-2 protein families. Abnormally activated apoptosis highly impacts the growth of neurons and readily reduces the neuronal count which leads to cognitive impairment (Fig. 4) (Walczak, 2013).

4.6 MAPK signalling pathway

Chemotherapy is also known to affect a few key cognitive and memory-related kinase pathways (Das et al., 2020). For example, doxorubicin activates p38 MAPK and ERK (extracellular signal-regulated kinase) where two different kinases perform two different roles in that synaptic inhibition is mediated by MAPK and synaptic facilitation is promoted by ERK (Du et al., 2021).

The serotonin inhibition in the sensory neurons of the hippocampus by chemotherapy facilitates and promotes Phe-Met-Arg-Phe-NH2 (FMRFa)-mediated chronic depression (Liu, Zhang, Coughlin, Cleary, & Byrne, 2014). This suggests that the chemotherapy drugs have a role in blocking learning-related activities in the hippocampal region. The long-term

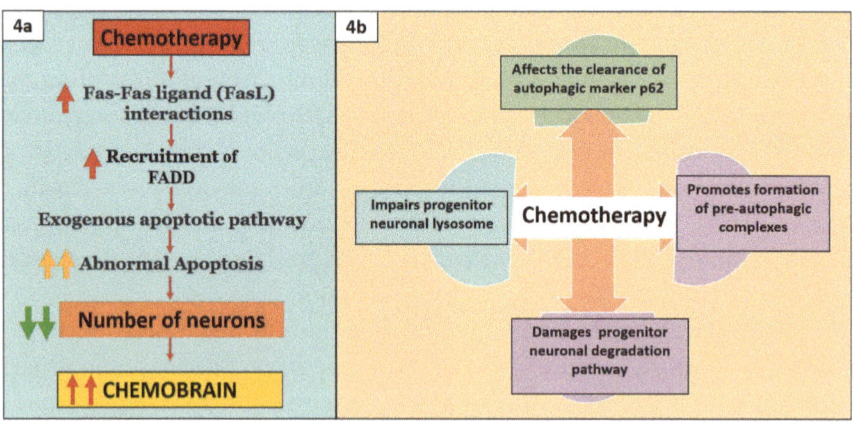

Fig. 4 Apoptosis and autophagy. Dysregulated activation of apoptosis and autophagic pathways resulting due to continuous chemotherapy greatly affect neurons and cause chemobrain.

facilitation (LTF) and its deviation were noted to be corrected by inhibitors of p38 (El-Agamy et al., 2019). This inhibition was superior to ERK effects. All of these studies support that the dominant activation of the p38 MAPK pathway induced by chemotherapy drugs has detrimental effects on long-term memory.

These chemotherapy drugs are also found to inhibit CREB2 (cAMP response element-binding protein 2) which is known to promote chronic depression. The long-term memory function and plasticity are dependent on ARC protein synthesis that is modulated by the ERK pathway. NMDA (n -methyl-d-aspartate) receptor antagonists and α-amino-3-hydroxy-5-methyl-4-isoxazolepropionic acid (AMPA) receptor antagonists inhibit neuronal activity. Studies show that these AMPA and NMDA receptor antagonists play a role in eliminating the induction of DNA double-strand breaks caused by chemotherapy like doxorubicin (Liu et al., 2014). This supports the contention that the neurotoxicity induced by chemotherapy like doxorubicin is highly dependent on an individual's neuronal activity (El-Agamy et al., 2019; Liu et al., 2014).

4.7 Genetic factors and epigenetic modulations

Several studies have shown long-term cognitive and memory impairment in cancer survivors after chemotherapy. The main factors responsible for this include genetic predisposition and decreases in dopamine and serotonin levels. In the case of breast cancer survivors, it was found out that the decline in neurocognitive functioning is related to or caused by single nucleotide polymorphism of three factors/genes (Single nucleotide polymorphisms (SNP) of apolipoprotein E (APOE), BDNF and catechol-O-methyltransferase (COMT)) (Cheng et al., 2016). The gene encoding for APOE protein is responsible for cholesterol transport in the blood. APOE e4 variation is associated with a decline in cognitive function and induction of Alzheimer's disease. Cancer survivors who are APOE e4 carriers showed a reduction in spatial orientation and visual memory along with a reduction in hippocampal volume directly impacting the memory function (Koffie et al., 2012).

BDNF that is present in the hippocampus and cerebral cortex helps in neuronal repair, and growth of axons along with maintaining synaptic plasticity. BDNF plays a significant role in the reduction of hippocampal volume causing a decline in memory function with increment in age. Val 66 mutation of BDNF is linked to cognitive depression in breast cancer survivors

(Ahles et al., 2003). In contrast, survivors with BDNF met allele were likely to be protected from cognitive impairment. In cancer survivors with Val 66 mutation, it was noted that there was significant plasma BDNF level reduction correlating directly with the level of cognitive impairment, in contrast to Met genotype survivors, with no changes in BDNF plasma levels (Savitz, Solms, & Ramesar, 2006).

The COMT gene helps in regulating the degradation of catecholamines. Polymorphism of this gene is associated with chemotherapy-induced cognitive impairment. COMT is the main modulator of dopamine, an important neurotransmitter involved in neurotransmission. Val 158 Met polymorphism is known to decrease dopamine availability. The patient's COMT Val allele showed decreased attention and motor functions compared to COMT-Met allele patients (Dooley, Ganz, Cole, Crespi, & Bower, 2016; Erickson et al., 2010).

Recent research has shown mediation of cellular memory by global DNA methylation and the phenomena of epigenesis by directly changing the homeostasis of the cytokine milieu. Epigenesis causes modifications of DNA without altering the nucleotide sequence, thereby altering the gene expression and not the sequence. One such array of modifications is DNA methylation and hydroxymethylation (Stenvinkel et al., 2007). Epigenetic modification of genes is involved in depression and Alzheimer's disease. DNA methylation plays a critical role in learning and memory formation. One such example is in patients with drug abuse, treated with conventional therapy, who have demonstrated long-term epigenetic modifications that affect the limbic system making them refractive to conventional therapy and prone to relapse (Heim & Binder, 2012). This study shows that cancer survivors face a wide range of biological deficits after chemotherapy. Even though various drugs with different mechanisms are used as chemotherapeutic agents to treat cancer, a stereotype is noted with respect to cognitive dysfunction post-chemotherapy. It should also be noted that the short-term alterations in the cytokine milieu result in chronic cognitive decline (Wang et al., 2015). All the above factors prove the involvement of a variety of triggering factors, culminating in a common upstream pathway causing cognitive impairment. It can also be speculated, that epigenome reprogramming by chemotherapy leading to cognitive impairment is a common pathway that is modulated by chemotherapeutic agents such as cyclophosphamide, methotrexate, 5-Fluorouracil increased histone H3 acetylation and decrease histone deacetylase activity (Reiser, 2019). This leads to cognitive impairment due to a reduction in hippocampal neuronal cell proliferation in the hippocampus (Reiser, 2019). In a recent study, it was noted that a subset of genes that have undergone transient methylation after chemotherapy persisted only in patients with persistent neuropsychological symptoms (Ahles & Saykin, 2007). Further research is required to know how exactly chemotherapy affects epigenetic stages (Adwan & Zawia, 2013).

5. Targets and treatment options

5.1 Targeting oxidative stress

Increased alteration of oxidative balance is the most critical pathway for the onset of chemobrain.

In determining DOX-induced cognitive deficits, oxidative stress has been found to be a crucial factor in facilitating its anticancer action. DOX contains a quinone functional group that experiences a redox cycling shift, in which quinone receives a single electron and converts into an unstable semi-quinone intermediate, which is then reconverted to parent quinone chaperoned with enormous amounts of ROS, causing numerous vital biomolecules to be disrupted (Joshi et al., 2010). DOX-mediated neurotoxicity is thought to be caused by increased protein oxidation, LPO, reduced GSH levels, and changes in antioxidant enzyme levels, along with a drop in manganese superoxide dismutase (MnSOD), one of the most important mitochondrial antioxidant enzymes MnSOD. It has been reported that nitration of MnSOD reduces its enzymatic activity and as a consequence, mitochondrial respiration (Daiber et al., 2013). Furthermore, generated ROS interacts with nitric oxide to produce peroxynitrite, a highly toxic RNS that nitrates tyrosine residues in proteins, preventing post-translational modifications and, as a result, flexible signaling pathways (Du et al., 2021).

In brain cells, autophagy has a protective effect, and its deregulation has been linked to several neurological conditions. Previous studies revealed that DOX therapy disrupted the autophagy-lysosome system in cultured neurons, resulting in an increase of autophagosomes and disrupted mitochondria (Ongnok et al., 2020). Furthermore, animals administered with the liposomal formulation of DOX had a significant accumulation of lipofuscin, a diagnostic of neuronal senescence that implies past exposure to significant oxidative disruption (Li & Caeyenberghs, 2018). Additionally, current findings have shown that pexophagy, selective autophagy of peroxisomes, is downregulated, resulting in the build-up of peroxisomes and overproduction of ROS. Thus, the foregoing findings emphasize that, despite

DOX's limited ability to penetrate BBB, a small quantity that does permeate may be sufficient to cause neurotoxic damage (Allegra et al., 2020).

Moreover, DOX-induced oxidative stress increased ROS-mediated acetylcholinesterase activity (AChE). The brain consumes 20% of overall body oxygen uptake for ATP generation through mitochondria while having only 2% of body weight. As a result, the brain is particularly sensitive to oxidative stress (Ongnok et al., 2021). Mitochondria are primary organelles that govern calcium absorption and redox signaling under the biological system, as well as being a supplier of ATP. Implications of doxorubicin administration on mitochondrial energy metabolism in rat hippocampus were discovered, as demonstrated by increased mitochondrial ROS levels and calcium retaining imbalance. (Lomeli, Di, Czerniawski, Guzowski, & Bota, 2017). In addition, DOX injection reduced energy metabolism in both bilateral cortices and hippocampus in vivo. These issues could be caused by the emergence of mitochondrial permeability transition pores (mPTP). This mPTP is a pore formed by the recruitment of three protein subunits in mitochondrial intermembrane space: VDAC, adenine nucleotide translocase (ANT), and cyclophilin D (CyP-D). Glycogen synthase kinase-3 (GSK-3) is triggered by ROS, which causes CyP-D to be phosphorylated and become functional. (C. Wang et al., 2021). Calcium is required for ANT conformational change and mitochondrial membrane depolarization. These events culminated in the formation of mPTP, which caused upregulated solute leakage and mitochondrial enlargement, which is a phenotypic characteristic of damaged mitochondria. (Gaman et al., 2016).

5.2 Targeting neurogenesis

Adult neurogenesis takes place largely in specialized areas such as the sub-granular zone (SGZ) of the hippocampus's dentate gyrus, subventricular zone bordering lateral ventricles, and striatum (Liao et al., 2018). In SGZ, neural precursor cells (NPC) divide to replenish themselves or to produce immature cells that can develop into neurons and glial cells. Neurogenesis acts as a buffer for the restoration of neurons lost owing to regular wear and strain. Neurogenesis subsequent to trauma is critical for the recovery of cognitive skills in situations of acute insults, such as a stroke. Neurogenesis declines with age and is a common feature of neurodegenerative diseases (Ramalingayya et al., 2017). It slows as people get older due to a decrease in NPCs. Reduced neurogenesis raises the risk of developing new cognitive problems or aggravating existing ones. Considering memory difficulties are a typical chemobrain sign, it is not surprising that decreased neurogenesis is the most researched chemobrain etiology (Reiser, 2019).

Since the hippocampus is involved in recollection, construction, and spatial processing, adult-born neural connection into circuitry is a crucial role in hippocampal neurogenesis and therefore, adequate control of cognitive abilities. (Nguyen & Ehrlich, 2020). Despite fact that DOX is limited to the peripheral nervous system, treated mice demonstrated a substantial reduction in neurogenesis, as shown by a large drop in the number of cells expressing neuronal markers. (Christie et al., 2012). DOX in conjunction with cyclophosphamide was also shown to decrease cell viability in the dentate gyrus and subgranular zones of rats. According to a recent study, astrocytic cell activation and the concurrent release of inflammatory mediators is a key factor in creating an environment that is not conducive to neuron longevity (El-Agamy et al., 2019). Furthermore, animals missing TNF receptor-1 (TNFR1) displayed enhanced subgranular zone proliferation, indicating that the anti-neurogenic effects of TNF-α are mediated by TNFR1. Scientists reported chronic inflammation hindered the recruitment of adult-born neurons into the hippocampal network that encodes contextual data, implying that neuroinflammation affects not only hippocampal proliferation, differentiation, and survival but also the integration of new neurons into pre-existing networks. (Liao et al., 2018).

IP/IV injections of a variety of drugs, including methotrexate, 5-FU, cyclophosphamide, dox, docetaxel, paclitaxel, cisplatin, and thioTEPA, were found to cause cognitive deficits from a few days to 20 weeks after injection. Diverse neurogenesis-related protein markers were diminished, though not in all investigations (Chiu et al., 2017). BrdU and Ki-67 are markers for proliferating cells; doublecortin (DCX) is a marker for NPCs and immature new neurons, and NeuN is a marker for mature neurons. Several studies discovered a decrease in the number of BrdU- or Ki-67-positive cells, implying that proliferating precursor cells were directly affected (Sobecki et al., 2016). 5-FU treatment; on either hand, did not impact the number of Ki-67-positive cells but did decrease the number of DCX-positive cells, revealing that the early maturation phase was disrupted (Christie et al., 2012). Likewise, cyclophosphamide or dox administration had no effect on the number of BrdU-positive cells, but it did lower the number of DCX-positive and BrdU/NeuN cells that were doubly tagged, suggesting that both early and late maturation stages were impacted. A variety of protein markers will be evaluated in future research to identify which stages of neurogenesis are impacted (Allegra et al., 2020).

In addition, to promote neurogenesis, BDNF, is secreted. Decreased levels of BDNF in blood have been linked to cognitive decline in cancer individuals (Butterfield, 2014). Following injections of 5-FU or DOX into the hippocampus of a rat model, BDNF levels in the hippocampus dropped. Processes behind BDNF depletion and how this impacts neurogenesis are unknown. Moreover, methotrexate therapy has been shown to reduce cortical BDNF mRNA and protein expression, suggesting that BDNF transcriptional regulation is a component. While the majority of research has not focused on other neurogenic areas, except in the hippocampus, which might be at risk (Cardoso et al., 2020). Cell division was shown to be reduced in SGZ, SVZ, and corpus callosum after systemic treatment by cisplatin, cytarabine, or 5-FU. Impaired neurogenesis in many areas might cause symptoms other than memory lapses. Olfactory impairment due to decreased SVZ neurogenesis, for example, is an early sign of Alzheimer's disease that precedes the onset of full dementia. However, neurogenesis can be impacted gradually, resulting in no apparent symptoms but an increased risk of cognitive decline later in life for survivors (Chiu et al., 2017).

Chemotherapy has been linked to a higher risk of dementia in older age. The adult hippocampus is still flexible and sensitive to changes in the environment, making it a suitable candidate for therapy. In rat models of ageing, AD, depression, and chemobrain, physical activity and environmental diversity improve neurogenesis and relieve symptoms. Fluoxetine, reboxetine, and tranylcypromine, among other antidepressants, promote neurogenesis in the adult hippocampus (Christie et al., 2012). Lithium, a mood stabilizer used to treat bipolar illness and as a supplement to depression therapies, also boosts neurogenesis. In rat models of chemobrain, fluoxetine and lithium both decrease cognitive impairment (Daiber et al., 2013). As a result, antidepressant medications may be beneficial in treating both chemobrain's cellular abnormalities and behavioral symptoms. Transplanting neural stem cells into various brain areas, such as the hippocampus, frontal cortex, and striatum is now being researched as a way to restore neurons lost in neurological conditions (Doll, Rellick, Barr, Ren, & Simpkins, 2015). Cells injected into mouse models, either of human or rodent origin, survive, integrate and develop into neurons and glia in the recipient's brain, alleviating cognitive deficits (Du et al., 2021). In a novel discovery, brain residual mesenchymal stem cells have the ability to transmit healthy mitochondria to damaged DCX^+ neuronal progenitors by forming actin-based tunneling nanotubes (TNTs) across cells. (El-Agamy et al., 2019). After cisplatin treatment, injured mtDNA and mitochondrial DAMPs, were released from NSCs and recognized by MSCs via toll-like receptors (TLRs), suggesting that damaged mtDNA and mitochondria, which are DAMPs, were released from NSCs and recognized by MSCs via TLRs. When MSCs were pre-incubated with an inhibitor of f-actin polymerization, protection against MSCs was lost, resulting in an increase in NSC mortality equal to the group given cisplatin monotherapy (El-Agamy et al., 2018).

5.3 Targeting neuroinflammation

Despite the fact that ROS plays a critical role in causing DOX neurotoxicity, the aforementioned method is effective for delivering ROS into all bodily organs except the brain since DOX is nearly impossible to penetrate the BBB (Fernández-Rojas et al., 2015). As a consequence, this suggests that CNS has a mechanism that does not require the cycling of redox. In particular respect, it has been discovered that ROS produced by the redox cycling process activates NF-κB, which then increases the transcription of multiple target genes, including inflammatory cytokines such as TNF-α, IL-1, and IL-6 (Gräff et al., 2012). Across multi-agent chemotherapy regimens, scientists found a significant increase in TNF-α plasma levels in cancer patients 6 hours after DOX delivery. Furthermore, DOX administration leads to a rise in plasma and brain TNF- levels in wild-type mice and animals deficient in the inducible form of nitric oxide (iNOSKO) (Gu et al., 2018).

Inflammatory insult might also be related to DOX-induced oxidative modification of apolipoprotein A1 (Apo-A1). Apo-A1 A1 plays a variety of roles in normal body function, including regulating inflammatory response by suppressing the production of inflammatory cytokines, primarily TNF-α, by upregulating the production of tristetraprolin, an mRNA-destabilizing protein that causes TNF-mRNA to disintegrate, limiting TNF-translation (Joshi et al., 2010).

Inflammation appears to be a ubiquitous feature, with cytokines being the primary culprit in DOX-induced cognitive deficits. Significant evidence links astrocytes to a continuous feedback loop of inflammation between the periphery and CNS (Li et al., 2018). Astrocytes are neuroglial cells that regulate synaptic transmission, neurovascular coupling, and cerebral microcirculation, as well as establishing and maintaining BBB's integrity. According to evidence, astrocytes are a critical part of neuroinflammation. Reactive astrogliosis, a reaction of astrocytic cells to persistent neuroinflammation, turns the beneficial astrocytic activity into a harmful influence, resulting in excess release of inflammatory mediators. This results in a vicious, never-ending circle of inflammation, which promotes oxidative and nitrosative damage to brain cells and decreases neuroplasticity (Li et al., 2018).

Distantly produced cytokines, on other hand, can enter the brain and cause local cytokine release (Liao et al., 2018). Cytokines can also weaken the protective BBB, allowing added cytokines and chemotherapeutic agents to enter the body. In fact, the peripheral and central cytokine levels can be regularly evaluated from blood or CSF making them useful as prospective diagnostics of translational importance.

Immunosenescence, or progressive weakening of the immune system as people age, underpins a persistent state of low-grade inflammation (Lim, Joung, Yu, Shim, & Kim, 2016). Increased levels of proinflammatory cytokines are also linked to Alzheimer's disease and traumatic brain injury. Increased levels of inflammatory cytokines are linked to worse cognitive function, as well as higher morbidity and death, under all circumstances (Lomeli et al., 2017). Surprisingly, increased peripheral cytokines were also found in cancer survivors who had received diverse chemotherapy treatment regimens. Researchers have looked at how activated microglia and astrocytes in the CNS produce cytokines directly (Marques-Aleixo et al., 2016). In the hippocampus, 5-FU and a combination of docetaxel, DOX, and cyclophosphamide elevate cytokines. Methotrexate stimulated microglia but had an insignificant effect on CNS cytokine levels. Certain chemotherapy medications, such as methotrexate, cisplatin, oxaliplatin, paclitaxel, and vincristine, on other hand, increased peripheral inflammatory cytokines to cause chronic discomfort (Marques-Aleixo et al., 2018). Increased peripheral cytokines may also cross the BBB, acting directly on the CNS and activating microglia and astrocytes to produce more cytokines. The relationship between increased peripheral cytokines and CNS inflammation on other hand is not well studied (Mitra, Siddiqui, & Khandelwal, 2015).

Neuroinflammation is still a major cause of neurodegeneration, and it may be controlled at both peripheral and central levels. NSAIDs like aspirin, ibuprofen, and naproxen, which are available over the counter, have been studied extensively to see whether they might help prevent and control Alzheimer's disease (Nguyen & Ehrlich, 2020). Outcomes were quite varied, ranging from good to neutral to detrimental. Natalizumab, which inhibits immune cell migration across the BBB, is among the drugs authorized to treat multiple sclerosis, a condition marked by severe inflammation and disruption of BBB (Nguyen, Fischer, & Ehrlich, 2021). Because these medicines have significant adverse effects, these may exceed the advantages of decreasing chemotherapy-mediated cognitive impairment.

Directly targeting astrocytes and microglia can also be used to combat neuroinflammation (Ongnok et al., 2020). According to new research, CSF1R (colony-stimulating factor 1 receptor) inhibitor PLX5622 can prevent methotrexate-induced cognitive loss. In a mouse model of Alzheimer's disease, PLX5622 decreased inflammation and restored cognitive impairments. Microglial activation is inhibited by minocycline, a popular antibiotic (Ongnok et al., 2021). Beyond two years, minocycline may not slow the course of cognitive deterioration in patients with moderate Alzheimer's disease. Such findings imply that even more precise microglia or astrocyte targeting is needed to relieve cognitive impairment without causing undesirable consequences (Ramalingayya et al., 2017). Since chemotherapy typically compromises white matter tracts, increasing neurogenesis and myelination is another treatment option. Methotrexate-induced myelin destruction and intellectual disability were shown to be rescued by LM22A-4, a TrkB agonist that increases OPC differentiation and neurogenesis (Reiser, 2019).

5.4 Targeting arborization

Synaptic plasticity relating to memory, and executive functions are regulated by spines and dendrites (Seo, Klauck, Efferth, & Panossian, 2019). Hippocampus and its surrounding areas are the focus of most current research (Wahdan et al., 2020). In chemobrain animal models, for example, behavioral tests to evaluate cortical-based performance, such as attention and executive skills, are absent. Only one research program examined prefrontal brain damage induced by cisplatin using the 5-choice serial response time measure. In mouse models of mental illnesses including schizophrenia and bipolar disorder, cortical-based tasks have been designed, updated, and used (Srivastava & Singh, 2020).

In adulthood, spine development and stability are still extremely dynamic and susceptible to environmental changes. Glutamate receptors, notably NMDARs and AMPARs, are important in the development and stability of the spine. For example, NMDAR inhibitor memantine and non-competitive NMDAR antagonist dextromethorphan restored cognitive impairment. Other NDMAR and AMPAR regulators, including ketamine and benzamides, can stimulate spine formation or modify dendritic arborization to relieve depressive symptoms and cognitive impairment (Suuberg, 2018). In the context of chemobrain, restoring a loss in spines might be a desirable goal, but glutamate modulators will need to be thoroughly investigated. Because glutamate excess, which is common in TBI, can disrupt neurons and decrease spinal and dendritic complexity, special care is required.

Calcium signaling is also required for appropriate spine and dendritic integrity(Vichaya et al., 2015). Dysregulated calcium/cAMP signaling, for example, during times of stress, might result in spine instability and atrophy. Phosphodiesterase inhibitors, such as rolipram and ibudilast, which control cAMP levels, restored cognitive decline caused by docetaxel and oxaliplatin, respectively. Protein kinase C hyperactivation is linked to dendritic integrity loss and cognitive loss in ageing and chronic psychological stress, implying that PKC suppression might be a viable treatment approach (Wahdan et al., 2020).

5.5 Targeting genetic factors

Chemotherapeutic medicines have been linked to epigenetic alterations (Roberti, Valdes, Torrecillas, Fraga, & Fernandez, 2019). Agents such as daunorubicin and etoposide increased the expression of the MDR1 gene in CEM-Bcl2 and SW620 cell lines. Chemotherapy resistance in cancer patients is frequently caused by the MDR1 gene. MDR1 promoter DNA methylation was reduced in cancer cell lines treated with these chemotherapeutic agents. Since DNA methylation is linked to the silencing of related genes, a decrease in methylation is linked to stimulation (Zimmer et al., 2015). There was also an elevation in H3K4 methylation together with hyperacetylation at the MDR1 promoter, indicating stimulation. These changes resulted in abnormally high MDR1 gene expression in cancer cell lines. In another study, it was found that chemotherapeutic agents such as bleomycin, etoposide, and cisplatin can effectively treat testicular cancer by causing epigenetic changes in sperm DNA. In 11 germ cells, rats given BEP demonstrated dose-dependent hypermethylation of 20 loci (Reiser, 2019; Yao et al., 2019). Even with low dosages of chemotherapy, several of these loci are hypermethylated, indicating that some loci are more susceptible to chemotherapy than others. The majority of these genes are engaged in key cell signaling pathways, such as RefSeq gene Cpne8, which is involved in calcium-mediated intracellular activities, and RefSeq gene Slc16a2, which is involved in thyroid hormone transport (Gaman et al., 2016). When rats were given BEP, their methylation levels increased, indicating that these genes were inappropriately suppressed. Some of these genes are also implicated in the development and invasion of tumors. Changes in methylation patterns of these genes might alter the father's germ cell development and as a consequence, future offspring. According to current research, it has been seen that chemotherapy causes epigenetic changes that seem to be important for normal biological processes (Du et al., 2021; Suuberg, 2018). Recent chemotherapy treatment was linked to alterations in DNA methylation patterns, which might explain chronic inflammation and tiredness. DNA was collected from 61 stages 0 IIIA breast cancer patients' peripheral blood mononuclear cells (PBMCs) before radiation therapy. Partial mastectomies, with or without chemotherapy, were performed on these individuals (Gu et al., 2018). At 485,000 CpG sites across the genome, DNA methylation was examined. There were indicators that have previously been linked to inflammation and tiredness. Chemotherapy patients had lower methylation levels in 8 CpG sites than their non-treated counterparts (Gräff et al., 2012). Exon 11 of transmembrane protein 49 (TMEM49), which allows particular molecules to pass through, has four differently methylated CpG sites. The reduction in methylation of these CpG sites was related to an increase in IL-6, a pro-inflammatory cytokine, and soluble tumor necrosis factor receptor (Reiser, 2019). According to the current findings, chemotherapy treatment was associated with an increment in inflammatory biomarkers. Pro-inflammatory indicators were examined again six months after radiation treatment. There was still decreased methylation of four of the initial eight CpG sites, which was linked to an increase in pro-inflammatory cytokines stated earlier. This study demonstrates epigenetic modifications that occur as a result of chemotherapy treatment, as well as the reversibility of certain of these alterations (Li et al., 2018).

Effects of cytotoxic chemotherapy medicines mitomycin C (MMC) and cyclophosphamide have been investigated In mouse prefrontal cortex and hippocampus, changes in the transcriptome and epigenetic alterations were examined. Three weeks following therapy, gene expression alterations in response to treatment were found to be mostly in females and unique to the prefrontal cortex of the brain (Sequeira et al., 2012). MMC also appeared to have a greater impact on gene expression than cyclophosphamide. MMC therapy caused oxidative DNA damage, as shown by a rise in 8-oxo-2′-deoxyguanosine (8-OxodG) 3 weeks later. A protein called 8-OxodG repair was also reduced. In the female mouse prefrontal cortex, MMC treatment reduced global DNA methylation and increased global DNA hydroxymethylation (Lim et al., 2016). The majority of alterations that occurred in PFC of female mice as a result of chemotherapy were comparable to those that arise with natural ageing. Chemotherapy's influence on epigenetic processes in the brain mimics brain ageing.

Certain neurodegenerative disorders, such as Alzheimer's, have been found to contain epigenetic pathways that influence cognitive decline. A blockage in gene regulation was discovered in the neurodegenerating brain.

This was mostly due to 14 histone deacetylase 2, which deacetylates histones and therefore inactivates gene expression, as its name implies. Histone deacetylase 2 levels were shown to be substantially higher in Alzheimer's patients and in two animal models of neurodegeneration (Li et al., 2018). Learning and memory-related genes are suppressed as a result of histone deacetylase 2 activity. Notably, inhibiting histone deacetylase activity utilizing short hairpin RNA causes these genes to become active again, reversing their suppression. This improves synaptic plasticity and reduces memory loss associated with neurodegeneration. Subsequently, in Drosophila, the neuronal function of euchromatin histone methyltransferase (EHMT) was investigated. EHMT is a conserved protein that silences transcription by methylating histone 3 lysine 9 as a repressive mark (Liao et al., 2018). The two EHMT mutant flies are alive even though EHMT is extensively expressed in the nervous system. EHMT plays a significant role in courting memory and associative learning, according to a study of neurodevelopment and behavior. When expression of EHMT was restored in adulthood, memory was restored. This suggests that memory loss induced by EHMT deficiency can be reversed. Researchers look at H3K9 demethylation patterns using ChIP-seq analysis. In EHMT mutants, 5% of euchromatin was lost due to H3K9 demethylation loss. This loss is similarly linked to the loss of genes involved in neural and behavioral processes in EHMT mutant flies. Hence, the involvement of EHMT in cognitive functions such as learning and memory has been reported (Nguyen et al., 2021).

Epigenetic changes have been demonstrated to have a role in the development of chemobrain. Chemotherapy medications cyclophosphamide, methotrexate, and 5-FU were given to rats. According to the findings, these medications impair spatial learning and long-term memory (Ongnok et al., 2020). When scientists looked at histone modification, they observed that histone acetylation on H3 in the hippocampus and prefrontal brain was considerably enhanced in chemotherapy-treated rats, whereas HDAC activity was reduced. These are unexpected findings because histone acetylation is generally linked to improved learning and memory (El-Agamy et al., 2019). These findings suggest that chemotherapy causes epigenetic alterations in the brain. Current studies revealed that chemotherapeutic medications produce an increase in proinflammatory cytokines, which leads to hypermethylation of the brain. A study was conducted along similar lines to evaluate the impact of a father's chemotherapy exposure on his transgenerational progeny's brains. Three chemotherapy medicines were given to male mice. In the progeny brain's frontal cortex, DNMT1 and Mecp2 levels were observed to be lower. In the transgenerational chemotherapy-exposed brain, decreased levels of DNMT1 indicate reduced levels of essential methylation (Gu et al., 2018). Mecp2 has been demonstrated to be necessary for genome stability and neural network maintenance. These findings suggest that aberrant methylation in chemotherapy-exposed brains can result in cognitive impairment in the offspring of chemotherapy patients. Earlier research has shown that chemotherapy treatment after cancer causes epigenetic changes that can lead to cognitive impairment. Because of the small volume of research that has been done in this field, as well as emerging knowledge of epigenetics, this topic is a viable research target (Joshi et al., 2010)

6. Current status and personalized therapy

Despite growing awareness, no proven or authorized diagnostics for chemobrain diagnosis exist currently. Indeed, numerous studies show that cancer survivors score within a normal range of the population, although at the lower end (Butterfield, 2014). This restriction is most likely related to evaluation instruments' lack of sensitivity. Chemobrain is extremely variable, with numerous confounding factors such as genetic variability, treatment regimen, and comorbidity with other neurological disorders. Also, chemobrain has no well-defined pathophysiological mechanisms. Each chemotherapy medication is likely to have a variety of side effects, which will vary even more when coupled with other treatments and treatment options (Wang et al., 2021). The present therapeutic strategy is cognitive rehabilitation, lifestyle modifications, mind-training exercises, intellectual therapy, and general coping methods due to the intricacy and unknown processes. Neuropsychiatric medications may also be administered to help with symptoms (Ongnok et al., 2020). Scientists believe that repurposing current medicines may prove to be a viable method for successfully treating chemobrain in near future, given the difficulty of finding, fine-tuning, and approving novel therapeutic molecules for CNS. Despite the diversity of molecular processes, there are cellular systems that may be addressed that are convergent. Of course, this isn't to downplay the importance of ongoing research into the molecular processes of each chemotherapy medication (Nguyen et al., 2021). Effective prevention or treatment alternatives can be created if enough information on the effects of chemotherapy is available at all levels—molecular, cellular, and behavioral levels. Ultimately, increasingly effective medicines will not only cure symptoms of chemobrain, but will also directly alter the course of the disease, either by reducing the first toxic effects or by improving recovery following chemotherapy (Nguyen & Ehrlich, 2020).

Undeniably, abundant research has already claimed cognitive debility even in the absence of chemotherapy. Considering the need for pre-chemotherapy evaluation of cognitive decline, the first-ever longitudinal study was conducted in 2004 that compared the baseline cognitive decline in breast cancer patients due to cancer and cognitive decline post-chemotherapy (FAC) in short-term and long-term. These studies are highly valued for their importance in evaluating cognitive decline during different timeframes. About 61% of patients showed a cognitive decline in short and 50% at long-term post-chemotherapy respectively, while 33% of women experienced cognitive decline pre-chemotherapy. Though the overall results were not momentous, alterations in the cognitive deterioration pre- and post-chemotherapy confirmed that CRCI should account for clinical significance. Although several longitudinal studies have suggested cognitive debility during treatment, the correct correlation of chemotherapy to cognitive decline is still a question. Firstly, studies evaluate both self-reported CRCI and diminished cognitive performance objectively through neuroimaging and neuropsychological tests. The lack of enough objective evidence to associate with self-reported data raises further inquiry into the tangibility of CRCI assessment. Regardless of the type of data pool, CRCI is extant and continues to affect the QoL of patients and survivors. Secondly, the documentation of CRCI in the absence of chemotherapy has speculated its relationship to the disease itself. A systemic review conducted in 2020 evaluated the same in cancer survivors with the inclusion of cancer as a factor to compromise cognitive functioning through neuroinflammatory mechanisms (Ahles et al., 2010; Bender et al., 2006; Hardy, Krull, Wefel, & Janelsins, 2018; Klemp et al., 2018; Sousa et al., 2020; Spooner & Pachana, 2006; Wefel, Lenzi, Theriault, Davis, & Meyers, 2004).

It should be acknowledged that there are other potential factors such as age, pre- or post-menopausal status, psychological issues such as fatigue, depression, and anxiety that could play a significant role in escalating the risk of chemotherapy-induced cognitive decline (Sousa et al., 2020). Ahles *et al.* proved that factors such as age and pre-cognitive reserve in cancer survivors may contribute to post-cognitive debilities. Older patients showed decreased cognitive reserve in comparison to younger cancer survivors and showed worsened cognitive performance upon exposure to chemotherapy. It was ascertained that in the older population, the speed of processing and verbal competence was affected the most. However, further discontinuation of chemotherapy restored verbal ability (Ahles et al., 2010).

Likewise, a study was conducted on chemotherapy-induced amenorrhea wherein a neuropsychological assessment of breast cancer survivors revealed the effect of menopause on CRCI. The study examined the effect of chemotherapy in pre- and post-menopausal breast cancer patients against healthy controls and proved that post-menopausal conditions failed to maintain visual and verbal cognitive function in cancer survivors. Additionally, the menopausal condition is also related to age which contributes to much lower cognitive performance (Conroy et al., 2013). Nevertheless, long-term follow-up evaluations of cognitive function suggest plausible partial or complete recovery from these deficits post-chemotherapy. However, which group of patients recuperates, and which subset of patients continue to endure through is still ambiguous. Contemplating the need to address these concerns, several randomized controlled trials have verified and recommended interventions. This includes mental stimulation by learning new skills like a musical instrument, cognitive rehabilitation delivered by web-based brain training programs, adaptation strategies such as journaling, exercise, and meditation therapy (Vardy et al., 2017). Nonetheless, the effect of these interventions on CRCI is still unclear and needs more trials in the future that enables improved QoL in cancer survivors.

Chemobrain has garnered a great deal of attention in recent years as a significant side effect of chemotherapy and numerous research has helped us grasp underlying processes and variables. Addressing chemobrain in the future will need comprehensive work on several fronts, guided by comparable efforts done for ageing, Alzheimer's disease, and TBI. On the clinical front, additional work is needed to raise the concern of the danger of chemobrain and better inform doctors, chemotherapy patients, and their caregivers, allowing for a more careful watch for subtle impairments like loss of memory that may otherwise go unnoticed (Srivastava & Singh, 2020). On the other hand, cancer survivors who do not experience chemobrain symptoms right after treatment may be at a higher risk of acquiring neurodegenerative disorders later in life. On a fundamental science level, further work is needed to develop animal models that more closely recapitulate the complexities and nuances of chemobrain in humans. Animal models that are old or have tumors and are given a common mix of agents rather than a single treatment are examples. In addition to memory acquisition tests, chemobrain research requires measurements that assess working memory, attention, and executive functioning (Wahdan et al., 2020). Deciding if certain cognitive modes, anatomical locations, or cell populations are more sensitive can also help in the development of therapeutic drugs. All the available targets and drug therapies are listed in Table 1.

Table 1 Targets and treatment options for chemobrain.

S·No	Parameters	Targets	Chemotherapy inducing chemobrain	Treatment drugs	References
1.	Oxidative stress	SOD Apolipoprotein A1 Oxidation LPO BDNF Lipofuscin ROS	Doxorubicin cyclophosphamide,	*Origanum majorana, Zingiber officinale,* Apigenin and Paclitaxel combination, Cysteamine, MicroRNA-497	Soliman, Desouky, Marzouk, and Sayed (2016), Ajith, Aswathy, and Hema (2008), Xu et al. (2011), Besouw, Masereeuw, van den Heuvel, and Levtchenko (2013), Wang et al. (2017)
2.	Neurogenesis	Astrocytic cell activation TNF receptor-1	Methotrexate, 5-FU, cyclophosphamide, docetaxel, paclitaxel, cisplatin, and thioTEPA	Fenofibrate, Metformin, Anti-depressant drugs	Ramanan et al. (2009), Sritawan et al. (2020), Drew, Hen, and Targets (2007)
3.	Neuroinflammation	NF-κB Microglial activation	5-FU and a combination of docetaxel, dox and cyclophosphamide, methotrexate, cisplatin, oxaliplatin, paclitaxel, and vincristine	MS-275 and resveratrol, *Ganoderma lucidum,* Fenofibrate, Green tea catechins, vasoactive intestinal peptide (VIP), NSAIDs like aspirin, ibuprofen, and naproxen	Schiaffino et al. (2018), Zhang et al. (2011), Ramanan et al. (2009), Farkhondeh et al. (2020), Delgado and Ganea (2003), Nguyen and Ehrlich (2020)

4.	Arborization	Glutamate ubiquitin-proteasome system disruption Pkc hyperactivation Dysregulated cAMP signaling	cisplatin, fluorouracil, dox, cyclophosphamide, paclitaxel	Memantine and vitamin-D combination, Atorvastatin and Fluoxetine combination, PDE4, PDE7 and PDE8 inhibitors, dextromethorphan, ketamine and benzamides	Annweiler, Brugg, Peyrin, Bartha, and Beauchet (2014), Ludka et al. (2017), Martinez and Gil (2014), Ajith et al. (2008); Kim et al. (2017); Ramanan et al. (2009); Suuberg (2018); Wang et al. (2017); Zhang et al. (2011)
5.	Genetic factors	MDR1 gene expression H3K4 methylation CpG sites methylation oxidative DNA damage Euchromatin histone methyltransferase (EHMT) HDAC	Daunorubicin, etoposide, as bleomycin, cis-platinum, mitomycin C, cyclophosphamide, methotrexate, and 5-FU, procarbazine	UNC0638 and UNC0642, amide procainamide, an antiarrhythmic drug, Belinostat, Panobinostat Quisinostat, Trichostatin A (TSA), Vorinostat (suberoylanilide hydroxamic acid, Abexinostat, Givinostat, Resminostat, Phenylbutyrate, Sodium Butyrate, Valproic Acid, Benzamide Entinostat, Mocetinostat, Tubastatin A, Sodium 4-phenylbutyrate	Kim et al. (2017), Castillo-Aguilera, Depreux, Halby, Arimondo, and Goossens (2017), Pasyukova and Vaiserman (2017), Gu et al. (2018)

Prospects of better treatments seem bright with these targeted methods (Nguyen et al., 2021).

7. Future perspectives

In summary, cognitive impairment post-chemotherapy remains a major challenge for oncologists as no prevention or treatment is available, despite pharmacological and non-pharmacological strategies being investigated clinically. Unlike other neurological disorders, the onset of chemobrain can be both scientifically and ethically studied with the administration of physiological stress, thereby observing the pattern of biological ramifications. This makes it possible to gain adequate knowledge on the mechanism underlying cognitive impairment in post-chemotherapy patients paving the way to effectively identify well-defined methods to prevent and treat chemobrain.

8. Conclusion

Based on the data obtained from various preclinical and clinical studies, it is evident that the presence of chemobrain situations pre-and post-chemotherapy is possible and corresponding effects like neuronal damage, memory impairment, and decrease in quality of life are inevitable. Many proposed mechanisms for this process include oxidative stress, loss of dendritic arborization, aberrant activation of apoptotic and autophagic pathways, dysregulated neurotransmitter release, unplanned activation of key kinases, and epigenetic factors. All of the above-mentioned pathways are notably interlinked and make a collective contribution to causing chemobrain. To date, no approved treatment is available for chemotherapy-induced cognitive impairment but few medications and practices like behavioral therapy and rehabilitation cognitive training can ensure improving the quality of life in patients.

Reference

Acharya, M. M., Martirosian, V., Chmielewski, N. N., Hanna, N., Tran, K. K., Liao, A. C., et al. (2015). Stem cell transplantation reverses chemotherapy-induced cognitive dysfunction. *Cancer Research*, *75*(4), 676–686.

Adwan, L., & Zawia, N. H. (2013). Epigenetics: A novel therapeutic approach for the treatment of Alzheimer's disease. *Pharmacology & Therapeutics*, *139*(1), 41–50.

Ahles, T. A., & Saykin, A. J. (2002). Breast cancer chemotherapy-related cognitive dysfunction. *Clinical Breast Cancer*, *3*, S84–S90.

Ahles, T. A., & Saykin, A. J. (2007). Candidate mechanisms for chemotherapy-induced cognitive changes. *Nature Reviews. Cancer*, 7(3), 192–201. https://doi.org/10.1038/nrc2073.

Ahles, T. A., Saykin, A. J., Furstenberg, C. T., Cole, B., Mott, L. A., Skalla, K., et al. (2002). Neuropsychologic impact of standard-dose systemic chemotherapy in long-term survivors of breast cancer and lymphoma. *Journal of Clinical Oncology*, *20*(2), 485–493.

Ahles, T. A., Saykin, A. J., McDonald, B. C., Li, Y., Furstenberg, C. T., Hanscom, B. S., et al. (2010). Longitudinal assessment of cognitive changes associated with adjuvant treatment for breast cancer: Impact of age and cognitive reserve. *Journal of Clinical Oncology*, *28*(29), 4434.

Ahles, T. A., Saykin, A. J., Noll, W. W., Furstenberg, C. T., Guerin, S., Cole, B., et al. (2003). The relationship of APOE genotype to neuropsychological performance in long-term cancer survivors treated with standard dose chemotherapy. *Psycho-Oncology: Journal of the Psychological, Social and Behavioral Dimensions of Cancer*, *12*(6), 612–619.

Ajith, T., Aswathy, M., & Hema, U. (2008). Protective effect of *Zingiber officinale* roscoe against anticancer drug doxorubicin-induced acute nephrotoxicity. *Food and Chemical Toxicology*, *46*(9), 3178–3181.

Ali, F., Gerhard, D. M., Sweasy, K., Pothula, S., Pittenger, C., Duman, R. S., et al. (2020). Ketamine disinhibits dendrites and enhances calcium signals in prefrontal dendritic spines. *Nature Communications*, *11*(1), 1–15.

Allegra, A., Innao, V., Basile, G., Pugliese, M., Allegra, A. G., Pulvirenti, N., et al. (2020). Post-chemotherapy cognitive impairment in hematological patients: Current understanding of chemobrain in hematology. *Expert Review of Hematology*, *13*(4), 393–404.

Andres, A. L., Gong, X., Di, K., & Bota, D. A. (2014). Low-doses of cisplatin injure hippocampal synapses: A mechanism for 'chemo'brain? *Experimental Neurology*, *255*, 137–144.

Annweiler, C., Brugg, B., Peyrin, J.-M., Bartha, R., & Beauchet, O. (2014). Combination of memantine and vitamin D prevents axon degeneration induced by amyloid-beta and glutamate. *Neurobiology of Aging*, *35*(2), 331–335.

Arafa, M. H., & Atteia, H. H. (2020). Protective role of epigallocatechin gallate in a rat model of cisplatin-induced cerebral inflammation and oxidative damage: Impact of modulating NF-κB and Nrf2. *Neurotoxicity Research*, *37*(2), 380–396.

Banks, W. A., Farr, S. A., & Morley, J. E. (2002). Entry of blood-borne cytokines into the central nervous system: Effects on cognitive processes. *Neuroimmunomodulation*, *10*(6), 319–327.

Bender, C. M., Sereika, S. M., Berga, S. L., Vogel, V. G., Brufsky, A. M., Paraska, K. K., et al. (2006). Cognitive impairment associated with adjuvant therapy in breast cancer. *Psycho-Oncology: Journal of the Psychological, Social and Behavioral Dimensions of Cancer*, *15*(5), 422–430.

Besouw, M., Masereeuw, R., van den Heuvel, L., & Levtchenko, E. (2013). Cysteamine: An old drug with new potential. *Drug Discovery Today*, *18*(15–16), 785–792.

Butterfield, D. A. (2014). The 2013 SFRBM discovery award: Selected discoveries from the butterfield laboratory of oxidative stress and its sequela in brain in cognitive disorders exemplified by Alzheimer disease and chemotherapy induced cognitive impairment. *Free Radical Biology & Medicine*, *74*, 157–174.

Cardoso, C. V., de Barros, M. P., Bachi, A. L. L., Bernardi, M. M., Kirsten, T. B., Martins, M.d. F. M., et al. (2020). Chemobrain in rats: Behavioral, morphological, oxidative and inflammatory effects of doxorubicin administration. *Behavioural Brain Research*, *378*, 112233.

Castillo-Aguilera, O., Depreux, P., Halby, L., Arimondo, P. B., & Goossens, L. (2017). DNA methylation targeting: The DNMT/HMT crosstalk challenge. *Biomolecules*, 7(1), 3.

Cheng, H., Li, W., Gan, C., Zhang, B., Jia, Q., & Wang, K. (2016). The COMT (rs165599) gene polymorphism contributes to chemotherapy-induced cognitive impairment in breast cancer patients. *American Journal of Translational Research*, *8*(11), 5087–5097.

Cheung, Y. T., Lim, S. R., Ho, H. K., & Chan, A. (2013). Cytokines as mediators of chemotherapy-associated cognitive changes: Current evidence, limitations and directions for future research. *PLoS One*, *8*(12), e81234.

Cheung, Y. T., Ng, T., Shwe, M., Ho, H. K., Foo, K. M., Cham, M. T., et al. (2015). Association of proinflammatory cytokines and chemotherapy-associated cognitive impairment in breast cancer patients: A multi-centered, prospective, cohort study. *Annals of Oncology*, *26*(7), 1446–1451. https://doi.org/10.1093/annonc/mdv206.

Chidambaram, S. B., Rathipriya, A., Bolla, S. R., Bhat, A., Ray, B., Mahalakshmi, A. M., et al. (2019). Dendritic spines: Revisiting the physiological role. *Progress in Neuro-Psychopharmacology and Biological Psychiatry*, *92*, 161–193.

Chikara, S., Nagaprashantha, L. D., Singhal, J., Horne, D., Awasthi, S., & Singhal, S. S. (2018). Oxidative stress and dietary phytochemicals: Role in cancer chemoprevention and treatment. *Cancer Letters*, *413*, 122–134.

Chiu, G. S., Maj, M. A., Rizvi, S., Dantzer, R., Vichaya, E. G., Laumet, G., et al. (2017). Pifithrin-μ prevents cisplatin-induced chemobrain by preserving neuronal mitochondrial function. *Cancer Research*, 77(3), 742–752.

Christie, L.-A., Acharya, M. M., Parihar, V. K., Nguyen, A., Martirosian, V., & Limoli, C. L. (2012). Impaired cognitive function and hippocampal neurogenesis following cancer chemotherapy. *Clinical Cancer Research*, *18*(7), 1954–1965.

Conroy, S. K., McDonald, B. C., Ahles, T. A., West, J. D., & Saykin, A. J. (2013). Chemotherapy-induced amenorrhea: A prospective study of brain activation changes and neurocognitive correlates. *Brain Imaging and Behavior*, 7(4), 491–500.

Couch, L. S., Groteluschen, D. L., Stewart, J. A., & Mulkerin, D. L. (2003). Capecitabine-related neurotoxicity presenting as trismus. *Clinical Colorectal Cancer*, *3*(2), 121–123.

Cranmer, D. (2019). CanCer SurvivorShip. *Cancer: Prevention, Early Detection, Treatment and Recovery*, 617–620.

Daiber, A., Daub, S., Bachschmid, M., Schildknecht, S., Oelze, M., Steven, S., et al. (2013). Protein tyrosine nitration and thiol oxidation by peroxynitrite—Strategies to prevent these oxidative modifications. *International Journal of Molecular Sciences*, *14*(4), 7542–7570.

Das, A., Ranadive, N., Kinra, M., Nampoothiri, M., Arora, D., & Mudgal, J. (2020). An overview on chemotherapy-induced cognitive impairment and potential role of antidepressants. *Current Neuropharmacology*, *18*(9), 838–851. https://doi.org/10.2174/1570159x18666200221113842.

Delgado, M., & Ganea, D. (2003). Neuroprotective effect of vasoactive intestinal peptide (VIP) in a mouse model of Parkinson's disease by blocking microglial activation. *The FASEB Journal*, *17*(8), 1–18.

Doll, D. N., Rellick, S. L., Barr, T. L., Ren, X., & Simpkins, J. W. (2015). Rapid mitochondrial dysfunction mediates TNF-alpha-induced neurotoxicity. *Journal of Neurochemistry*, *132*(4), 443–451.

Dong, H., & Czaja, M. J. (2011). Regulation of lipid droplets by autophagy. *Trends in Endocrinology and Metabolism*, *22*(6), 234–240. https://doi.org/10.1016/j.tem.2011.02.003.

Dooley, L. N., Ganz, P. A., Cole, S. W., Crespi, C. M., & Bower, J. E. (2016). Val66Met BDNF polymorphism as a vulnerability factor for inflammation-associated depressive symptoms in women with breast cancer. *Journal of Affective Disorders*, *197*, 43–50. https://doi.org/10.1016/j.jad.2016.02.059.

Drew, M. R., Hen, R. J. C., & Targets, N. D.-D. (2007). Adult hippocampal neurogenesis as target for the treatment of depression. *CNS & Neurological Disorders Drug Targets*, *6*(3), 205–218.

Du, J., Zhang, A., Li, J., Liu, X., Wu, S., Wang, B., et al. (2021). Doxorubicin-induced cognitive impairment: The mechanistic insights. *Frontiers in Oncology*, *11*, 1461.

Edwardson, D. W., Boudreau, J., Mapletoft, J., Lanner, C., Kovala, A. T., & Parissenti, A. M. (2017). Inflammatory cytokine production in tumor cells upon chemotherapy drug exposure or upon selection for drug resistance. *PLoS One*, *12*(9), e0183662.

El-Agamy, S. E., Abdel-Aziz, A. K., Ahmed, E., & Azab, S. S. (2019). Chemotherapy and cognition: Comprehensive review on doxorubicin-induced chemobrain. *Cancer Chemotherapy and Pharmacology*, *84*(1), 1–14.

El-Agamy, S. E., Abdel-Aziz, A. K., Wahdan, S., Esmat, A., & Azab, S. S. (2018). Astaxanthin ameliorates doxorubicin-induced cognitive impairment (chemobrain) in experimental rat model: Impact on oxidative, inflammatory, and apoptotic machineries. *Molecular Neurobiology*, *55*(7), 5727–5740.

Erickson, K. I., Prakash, R. S., Voss, M. W., Chaddock, L., Heo, S., McLaren, M., et al. (2010). Brain-derived neurotrophic factor is associated with age-related decline in hippocampal volume. *The Journal of Neuroscience*, *30*(15), 5368–5375. https://doi.org/10.1523/jneurosci.6251-09.2010.

Evens, K., & Eschiti, V. S. (2009). Cognitive effects of Cancer treatment:" chemo brain" explained. *Clinical Journal of Oncology Nursing*, *13*(6).

Fardell, J. E., Vardy, J., Johnston, I. N., & Winocur, G. (2011). Chemotherapy and cognitive impairment: Treatment options. *Clinical Pharmacology and Therapeutics*, *90*(3), 366–376. https://doi.org/10.1038/clpt.2011.112.

Farkhondeh, T., Pourbagher-Shahri, A. M., Ashrafizadeh, M., Folgado, S. L., Rajabpour-Sanati, A., Khazdair, M. R., et al. (2020). Green tea catechins inhibit microglial activation which prevents the development of neurological disorders. *Neural Regeneration Research*, *15*(10), 1792.

Fernández-Rojas, B., Rodríguez-Rangel, D. S., Granados-Castro, L. F., Negrette-Guzmán, M., León-Contreras, J. C., Hernández-Pando, R., et al. (2015). C-phycocyanin prevents cisplatin-induced mitochondrial dysfunction and oxidative stress. *Molecular and Cellular Biochemistry*, *406*(1), 183–197.

Fitzgerald, P. J. (2021). Norepinephrine may be an underlying factor in chemobrain. *International Journal of Cancer*, *148*(12), 3119–3120. https://doi.org/10.1002/ijc.33548.

Gaman, A. M., Uzoni, A., Popa-Wagner, A., Andrei, A., & Petcu, E.-B. (2016). The role of oxidative stress in etiopathogenesis of chemotherapy induced cognitive impairment (CICI)-"chemobrain". *Aging and Disease*, 7(3), 307.

Gorrini, C., Harris, I. S., & Mak, T. W. (2013). Modulation of oxidative stress as an anticancer strategy. *Nature Reviews Drug Discovery*, *12*(12), 931–947.

Gräff, J., Rei, D., Guan, J.-S., Wang, W.-Y., Seo, J., Hennig, K. M., et al. (2012). An epigenetic blockade of cognitive functions in the neurodegenerating brain. *Nature*, *483*(7388), 222–226.

Grisold, W., Cavaletti, G., & Windebank, A. J. (2012). Peripheral neuropathies from chemotherapeutics and targeted agents: Diagnosis, treatment, and prevention. *Neuro-Oncology*, *14*(suppl_4), iv45-iv54.

Groves, T. R., Farris, R., Anderson, J. E., Alexander, T. C., Kiffer, F., Carter, G., et al. (2017). 5-fluorouracil chemotherapy upregulates cytokines and alters hippocampal dendritic complexity in aged mice. *Behavioural Brain Research*, *316*, 215–224.

Gu, T., Lin, X., Cullen, S. M., Luo, M., Jeong, M., Estecio, M., et al. (2018). DNMT3A and TET1 cooperate to regulate promoter epigenetic landscapes in mouse embryonic stem cells. *Genome Biology*, *19*(1), 1–15.

Gutmann, D. H. (2019). Clearing the fog surrounding chemobrain. *Cell*, *176*(1–2), 2–4.

Hafner, D. L. (2009). Lost in the fog: Understanding" chemo brain". *Nursing 2020*, *39*(8), 42–45.

Haines, L., & Shapiro, C. L. (2019). *Cancer survivorship* (pp. 535–542). Mount Sinai Expert Guides: Oncology.

Hardy, S. J., Krull, K. R., Wefel, J. S., & Janelsins, M. (2018). Cognitive changes in cancer survivors. *American Society of Clinical Oncology Educational Book, 38*, 795–806.

Harry, G. J., & Kraft, A. D. (2008). Neuroinflammation and microglia: Considerations and approaches for neurotoxicity assessment. *Expert Opinion on Drug Metabolism & Toxicology, 4*(10), 1265–1277. https://doi.org/10.1517/17425255.4.10.1265.

Heim, C., & Binder, E. B. (2012). Current research trends in early life stress and depression: Review of human studies on sensitive periods, gene–environment interactions, and epigenetics. *Experimental Neurology, 233*(1), 102–111.

Henderson, F. M., Cross, A. J., & Baraniak, A. R. (2019). 'A new normal with chemobrain': Experiences of the impact of chemotherapy-related cognitive deficits in long-term breast cancer survivors. *Health Psychology Open, 6*(1). 2055102919832234.

Ho, K.-H. (2015). Insights into the mechanism of'chemobrain': Deriving a multi-factorial model of pathogenesis. *Australian Medical Student Journal, 6*(1).

Horowitz, T. S., Suls, J., & Treviño, M. (2018). A call for a neuroscience approach to cancer-related cognitive impairment. *Trends in Neurosciences, 41*(8), 493–496.

Huo, X., Reyes, T. M., Heijnen, C. J., & Kavelaars, A. (2018). Cisplatin treatment induces attention deficits and impairs synaptic integrity in the prefrontal cortex in mice. *Scientific Reports, 8*(1), 1–11.

Janelsins, M. C., Kohli, S., Mohile, S. G., Usuki, K., Ahles, T. A., & Morrow, G. R. (2011). An update on cancer-and chemotherapy-related cognitive dysfunction: Current status. In *Paper presented at the Seminars in oncology*, Elsevier.

Jarmolowicz, D. P., Gehringer, R., Lemley, S. M., Sofis, M. J., Kaplan, S., & Johnson, M. A. (2019). 5-fluorouracil impairs attention and dopamine release in rats. *Behavioural Brain Research, 362*, 319–322.

Jim, H. S., Phillips, K. M., Chait, S., Faul, L. A., Popa, M. A., Lee, Y.-H., et al. (2012). Meta-analysis of cognitive functioning in breast cancer survivors previously treated with standard-dose chemotherapy. *Journal of Clinical Oncology, 30*(29), 3578.

Joshi, G., Aluise, C. D., Cole, M. P., Sultana, R., Pierce, W., Vore, M., et al. (2010). Alterations in brain antioxidant enzymes and redox proteomic identification of oxidized brain proteins induced by the anti-cancer drug adriamycin: Implications for oxidative stress-mediated chemobrain. *Neuroscience, 166*(3), 796–807.

Kang, S., Lee, S., Kim, J., Kim, J.-C., Kim, S.-H., Son, Y., et al. (2018). Chronic treatment with combined chemotherapeutic agents affects hippocampal micromorphometry and function in mice, independently of neuroinflammation. *Experimental Neurobiology, 27*(5), 419.

Kannarkat, G., Lasher, E. E., & Schiff, D. (2007). Neurologic complications of chemotherapy agents. *Current Opinion in Neurology, 20*(6), 719–725.

Kaplan, S. V., Limbocker, R. A., Gehringer, R. C., Divis, J. L., Osterhaus, G. L., Newby, M. D., et al. (2016). Impaired brain dopamine and serotonin release and uptake in Wistar rats following treatment with carboplatin. *ACS Chemical Neuroscience*, 7(6), 689–699. https://doi.org/10.1021/acschemneuro.5b00029.

Kasai, H., Hayama, T., Ishikawa, M., Watanabe, S., Yagishita, S., & Noguchi, J. (2010). Learning rules and persistence of dendritic spines. *European Journal of Neuroscience, 32*(2), 241–249.

Keeney, J. T. R., Ren, X., Warrier, G., Noel, T., Powell, D. K., Brelsfoard, J. M., et al. (2018). Doxorubicin-induced elevated oxidative stress and neurochemical alterations in brain and cognitive decline: Protection by MESNA and insights into mechanisms of chemotherapy-induced cognitive impairment ("chemobrain"). *Oncotarget, 9*(54), 30324–30339. https://doi.org/10.18632/oncotarget.25718.

Kennedy, M. B. (2016). Synaptic signaling in learning and memory. *Cold Spring Harbor Perspectives in Biology, 8*(2), a016824.

Kiffer, F., Howe, A. K., Carr, H., Wang, J., Alexander, T., Anderson, J. E., et al. (2018). Late effects of 1H irradiation on hippocampal physiology. *Life Sciences and Space Research*, *17*, 51–62.

Kim, Y., Lee, H.-M., Xiong, Y., Sciaky, N., Hulbert, S. W., Cao, X., et al. (2017). Targeting the histone methyltransferase G9a activates imprinted genes and improves survival of a mouse model of Prader–Willi syndrome. *Nature Medicine*, *23*(2), 213–222.

Klemp, J. R., Myers, J. S., Fabian, C. J., Kimler, B. F., Khan, Q. J., Sereika, S. M., et al. (2018). Cognitive functioning and quality of life following chemotherapy in pre-and peri-menopausal women with breast cancer. *Supportive Care in Cancer*, *26*(2), 575–583.

Koffie, R. M., Hashimoto, T., Tai, H.-C., Kay, K. R., Serrano-Pozo, A., Joyner, D., et al. (2012). Apolipoprotein E4 effects in Alzheimer's disease are mediated by synaptotoxic oligomeric amyloid-β. *Brain*, *135*(7), 2155–2168.

Lee, B.-N., Dantzer, R., Langley, K. E., Bennett, G. J., Dougherty, P. M., Dunn, A. J., et al. (2004). A cytokine-based neuroimmunologic mechanism of cancer-related symptoms. *Neuroimmunomodulation*, *11*(5), 279–292.

Li, M., & Caeyenberghs, K. J. N. (2018). Longitudinal assessment of chemotherapy-induced changes in brain and cognitive functioning: A systematic review. *Neuroscience and Biobehavioral Reviews*, *92*, 304–317.

Li, Z., Zhao, S., Zhang, H.-L., Liu, P., Liu, F.-F., Guo, Y.-X., et al. (2018). Proinflammatory factors mediate paclitaxel-induced impairment of learning and memory. *Mediators of Inflammation*, 2018.

Liao, D., Xiang, D., Dang, R., Xu, P., Wang, J., Han, W., et al. (2018). Neuroprotective effects of dl-3-n-butylphthalide against doxorubicin-induced neuroinflammation, oxidative stress, endoplasmic reticulum stress, and behavioral changes. *Oxidative Medicine and Cellular Longevity*, *2018*.

Licinio, J., Kling, M. A., & Hauser, P. (1998). Cytokines and brain function: Relevance to interferon-alpha-induced mood and cognitive changes. In *Paper presented at the Seminars in oncology*, Europe PMC.

Lim, I., Joung, H.-Y., Yu, A. R., Shim, I., & Kim, J. S. (2016). PET evidence of the effect of donepezil on cognitive performance in an animal model of chemobrain. *BioMed Research International*, 2016.

Liu, R. Y., Zhang, Y., Coughlin, B. L., Cleary, L. J., & Byrne, J. H. (2014). Doxorubicin attenuates serotonin-induced long-term synaptic facilitation by phosphorylation of p38 mitogen-activated protein kinase. *The Journal of Neuroscience*, *34*(40), 13289–13300. https://doi.org/10.1523/jneurosci.0538-14.2014.

Lomeli, N., Di, K., Czerniawski, J., Guzowski, J. F., & Bota, D. A. (2017). Cisplatin-induced mitochondrial dysfunction is associated with impaired cognitive function in rats. *Free Radical Biology & Medicine*, *102*, 274–286.

Ludka, F. K., Dal-Cim, T., Binder, L. B., Constantino, L. C., Massari, C., & Tasca, C. I. (2017). Atorvastatin and fluoxetine prevent oxidative stress and mitochondrial dysfunction evoked by glutamate toxicity in hippocampal slices. *Molecular Neurobiology*, *54*(5), 3149–3161.

Manchon, J. F. M., Uzor, N.-E., Kesler, S. R., Wefel, J. S., Townley, D. M., Nagaraja, A. S., et al. (2016). TFEB ameliorates the impairment of the autophagy-lysosome pathway in neurons induced by doxorubicin. *Aging (Albany NY)*, *8*(12), 3507.

Marques-Aleixo, I., Santos-Alves, E., Balca, M., Moreira, P., Oliveira, P., Magalhães, J., et al. (2016). Physical exercise mitigates doxorubicin-induced brain cortex and cerebellum mitochondrial alterations and cellular quality control signaling. *Mitochondrion*, *26*, 43–57.

Marques-Aleixo, I., Santos-Alves, E., Oliveira, P., Moreira, P., Magalhaes, J., & Ascensao, A. (2018). The beneficial role of exercise in mitigating doxorubicin-induced Mitochondrionopathy. *Biochimica Et Biophysica Acta. Reviews on Cancer*, *1869*(2), 189–199.

Martinez, A., & Gil, C. (2014). cAMP-specific phosphodiesterase inhibitors: Promising drugs for inflammatory and neurological diseases. *Expert Opinion on Therapeutic Patents*, *24*(12), 1311–1321.

McDonald, B. C., Conroy, S. K., Ahles, T. A., West, J. D., & Saykin, A. J. (2010). Gray matter reduction associated with systemic chemotherapy for breast cancer: A prospective MRI study. *Breast Cancer Research and Treatment*, *123*(3), 819–828.

Meyers, C. A., Albitar, M., & Estey, E. (2005). Cognitive impairment, fatigue, and cytokine levels in patients with acute myelogenous leukemia or myelodysplastic syndrome. *Cancer*, *104*(4), 788–793.

Milne, G. W., Nicklaus, M. C., Driscoll, J. S., Wang, S., & Zaharevitz, D. (1994). National Cancer Institute drug information system 3D database. *Journal of Chemical Information and Computer Sciences*, *34*(5), 1219–1224.

Mitra, S., Siddiqui, W. A., & Khandelwal, S. (2015). C-Phycocyanin protects against acute tributyltin chloride neurotoxicity by modulating glial cell activity along with its anti-oxidant and anti-inflammatory property: A comparative efficacy evaluation with N-acetyl cysteine in adult rat brain. *Chemico-Biological Interactions*, *238*, 138–150.

Moore, H. C. (2014). An overview of chemotherapy-related cognitive dysfunction, or'chemobrain'. *Oncology*, *28*(9), 797.

Murillo Espitia, L.C. An update on neurobiological mechanisms involved in the development of chemotherapy-induced cognitive impairment (CICI).

Myers, J. S. (2010). The possible role of cytokines in chemotherapy-induced cognitive deficits. In *Chemo Fog* (pp. 119–123). Springer.

Nguyen, L. D., & Ehrlich, B. E. (2020). Cellular mechanisms and treatments for chemobrain: Insight from aging and neurodegenerative diseases. *EMBO Molecular Medicine*, *12*(6), e12075.

Nguyen, L. D., Fischer, T. T., & Ehrlich, B. E. (2021). Pharmacological rescue of cognitive function in a mouse model of chemobrain. *Molecular Neurodegeneration*, *16*(1), 1–16.

Nikolic, L., Shen, W., Nobili, P., Virenque, A., Ulmann, L., & Audinat, E. (2018). Blocking TNFα-driven astrocyte purinergic signaling restores normal synaptic activity during epileptogenesis. *Glia*, *66*(12), 2673–2683. https://doi.org/10.1002/glia.23519.

Ongnok, B., Chattipakorn, N., & Chattipakorn, S. C. (2020). Doxorubicin and cisplatin induced cognitive impairment: The possible mechanisms and interventions. *Experimental Neurology*, *324*, 113118.

Ongnok, B., Khuanjing, T., Chunchai, T., Pantiya, P., Kerdphoo, S., Arunsak, B., et al. (2021). Donepezil protects against doxorubicin-induced Chemobrain in rats via attenuation of inflammation and oxidative stress without interfering with doxorubicin efficacy. *Neurotherapeutics*, 1–19.

Pasyukova, E. G., & Vaiserman, A. M. (2017). HDAC inhibitors: A new promising drug class in anti-aging research. *Mechanisms of Ageing and Development*, *166*, 6–15.

Pisoschi, A. M., & Pop, A. (2015). The role of antioxidants in the chemistry of oxidative stress: A review. *European Journal of Medicinal Chemistry*, *97*, 55–74.

Pistritto, G., Trisciuoglio, D., Ceci, C., Garufi, A., & D'Orazi, G. (2016). Apoptosis as anticancer mechanism: Function and dysfunction of its modulators and targeted therapeutic strategies. *Aging (Albany NY)*, *8*(4), 603–619. https://doi.org/10.18632/aging.100934.

Powell, C. M., & Miyakawa, T. (2006). Schizophrenia-relevant behavioral testing in rodent models: A uniquely human disorder? *Biological Psychiatry*, *59*(12), 1198–1207.

Ramalingayya, G. V., Cheruku, S. P., Nayak, P. G., Kishore, A., Shenoy, R., Rao, C. M., et al. (2017). Rutin protects against neuronal damage in vitro and ameliorates doxorubicin-induced memory deficits in vivo in Wistar rats. *Drug Design, Development and Therapy*, *11*, 1011.

Ramanan, S., Kooshki, M., Zhao, W., Hsu, F.-C., Riddle, D. R., & Robbins, M. E. (2009). The PPARα agonist fenofibrate preserves hippocampal neurogenesis and inhibits microglial activation after whole-brain irradiation. *International Journal of Radiation Oncology, Biology, Physics*, *75*(3), 870–877.

Reis, P. A., de Albuquerque, C. F. G., Gutierrez, T., Silva, A. R., & de Castro Faria Neto, H. (2017). Role of nitric oxide synthase in the function of the central nervous system under normal and infectious conditions. In *Nitric Oxide Synthase–Simple Enzyme-Complex Roles* (pp. 55–70). *London*: InTech.

Reiser, J. (2019). *Cyclophosphamide and doxorubicin contribute to onset of chemobrain by altering epigenetic factors DNMT1, DNMT3A, Tet1, Mecp2, and CBP*. City University of New York (CUNY).

Ren, X., Boriero, D., Chaiswing, L., Bondada, S., Clair, D. K. S., & Butterfield, D. A. (2019). Plausible biochemical mechanisms of chemotherapy-induced cognitive impairment ("chemobrain"), a condition that significantly impairs the quality of life of many cancer survivors. *Biochimica et Biophysica Acta (BBA) - Molecular Basis of Disease, 1865*(6), 1088–1097.

Ren, X., Keeney, J. T., Miriyala, S., Noel, T., Powell, D. K., Chaiswing, L., et al. (2019). The triangle of death of neurons: Oxidative damage, mitochondrial dysfunction, and loss of choline-containing biomolecules in brains of mice treated with doxorubicin. Advanced insights into mechanisms of chemotherapy induced cognitive impairment ("chemobrain") involving TNF-α. *Free Radical Biology and Medicine*, *134*, 1–8.

Renouf, D., & Gill, S. (2006). Capecitabine-induced cerebellar toxicity. *Clinical Colorectal Cancer*, *6*(1), 70–71.

Roberti, A., Valdes, A. F., Torrecillas, R., Fraga, M. F., & Fernandez, A. F. (2019). Epigenetics in cancer therapy and nanomedicine. *Clinical Epigenetics*, *11*(1), 81. https://doi.org/10.1186/s13148-019-0675-4.

Roggenhofer, E., Fidzinski, P., Bartsch, J., Kurz, F., Shor, O., & Behr, J. (2010). Activation of dopamine D1/D5 receptors facilitates the induction of presynaptic long-term potentiation at hippocampal output synapses. *The European Journal of Neuroscience*, *32*(4), 598–605. https://doi.org/10.1111/j.1460-9568.2010.07312.x.

Savitz, J., Solms, M., & Ramesar, R. (2006). The molecular genetics of cognition: Dopamine, COMT and BDNF. *Genes, Brain and Behavior*, *5*(4), 311–328.

Schiaffino, L., Bonafede, R., Scambi, I., Parrella, E., Pizzi, M., & Mariotti, R. (2018). Acetylation state of RelA modulated by epigenetic drugs prolongs survival and induces a neuroprotective effect on ALS murine model. *Scientific Reports*, *8*(1), 1–13.

Seigers, R., Timmermans, J., van der Horn, H. J., de Vries, E. F., Dierckx, R. A., Visser, L., et al. (2010). Methotrexate reduces hippocampal blood vessel density and activates microglia in rats but does not elevate central cytokine release. *Behavioural Brain Research*, *207*(2), 265–272. https://doi.org/10.1016/j.bbr.2009.10.009.

Seo, E.-J., Klauck, S. M., Efferth, T., & Panossian, A. (2019). Adaptogens in chemobrain (Part I): Plant extracts attenuate cancer chemotherapy-induced cognitive impairment–Transcriptome-wide microarray profiles of neuroglia cells. *Phytomedicine*, *55*, 80–91.

Sequeira, A., Morgan, L., Walsh, D. M., Cartagena, P. M., Choudary, P., Li, J., et al. (2012). Gene expression changes in the prefrontal cortex, anterior cingulate cortex and nucleus accumbens of mood disorders subjects that committed suicide. *PLoS One*, 7(4), e35367. https://doi.org/10.1371/journal.pone.0035367.

Sheldrick, A., Krug, A., Markov, V., Leube, D., Michel, T., Zerres, K., et al. (2008). Effect of COMT val158met genotype on cognition and personality. *European Psychiatry*, *23*(6), 385–389.

Silberfarb, P. M. (1983). Chemotherapy and cognitive defects in cancer patients. *Annual Review of Medicine*, *34*(1), 35–46.

Small, B. J., Rawson, K. S., Walsh, E., Jim, H. S., Hughes, T. F., Iser, L., et al. (2011). Catechol-O-methyltransferase genotype modulates cancer treatment-related cognitive deficits in breast cancer survivors. *Cancer*, *117*(7), 1369–1376.

Sobecki, M., Mrouj, K., Camasses, A., Parisis, N., Nicolas, E., Llères, D., et al. (2016). The cell proliferation antigen Ki-67 organises heterochromatin. *eLife*, *5*, e13722. https://doi.org/10.7554/eLife.13722.

Soliman, A. M., Desouky, S., Marzouk, M., & Sayed, A. A. (2016). *Origanum majorana* attenuates nephrotoxicity of cisplatin anticancer drug through ameliorating oxidative stress. *Nutrients*, *8*(5), 264.

Sousa, H., Almeida, S., Bessa, J., & Pereira, M. G. (2020). The developmental trajectory of cancer-related cognitive impairment in breast cancer patients: A systematic review of longitudinal neuroimaging studies. *Neuropsychology Review*, 1–23.

Spooner, D. M., & Pachana, N. A. (2006). Ecological validity in neuropsychological assessment: A case for greater consideration in research with neurologically intact populations. *Archives of Clinical Neuropsychology*, *21*(4), 327–337.

Sritawan, N., Prajit, R., Chaisawang, P., Sirichoat, A., Pannangrong, W., Wigmore, P., et al. (2020). Metformin alleviates memory and hippocampal neurogenesis decline induced by methotrexate chemotherapy in a rat model. *Biomedicine & Pharmacotherapy*, *131*, 110651.

Srivastava, R. K., & Singh, P. (2020). Stem cell therapies as a therapeutic option to counter chemo brain: A negative effect of cancer treatment. *Regenerative Medicine*, *15*(6), 1789–1800.

Stenvinkel, P., Karimi, M., Johansson, S., Axelsson, J., Suliman, M., Lindholm, B., et al. (2007). Impact of inflammation on epigenetic DNA methylation–a novel risk factor for cardiovascular disease? *Journal of Internal Medicine*, *261*(5), 488–499.

Stewart, A., Bielajew, C., Collins, B., Parkinson, M., & Tomiak, E. (2006). A meta-analysis of the neuropsychological effects of adjuvant chemotherapy treatment in women treated for breast cancer. *The Clinical Neuropsychologist*, *20*(1), 76–89.

Suuberg, A. (2018). Mechanisms and treatment of chemotherapy-induced cognitive impairment ('Chemobrain'). *SSRN Electronic Journal*.

Taillibert, S., Voillery, D., & Bernard-Marty, C. (2007). Chemobrain: Is systemic chemotherapy neurotoxic? *Current Opinion in Oncology*, *19*(6), 623–627.

Tangpong, J., Cole, M. P., Sultana, R., Estus, S., Vore, M., St. Clair, W., et al. (2007). Adriamycin-mediated nitration of manganese superoxide dismutase in the central nervous system: Insight into the mechanism of chemobrain. *Journal of Neurochemistry*, *100*(1), 191–201.

Thomas, T. C., Beitchman, J. A., Pomerleau, F., Noel, T., Jungsuwadee, P., Butterfield, D. A., et al. (2017). Acute treatment with doxorubicin affects glutamate neurotransmission in the mouse frontal cortex and hippocampus. *Brain Research*, *1672*, 10–17. https://doi.org/10.1016/j.brainres.2017.07.003.

ur Rashid, H., Rasool, S., Ali, Y., Khan, K., & Martines, M. A. U. (2020). Anti-cancer potential of sophoridine and its derivatives: Recent progress and future perspectives. *Bioorganic Chemistry*, *99*, 103863.

van Dam, F. S., Boogerd, W., Schagen, S. B., Muller, M. J., Droogleever Fortuyn, M. E., Wall, E. V., et al. (1998). Impairment of cognitive function in women receiving adjuvant treatment for high-risk breast cancer: High-dose versus standard-dose chemotherapy. *JNCI: Journal of the National Cancer Institute*, *90*(3), 210–218.

Vardy, J. L., Bray, V. J., & Dhillon, H. M. (2017). Cancer-induced cognitive impairment: Practical solutions to reduce and manage the challenge. *Future Medicine*.

Vichaya, E. G., Chiu, G. S., Krukowski, K., Lacourt, T. E., Kavelaars, A., Dantzer, R., et al. (2015). Mechanisms of chemotherapy-induced behavioral toxicities. *Frontiers in Neuroscience, 9*, 131.

Wahdan, S. A., El-Derany, M. O., Abdel-Maged, A. E., & Azab, S. S. (2020). Abrogating doxorubicin-induced chemobrain by immunomodulators IFN-beta 1a or infliximab: Insights to neuroimmune mechanistic hallmarks. *Neurochemistry International, 138*, 104777.

Walczak, H. (2013). Death receptor–ligand systems in cancer, cell death, and inflammation. *Cold Spring Harbor Perspectives in Biology, 5*(5), a008698.

Walker, K. R., & Tesco, G. (2013). Molecular mechanisms of cognitive dysfunction following traumatic brain injury. *Frontiers in Aging Neuroscience, 5*, 29.

Wang, J., Hodes, G. E., Zhang, H., Zhang, S., Zhao, W., Golden, S. A., et al. (2018). Epigenetic modulation of inflammation and synaptic plasticity promotes resilience against stress in mice. *Nature Communications, 9*(1), 1–14.

Wang, P., Meng, X., Huang, Y., Lv, Z., Liu, J., Wang, G., et al. (2017). MicroRNA-497 inhibits thyroid cancer tumor growth and invasion by suppressing BDNF. *Oncotarget, 8*(2), 2825.

Wang, X.-M., Walitt, B., Saligan, L., Tiwari, A. F., Cheung, C. W., & Zhang, Z.-J. (2015). Chemobrain: A critical review and causal hypothesis of link between cytokines and epigenetic reprogramming associated with chemotherapy. *Cytokine, 72*(1), 86–96.

Wang, C., Zhao, Y., Wang, L., Pan, S., Liu, Y., Li, S., et al. (2021). C-phycocyanin mitigates cognitive impairment in doxorubicin-induced chemobrain: Impact on neuroinflammation, oxidative stress, and brain mitochondrial and synaptic alterations. *Neurochemical Research, 46*(2), 149–158.

Wefel, J. S., Lenzi, R., Theriault, R. L., Davis, R. N., & Meyers, C. A. (2004). The cognitive sequelae of standard-dose adjuvant chemotherapy in women with breast carcinoma: Results of a prospective, randomized, longitudinal trial. *Cancer: Interdisciplinary International Journal of the American Cancer Society, 100*(11), 2292–2299.

Wefel, J. S., & Schagen, S. B. (2012). Chemotherapy-related cognitive dysfunction. *Current Neurology and Neuroscience Reports, 12*(3), 267–275.

Winocur, G., Berman, H., Nguyen, M., Binns, M. A., Henkelman, M., van Eede, M., et al. (2018). Neurobiological mechanisms of chemotherapy-induced cognitive impairment in a transgenic model of breast cancer. *Neuroscience, 369*, 51–65. https://doi.org/10.1016/j.neuroscience.2017.10.048.

Xu, Y., Xin, Y., Diao, Y., Lu, C., Fu, J., Luo, L., et al. (2011). Synergistic effects of apigenin and paclitaxel on apoptosis of cancer cells. *PLoS One, 6*(12), e29169.

Yao, S., Hu, Q., Kerns, S., Yan, L., Onitilo, A. A., Misleh, J., et al. (2019). Impact of chemotherapy for breast cancer on leukocyte DNA methylation landscape and cognitive function: A prospective study. *Clinical Epigenetics, 11*(1), 1–10.

Zhang, R., Xu, S., Cai, Y., Zhou, M., Zuo, X., & Chan, P. (2011). Ganoderma lucidum protects dopaminergic neuron degeneration through inhibition of microglial activation. *Evidence-Based Complementary and Alternative Medicine*, 2011.

Zimmer, P., Mierau, A., Bloch, W., Strüder, H. K., Hülsdünker, T., Schenk, A., et al. (2015). Post-chemotherapy cognitive impairment in patients with B-cell non-Hodgkin lymphoma: A first comprehensive approach to determine cognitive impairments after treatment with rituximab, cyclophosphamide, doxorubicin, vincristine and prednisone or rituximab and bendamustine. *Leukemia & Lymphoma, 56*(2), 347–352.

Zunini, R. A. L., Scherling, C., Wallis, N., Collins, B., MacKenzie, J., Bielajew, C., et al. (2013). Differences in verbal memory retrieval in breast cancer chemotherapy patients compared to healthy controls: A prospective fMRI study. *Brain Imaging and Behavior*, 7(4), 460–477.

Further reading

Gradishar, W. J., Tjulandin, S., Davidson, N., Shaw, H., Desai, N., Bhar, P., et al. (2005). Phase III trial of nanoparticle albumin-bound paclitaxel compared with polyethylated castor oil–based paclitaxel in women with breast cancer. *Journal of Clinical Oncology*, *23*(31), 7794–7803.

Kovalchuk, A., Rodriguez-Juarez, R., Ilnytskyy, Y., Byeon, B., Shpyleva, S., Melnyk, S., et al. (2016). Sex-specific effects of cytotoxic chemotherapy agents cyclophospha-mide and mitomycin C on gene expression, oxidative DNA damage, and epigenetic alterations in the prefrontal cortex and hippocampus–an aging connection. *Aging (Albany NY)*, *8*(4), 697.

Shi, D.-D., Dong, C. M., Ho, L. C., Lam, C., Zhou, X.-D., Wu, E. X., et al. (2018). Resveratrol, a natural polyphenol, prevents chemotherapy-induced cognitive impairment: Involvement of cytokine modulation and neuroprotection. *Neurobiology of Disease*, *114*, 164–173.

Shi, D.-D., Huang, Y.-H., Lai, C. S. W., Dong, C. M., Ho, L. C., Li, X.-Y., et al. (2019a). Ginsenoside Rg1 prevents chemotherapy-induced cognitive impairment: Associations with microglia-mediated cytokines, neuroinflammation, and neuroplasticity. *Molecular Neurobiology*, *56*(8), 5626–5642.

Shi, D.-D., Huang, Y.-H., Lai, C. S. W., Dong, C. M., Ho, L. C., Wu, E. X., et al. (2019b). Chemotherapy-induced cognitive impairment is associated with cytokine dysregulation and disruptions in neuroplasticity. *Molecular Neurobiology*, *56*(3), 2234–2243.

Videnovic, A., Semenov, I., Chua-Adajar, R., Baddi, L., Blumenthal, D., Beck, A., et al. (2005). Capecitabine-induced multifocal leukoencephalopathy: A report of five cases. *Neurology*, *65*(11), 1792–1794.

CHAPTER THREE

Nephrotoxicity in cancer treatment: An update

Chaoling Chen, Dengpiao Xie, David A. Gewirtz, and Ningjun Li*
Department of Pharmacology and Toxicology, Virginia Commonwealth University School of Medicine, Richmond, VA, United States
*Corresponding author: e-mail address: ningjun.li@vcuhealth.org

Contents

Advances in Cancer Research, Volume 155
ISSN 0065-230X
https://doi.org/10.1016/bs.acr.2022.03.005

Abstract

It has been estimated that nearly 80% of anticancer drug-treated patients receive potentially nephrotoxic drugs, while the kidneys play a central role in the excretion of anticancer drugs. Nephrotoxicity has long been a serious complication that hampers the effectiveness of cancer treatment and continues to influence both mortality and length of hospitalization among cancer patients exposed to either conventional cytotoxic agents or targeted therapies. Kidney injury arising from anticancer drugs tends to be associated with preexisting comorbidities, advanced cancer stage, and the use of concomitant non-chemotherapeutic nephrotoxic drugs. Despite the prevalence and impact of kidney injury on therapeutic outcomes, the field is sorely lacking in an understanding of the mechanisms driving cancer drug-induced renal pathophysiology, resulting in quite limited and largely ineffective management of anticancer drug-induced nephrotoxicity. Consequently, there is a clear imperative for understanding the basis for nephrotoxic manifestations of anticancer agents for the successful management of kidney injury by these drugs. This article provides an overview of current preclinical research on the nephrotoxicity of cancer treatments and highlights prospective approaches to mitigate cancer therapy-related renal toxicity.

Abbreviations

ABC transporters	ATP-binding cassette transporters
ACEi	angiotensin-converting enzyme inhibitors
AIN	acute interstitial nephritis
AKI	acute kidney injury
ARBs	angiotensin receptor blockers
CCB	calcium channel blocker
CECT	contrast-enhanced computed tomography
CIN	contrast-induced nephropathy
CKD	chronic kidney disease
CPIs	immune checkpoint inhibitors
CYP450	cytochrome P450
DPP-4	dipeptidyl peptidase-4
eGFR	estimated glomerular filtration rate
FSGF	focal segmental glomerulosclerosis
5-FU	5-fluorouracil
GCs	glucocorticoids
GGT	gamma-glutamyl transpeptidase
GSTs	glutathione S-transferases
HDACs	histone deacetylases
IL-2	interleukin-2
MCD	minimal change disease
mTOR	mammalian target of rapamycin
MTX	methotrexate
NAC	N-acetylcysteine or acetylcysteine
NSAIDs	nonsteroidal anti-inflammatory drugs
OAT	organic anion transporters

OCT	organic cation transporters
PDE5	phosphodiesterase type 5
PPI	proton pump inhibitors
SLC	renal solute carrier transporter
TLR	toll-like receptor
TMA	thrombotic microangiopathy
TNF-α	tumor necrosis factor-α

1. Introduction

The kidneys represent one of the major sites for the elimination of anticancer drugs. Approximately 80% of cancer patients are administered anticancer drugs that potentially have nephrotoxic effects (Launay-Vacher et al., 2007). The nephrotoxicity of anticancer drugs remains a serious complication of cancer therapy, leading to delayed treatment, the lengthening of hospital stays, as well as increased mortality rates for cancer patients (Kitchlu et al., 2019; Lameire, Flombaum, Moreau, & Ronco, 2005). Therefore, anticancer drug-induced renal toxicity is increasingly regarded as a principal factor that limits the efficacy of cancer treatments.

Nephrotoxicity is defined as the loss of the kidneys' ability to filtrate, detoxify and/or excrete harmful metabolites due to anticancer drugs-induced injuries to renal structure and/or function. Drug-induced renal toxicity can affect different cell types in the kidneys including glomerular, tubular, interstitial or vascular cells. Both conventional chemotherapy and targeted agents can cause a diversity of renal complications such as proteinuria, electrolyte imbalance, glomerulopathy, hypertension, interstitial nephritis, and tubulointerstitial damage. Consequently, in many cases, anticancer drugs require dose modification or discontinuation in the setting of renal impairment (Naughton, 2008).

It is also of critical importance to identify risk factors and associated pathogenic mechanisms for nonchemotherapeutic nephrotoxic therapies that could potentially contribute to the deterioration of renal function in cancer patients. In addition to radiation therapy-induced kidney injury (Klaus, Niyazi, & Lange-Sperandio, 2021), concomitant use of analgesics, bisphosphonates, or other chronic disease mediations are common factors leading to renal dysfunction, which need to be taken into consideration (Lameire et al., 2005). Moreover, cancer patients with underlying compromised renal function are prone to develop acute kidney injury (AKI) and subsequent long-term chronic kidney disease (CKD) during anticancer therapy (Kitchlu et al., 2019). Unfortunately, the biochemical, physiological and molecular mechanisms underlying anticancer drug-induced renal toxicities remain poorly understood. As nephrological management for most anticancer drugs largely depend on dosage adjustment, discontinuation of treatment and/or symptom management (Chiruvella, Annamaraju, & Guddati, 2020), a comprehensive understanding of nephrotoxic mechanisms of anticancer drugs is necessary to more effectively mitigate nephrotoxicity in cancer patients. This chapter presents an overview of fundamental risk factors associated with nephrotoxicity in cancer patients, findings during the last decade related to some novel molecular pathways putatively involved in the nephrotoxicity as well as efforts to minimize cancer therapeutics-induced nephrotoxicity.

2. Risk factors for renal toxicities in cancer therapy

2.1 General risk factors for nephrotoxicity of anticancer drugs and treatment

A variety of risk factors can exacerbate the course of renal dysfunction and are likely to enhance the nephrotoxicity of cancer therapeutic agents. Renal toxicities can be initiated by the antitumor drugs themselves, elevated doses or prolonged treatment, formation of toxic crystals within intratubular lumens or concomitant use of other nephrotoxic drugs (including aminoglycosides, nonsteroidal anti-inflammatory drugs, radiographic ionic contrast media or other anticancer drugs). In addition, gene variants in hepatic and renal cytochrome P450 (CYP) enzymes or renal transporters that result in decreased metabolism or reduced drug excretion can increase intracellular anticancer drug concentrations, thereby elevating nephrotoxic risk (Miteva-Marcheva, Ivanov, Dimitrov, & Stoyanova, 2020). Fluid deficits are also one of the most common factors that potentiate nephrotoxic effects due to vomiting or diarrhea caused by antineoplastic drugs (Izzedine & Perazella, 2017; Perazella, 2009, 2012; Perazella & Moeckel, 2010). A spectrum of risk factors for renal toxicity in cancer therapeutics are listed in Table 1.

2.2 Patient-specific factors

Several patient-specific risk factors contribute to increased risk of anticancer agent-induced nephrotoxicity, including underlying disease, advanced age

Table 1 Risk factors for renal toxicities in cancer therapy.

Patient-specific factors
Older age
Female sex
Preexisting AKI or CKD
Hepatic failure
Other comorbidities (multiple myeloma, lymphoma, leukemia, renal cancer, diabetes, sepsis, and acid-basedisturbances)
Kidney-specific factors
High rate of blood delivery (20–25% of cardiac output)
Proximal tubular uptake of toxins
apical tubular uptake by endocytosis/pinocytosis
basolateral tubular transport through OAT and OCT pathways
Relatively hypoxic renal environment
high metabolic rate of tubular cells in the loop of Henle
Increased drug/toxin concentration in renal medulla and interstitium
Biotransformation of substances to reactive oxygen species causing oxidative stress
Sex difference/species difference
Tumor cell-induced alternation in expressions of renal transporters
Drug-specific factors
Prolonged dosing periods, rapid infusion rates of drugs
Potent direct nephrotoxic effects of the drug or compound
Concomitant use of nephrotoxic drugs
Competition between endogenous and exogenous toxins for transporters, increasing toxin accumulation within the tubular cell
Insoluble parent compound and metabolite with intratubular crystal precipitation (urine pH/drug pKa)

AKI, acute kidney injury; CKD, chronic kidney disease; OAT, organic anion transporters; OCT, organic cation transporters.

Modified from (Perazella, M. A. (2009). Renal vulnerability to drug toxicity. Clinical Journal of the American Society of Nephrology, 4(7), 1275-1283. doi:10.2215/cjn.02050309).

and (female) gender. Those factors can be associated with decreased estimated glomerular filtration rates (eGFR), reduced body water or lower serum albumin levels, resulting in increases in drug concentration. However, there is still no consensus as to sex difference in acute kidney injury (AKI) (Schiffl, 2020). Additionally, preexisting AKI and chronic kidney disease (CKD) may raise the risk for nephrotoxicity. For instance, AKI resulting from intravascular volume depletion results in increased drug exposure to renal cells. Nephrotic syndrome with hypoalbuminemia can lead to decreased drug binding and increased active unbound drug concentrations (Perazella, 2018). Patients with advanced hepatic failure are especially at high risk for drug-induced nephrotoxicity, given that advanced hepatic failure exhibits hypoalbuminemia, intravascular volume depletion, and underlying AKI or CKD (Yeung, Yong, & Wong, 2004). In addition to these factors, comorbidities in cancer patients such as multiple myeloma, lymphoma, leukemia, renal cancer, diabetes, sepsis and acid-base disturbances can also enhance the risk for nephrotoxicity (Izzedine & Perazella, 2017).

2.3 Kidney-specific factors

The kidney and its segments are vulnerable to drugs-induced nephrotoxicity owing to the high renal blood flow rate (25% of cardiac output). The nephrotoxic potential of drugs is highly related to the renal microenvironment. High concentration of drugs or drug metabolites can present in the glomerular ultrafiltrate. Furthermore, the renal cortex is likely to be exposed to high concentrations of toxic metabolites as this portion of the kidney receives approximately 80% of total renal blood flow; consequently, many anticancer drugs can accumulate in proximal tubular cells through both apical uptake and basolateral transport (Drozdzik, Drozdzik, & Oswald, 2021; Enomoto & Endou, 2005; Hucke & Ciarimboli, 2016). In addition to the proximal tubules, epithelial membranes of distal and collecting tubules express several carriers and transporters that facilitate bidirectional movement of substrate molecules. In contrast to the cortex region, the loop of Henle and medullary collecting duct cells in the renal medulla require high metabolic rates for energy-consuming active transport of many solutes; these excess cellular workloads generate hypoxic environments, thereby increasing susceptibility to the nephrotoxicity of anticancer drug therapies (Izzedine & Perazella, 2017; Perazella, 2010).

Renal biotransformation of drugs also plays a critical role in drug-related nephrotoxicity. Renal biotransformation involves several enzyme systems, where the generation of oxidative stress and the formation of injurious reactive oxygen species are likely to damage the kidney. For instance, glutathione S-transferases (GSTs) are well-known to catalyze the conjugation of glutathione (GSH) to xenobiotics electrophilic compounds, a central step in drug detoxication and biotransformation (Townsend & Tew, 2003). On the other hand, this mechanism could also produce strong mutagens or carcinogens due to GST polymorphisms.

GST-α is primarily expressed in the convoluted proximal tubule with some existing in the thin loops of Henle (Bauchet, Masson, Guffroy, & Slaoui, 2011). GST-π is mainly localized in podocytes, parietal cells of the Bowman's capsule, distal convoluted tubules and collecting ducts, with significantly lower levels in proximal tubules (Harrison, Kharbanda, Cunningham, McLellan, & Hayes, 1989). Glutathione (GSH) has been found to protect against cisplatin-induced renal and systemic toxicity without affecting its antitumor activity (Xu, Jiang, Liu, Qu, & Wang, 2012; Zunino et al., 1989); however, biotransformation of trichloroethylene (TCE) conjugated GSH can result in nephrotoxic metabolites such as S-(1,2-dichlorovinyl)-L-cysteine (DCVC) through γ-glutamyltransferase and subsequent β-elimination reaction as well as the metabolite trichloroethylene-cysteine S-conjugate sulfoxide (DCVCS, a highly reactive Michael acceptor) via flavin-containing monooxygenase 3. N-acetylcysteine (NAC, or acetylcysteine) is a widely available nutritional supplement and a medication that can rescue the GSH depletion. NAC has been extensively studied for its antioxidant effects against both cancers and kidney injuries (Kwon, 2021; Mlejnek, Dolezel, Kriegova, & Pastvova, 2021; Zhang et al., 2021). NAC has been shown to prevent radiocontrast, cisplatin- and radiation-induced nephrotoxicity in preclinical studies (Güntürk et al., 2019; Huang et al., 2019; Mercantepe et al., 2019; Richter & Crannage, 2015). Moreover, a relatively recent study demonstrated the development of tumor-selective GSH-dependent Michael acceptor prodrugs of 6-mercaptopurine (6-MP) and 6-thioguanine (6-TG) which can have antitumor effects with reduced toxicity in mice (Zhang & Elfarra, 2018).

P450 isoenzymes and UDP-glucuronosyltransferase (UGT) also play significant roles in drug biotransformation and detoxification in the kidneys. Of note, some of these enzymes are particularly expressed in the kidneys and specific renal segments (Bauchet et al., 2011; Knights, Rowland, & Miners, 2013). Consequently, alternations in transporter activity such as loss of function mutations and/or competition for carrier proteins/enzymes can hinder parent drug/metabolite excretion and induce nephrotoxicity.

Alternation in the activity of these enzymes can be attributed to various factors, such as age, ethnicity, disease states and genetic polymorphisms.

2.3.1 *Genetic polymorphisms, sex and other kidney-related factors in renal enzymes*

The activity of metabolizing enzymes in the renal tubules can differ between individuals and sexes (Bozina, Bradamante, & Lovrić, 2009). Genetic polymorphisms may affect enzyme expression. Multiple single nucleotide polymorphisms (SNPs) for the organic cation transporter (OCT) genes (SLC22A1, SLC22A2 and SLC22A3) are found to alter their transport functions, where a gain-in-function mutation results in drug accumulation in renal proximal tubule cells (Huang et al., 2020; Yee et al., 2018; Zazuli et al., 2020). However, Fujita et al showed that a polymorphism in renal ATP-binding cassette transporters (ABC transporters) or renal solute carrier (SLC) transporter did not directly affect cisplatin-induced nephrotoxicity in patients with esophageal cancer (Fujita et al., 2016). A study of AKI outcome based on serum creatinine (SCr) showed associations between genetic variants of ERCC1, ERCC2, SLC22A2 and cisplatin-induced nephrotoxicity in adult testicular cancer, suggesting that genetic variations are involved in the inter-individual susceptibility to cisplatin-induced nephrotoxicity (Trendowski et al., 2019).

Interestingly, the expression levels of transporters in renal proximal tubules also differ across species and sexes. A recent study showed that the sex differences in protein abundance of 12 transporters were only significant in rodents and dogs but not in humans (Basit, Radi, Vaidya, Karasu, & Prasad, 2019), indicating that cross-species and sex-dependent protein abundance data should be taken into consideration when evaluating interspecies scaling in drug clearance and assessing kidney toxicity for new therapeutic candidates under development.

2.4 Drug-specific factors

Nephrotoxicity can be augmented and prolonged with repeated drug dosing. Hartmann et al showed that patients who received 2 cycles of a single 50-mg/m^2 dose of cisplatin had a 24% decrease in GFR after the second cycle (Hartmann, Kollmannsberger, Kanz, & Bokemeyer, 1999). Patients with higher doses and prolonged durations of treatment with aminoglycosides or amphotericin showed increased nephrotoxicity (Perazella, 2009). In addition, the cumulative dose, rate of administration, and increased frequency of anticancer drug administration are associated with an increased risk for nephrotoxicity (Caglar et al., 2002; Kobayashi et al., 2016).

Combinations of nephrotoxic drugs with anticancer drugs also contribute to enhanced nephrotoxicity. Many drugs, including angiotensin-spi1; converting enzyme (ACE) inhibitors, angiotensin receptor blockers (ARBs), immunosuppressants such as cyclosporins, tacrolimus and nonsteroidal anti-inflammatory drugs (NSAIDs), can influence intraglomerular hemodynamics and decrease glomerular filtration rates (Ghane Shahrbaf & Assadi, 2015; Patel & Sapra, 2021; Perazella, 2005). Furthermore, drugs such as aminoglycosides, amphotericin B, adefovir, foscarnet and cisplatin can potentiate tubular cell toxicity via the formation of free radicals, mitochondrial dysfunction, and transport systems damage (Perazella, 2018). Importantly, the use of drugs that cause drug-related acute interstitial nephritis (AIN) should not be ignored; these include the antituberculosis drug, rifampin, antibiotics, proton pump inhibitors (PPI), NSAIDs and many types of anticancer drugs (Martínez-Valenzuela et al., 2021; Mérida & Praga, 2019; Moledina & Perazella, 2016; Nagata, Ohji, & Iwata, 2019; Perazella & Markowitz, 2010). In addition, chronic interstitial nephritis can arise from medications, including analgesics, lithium, anticancer drugs, and calcineurin inhibitors (Patel & Sapra, 2021). Of note, despite the high frequency of NSAIDs-induced AIN, the occurrence of nephrotic syndrome does not appear to be similarly increased (González et al., 2008). This suggests that patients exposed to NSAIDs-related AIN rarely reach the nephrotic range; instead, they are prone to present increased proteinuria rather than other types of drug-induced AIN (Markowitz & Perazella, 2005). Therefore, monitoring of renal injury and function is crucial for patients receiving combinations of anticancer drugs and other nephrotoxic drugs.

Drugs or metabolites that are insoluble in the urine may cause acute crystalline nephropathy upon precipitation in distal tubular lumens and collecting ducts. This process can occur with or without an interstitial reaction and can result from the reduction of urinary flow rates, change of urine pH/drug pKa, excessive doses and rapid drug infusion rates (Markowitz & Perazella, 2005; Perazella, 2018). Agents such as the antiviral, acyclovir, and the antibiotic, ampicillin, have demonstrated common etiologies of insoluble crystal formation within renal tissue (Chávez-Iñiguez et al., 2018; Garnier et al., 2020; Yarlagadda & Perazella, 2008).

Certain drugs that interfere with renal transporters may also enhance the nephrotoxicity of anticancer agents. For instance, cimetidine, trimethoprim, pyrimethamine and salicylates block tubular secretion of creatinine with or without affecting their glomerular filtration (Andreev, Koopman, & Arisz, 1999). In addition, a variety of drugs can act as substrates competing for a renal transporter and further increase drug deposition in kidneys

(Drozdzik et al., 2021; Zhou, Zeng, & Shu, 2021). Moreover, Hong et al. demonstrated that repeated contrast-enhanced computed tomography (CECT) within 72h can predispose cancer patients to contrast-induced nephropathy (CIN) (Hong et al., 2016). Therefore, concomitant use of radiocontrast and NSAIDs should be avoided with nephrotoxic anticancer drugs (Ozkok & Ozkok, 2017; Perazella, 2009). The involvement of novel anticancer drugs in AIN and other kidney injuries has been discussed in previous review articles (Izzedine & Perazella, 2017; Martínez-Valenzuela et al., 2021). Renal manifestations associated with commonly used anticancer drugs and recent novel cancer therapies are listed in Table 2.

3. Nephrotoxicity of chemotherapeutic agents and management

As basic information describing anticancer drug-induced nephrotoxicity has been extensively reviewed and summarized elsewhere in the literature, this review will focus on the most recent findings related to the kidney damage and promising management for the nephrotoxicity of anticancer drugs. A summary of current and potential strategies for the management of cancer therapeutics-induced nephrotoxicity is provided in Table 3.

3.1 Platinum agents

Cisplatin has been used as an effective platinum-containing chemotherapeutic drug for several decades in a wide range of cancers (Brown, Kumar, & Tchounwou, 2019). Cisplatin elicits its antitumor effects through binding to DNA, generation of DNA cross links and damage, a blockade to cell division and, in many cases, the promotion of apoptosis (Tchounwou, Dasari, Noubissi, Ray, & Kumar, 2021). Nephrotoxicity has been recognized as the most critical side effect that limits the clinical use of cisplatin (Manohar & Leung, 2018). Current measures to reduce cisplatin-induced nephrotoxicity are limited to hydration and magnesium preloading (McKibbin et al., 2016; Workeneh, Uppal, Jhaveri, & Rondon-Berrios, 2021; Yoshida et al., 2014). Unfortunately, the effectiveness of these approaches is modest, at the best. The pathways that contribute to cisplatin-induced kidney damage are complex and include apoptosis, necrosis, necroptosis, oxidative stress, endoplasmic reticulum (ER) stress, mitochondrial dysfunction, autophagy, inflammatory responses, and cell cycle dysregulation (Duan, Cai, Li, & Chen, 2020; Sharp & Siskind, 2017; Xu et al., 2015; Zhang, Ye, Tew, & Townsend, 2021). In the course of the last decade, numerous studies have

Table 2 Renal manifestation associated with commonly used anticancer drugs and recent novel cancer therapies.

Renal manifestation						
Tubular injury	**Glomerular injury**	**TMA**	**Fanconi syndrome**	**Proteinuria**	**AKI**	**Tubulointerstitial disease**
Anticancer agents						
Ifosfamide	MitomycinC	Ipilimumab	Ifosfamide	Bevacizumab	Doxorubicin	Ipilimumab
Cisplatin	Doxorubicin	Bevacizumab	Lenalidomide	Carfilzomib	Nivobumab	Nivobumab
Doxorubicin	Actinomycin	Carfilzomib		Everolimus	Pembrolizumab	Pembrolizumab
Cetuximab	Bevacizumab	Bortezomib		Sirolimus	Cisplatin	Sorafenib
Panitumumab	Carfilzomib			Clofarabine	Carfilzomib	Sunitinib
Methotrexate	IFN-α			Vemurafenib	Sirolimus	Vemurafenib
Vemurafenib	Lenalidomide				Clofarabine	
Crizotinib					Dabrafenib	
Everolimus					Vemurafenib	

AKI, acute kidney injury; IFN-α, interferon alpha; TMA, thrombotic microangiopathy

Table 3 Summary of current and potential strategies for the management of cancer therapeutics-induced nephrotoxicity.

Agent	Nephrotoxicity	Management	Potential Strategies
Platinum agents (Cisplatin)	AKI; CKD; Fanconi-like syndrome (hypomagnesemia, fanconi-like syndrome, hypocalcemia, proteinuria, hyperuricemia)	Hydration; Mannitol; Furosemide; Magnesium supplement	N-acetylcysteine; Inhibition of renal uptake transporters; Reducing inflammation response and cellular death; Use of DPP-4 inhibitors and pioglitazone; Appropriate dose of ACEi, ARB, CCB, aprepitant, thrombin inhibitor or PPI; Targeting GGT enzyme and GGT-associated inflammatory pathway; Activation of Pregnane X receptor; Use of sodium thiosulfate or probiotics; Tempol (an antioxidant)
Alkylating agents (Ifosfamide, cyclophosphamide)	Tubular damage (glucosuria, aminoaciduria, polyuria, and proteinuria); Interstitial inflammation and renal fibrosis	Control total dose usage; Monitor and supplementation of electrolyte	N-acetylcysteine; Targeting oxidative stress, mitochondrial dysfunction, apoptosis; DPP-4 or vasopressin V2 receptor and use of acetylcysteine
Antitumor antibiotics (Mitomycin C, Doxorubicin, Actinomycin)	AKI; Nephrotic syndrome; Endothelial cell and podocyte injury	Control total dose usage	Combination of cisplatin with mitomycin C; Targeting oxidative stress, inflammatory and apoptosis; Inhibition of renal podocyte detachment (angiopoietin-like-3)

Antimetabolites cancer drugs (Methotrexate, Clofarabine)	Decreased GFR; Tubular injury; AKI; Proteinuria	Hydration; Alkalinization; Hemodialysis	Use of Glucarpidase; Targeting oxidative stress, inflammation, and apoptosis; Inhibiting miRNA-145-5p-mediated oxidative stress; Targeting ER stress signaling; Restoring OAT3 expression; Use of Rebamipide or HA-230 adsorber; Targeting miR-181a; Use of camel milk, hesperidin and curcumin
EGFR inhibitors (Cetuximab, Panitumumab	Tubular injury; Glomerulopathy; Nephritic syndrome; Electrolyte disorder	Glucocorticoid; Discontinuation; Monitor and supplementation of electrolyte	Sodium restriction and diuretics treatment
VEGF inhibitors (Bevacizumab, Sorafenib, Sunitinib)	Proteinuria; TMA; Hypertension; Glomerulopathy; Electrolyte disorder; Interstitial nephritis	ACEi or ARBs	Use of PDE5 inhibitor or selective ETA receptor antagonists; Targeting downstream of ET-1 signaling
Immune checkpoint inhibitors (Ipilimumab, Nivolumab, Pembrolizumab)	AKI; Acute tubulointerstitial nephritis; Electrolyte disturbance and TMA	Glucocorticoids; Discontinuation	Use of weak checkpoint blockade; TNF-α blocker
Proteasome inhibitors (Carfilzomib, Bortezomib)	TMA; Proteinuria; AKI; Podocyte injury	Anti-hypertension; Intravenous hydration; Reduction of dose; ACEI or ARBs	Use of acetylcysteine or glucocorticoid; Targeting oxidative stress and inflammation

Continued

Table 3 Summary of current and potential strategies for the management of cancer therapeutics-induced nephrotoxicity.—cont'd

Agent	Nephrotoxicity	Management	Potential Strategies
mTOR protein kinase inhibitors (Sirolimus, Everolimus)	AKI; Proteinuria	ACEI or ARBs; Discontinuation	NA
Biologic agents (IL-2, IFN-α)	Hypotension; Edema; Oliguria; MCD; FSGF; TMA	Fluid management; Discontinuation; Urine output monitoring	Low and intermediate-dose of vasopressors (dopamine); Targeting vascular endothelial protein tyrosine phosphatase (VE-PTP)
BRAF inhibitors (Dabrafenib, Vemurafenib)	Hyperkalemia; Fanconi syndrome; MCD; AKI; Crystal nephropathy	Routine monitoring of electrolytes and serum creatinine	Combined therapy of MEK inhibitor, cobimetinib and vemurafenib
ALK inhibitor (Crizotinib)	Electrolyte disorder; Renal cyst; Acute tubular necrosis; Glomerular mesangiolysis	Discontinuation	Monitoring of renal function and examination of kidney biopsy

ACEi, angiotensin-converting enzyme inhibitors; AKI, acute kidney injury; ALK, anaplastic lymphoma kinase; CCB, calcium channel blocker; CKD, chronic kidney disease; DPP-4, dipeptidyl peptidase-4; ET, endothelin; FSGF, focal segmental glomerulosclerosis; GGT, gamma-glutamyl transpeptidase; MCD, minimal change disease; NA, not available; OAT, organic anion transporters; PDE5, phosphodiesterase type 5; PPI, proton pump inhibitors; TMA, thrombotic microangiopathy.

explored potential therapeutic strategies involving inhibition of drug uptake transporters such as organic cation transporter 2 (OCT2) and copper transporter 1 (CTR1), and reducing the inflammation response (McSweeney et al., 2021). Nonetheless, clinical applications by targeting the above mechanistic pathways in cisplatin-induced kidney toxicity are challenging as many of these molecular pathways are at the same time crucially involved in the cytotoxic activity of cisplatin against tumor cells; consequently, alterations in their functions could possibly interfere with cisplatin-induced anti-tumor effects (Volarevic et al., 2019). Notable efforts have also been made in the potential application of over-the counter (OTC) drugs and prescription drugs against cisplatin-induced nephrotoxicity. The most recent findings on the potential applications of clinically used drugs in cisplatin-induced nephrotoxicity are discussed below.

3.1.1 Prospective use of diabetes drugs

A retrospective study revealed that the dipeptidyl peptidase-4 (DPP-4) inhibitors reduced cisplatin-induced AKI in cancer patients with diabetes mellitus by suppressing inflammation and promoting tubular regeneration (Iwakura et al., 2019, 2020). Another DPP-4 inhibitor, gemigliptin, was shown to protect against cisplatin-induced nephrotoxicity through the inhibition of apoptosis and inflammatory responses by increasing heme oxygenase-1 (HO-1) and NAD(P)H:quinone oxidoreductase 1 (NQO1) expression (Choi, Leem, & Lee, 2017). The outcome of a clinical trial evaluating the effect of gemigliptin on cisplatin-induced nephrotoxicity is not yet known (ClinicalTrials.gov Identifier: NCT02250872). Cisplatin-induced nephrotoxicity was also found to be prevented by the thiazolidinedione-type diabetes drugs, pioglitazone, via inhibition of the p53-mediated mitochondrial apoptotic pathway via SIRT1 activation (Zhang et al., 2020). These findings suggest the potential benefits of these drugs in cancer patient with diabetes, and possibly, non-diabetic cancer patients. However, more clinical data are needed for their use as a renoprotective strategy against cisplatin-induced nephrotoxicity.

3.1.2 Prospective use of antihypertensive agents

Several types of antihypertensive agents appear to represent potential strategies for mitigation of cisplatin nephrotoxicity. One study demonstrated protective effects of enalaprilat, an angiotensin-converting enzyme inhibitor (ACEi), for cisplatin-induced renal toxicity in mice not only by reducing the formation of angiotensin II, but also by reversing cisplatin-induced upregulation of the kinin B1 receptor and decreased aminopeptidase P activity, thereby preventing tubular cell apoptosis and inflammation (Estrela et al., 2020). In contrast, a recent single-center observational study revealed that patients exposed to target or above target dosage of ACEi or angiotensin II receptor blockers (ARBs) had a higher risk of AKI in emergency medical admissions (Feidakis et al., 2021). Hence, dosage adjustment of ACEi for cisplatin-treated patients should not be overlooked for future potential application in cisplatin-induced nephrotoxicity.

Amlodipine, a calcium channel blocker (CCB), was shown to inhibit the gamma-glutamyl transpeptidase (GGT) enzyme, which metabolizes platinum-glutathione (Pt-GSH)-conjugates to a reactive toxic thiol that causes tubular cell death. Amlodipine can also interfere with the GGT-associated inflammatory pathway (Azouz, Abdel-Nassir Abdel-Razek, & Abo-Youssef, 2020). Although the β-adrenoceptor blockers carvedilol and high-dose propranolol exhibited renoprotection against cisplatin renal dysfunction, they failed to improve cisplatin-induced electrolyte imbalance (Esmaeeli, Keshavarz, Dehdar, Assadi, & Seyedabadi, 2020). As a result, we should carefully interpret the effects of β-adrenoceptor blockers on electrolyte levels in cisplatin-treated patients. Overall, these findings indicate potential uses of these antihypertensive agents in cisplatin-induced nephrotoxicity.

3.1.3 Prospective use of other clinically available drugs

Some other types of clinically available drugs are also found to mitigate cisplatin nephrotoxicity. A thrombin inhibitor, dabigatran (Pradaxa), may offer beneficial effects against cisplatin-mediated nephrotoxicity through inhibiting the thrombin-associated apoptotic and oxidative effects (Ewees et al., 2021). In addition, proton pump inhibitors (PPIs), omeprazole and pantoprazole, were demonstrated to provide protection against cisplatin-induced kidney injury in both *in vivo* animal studies and randomized controlled clinical trial (ClinicalTrials.gov Identifier: NCT04217512) (Gao et al., 2020; Ghonaim, El-Haggar, & Gohar, 2021; Ismail, El-Awady, & Hassan, 2020). The underlying mechanism for the renoprotection of PPIs could be by suppressing the release and production of inflammatory cytokines stimulated by cisplatin. On the other hand, data from the US FDA adverse event reporting system showed a high correlation between PPIs use and the AKI and CKD events. Hence, monitoring of renal function and dosage modification of PPIs would be necessary in patients receiving cisplatin treatment. Finally, aprepitant is an antiemetic drug provided for cisplatin-treated patients. Interestingly, a novel role for aprepitant was revealed as an effective option for cisplatin-induced nephrotoxicity through anti-oxidative and anti-inflammatory effects (Un et al., 2020).

3.1.4 Prospective targets and other compounds

Several novel mechanisms underlying protective effects against cisplatin-induced renal injury were discovered recently. Pregnane X receptor (PXR), a master transcription factor of xenobiotic detoxification, was shown to play a protective role in cisplatin-AKI mediated by activating the PI3K/AKT pathway (Luan et al., 2021). The hydrogen sulfide metabolite, sodium thiosulfate (STS), was identified as a promising candidate molecule that could protect against renal toxicity following hyperthermic intraperitoneal chemotherapy (HIPEC) with cisplatin; this protection was possibly mediated by the binding of STS to free platinum (Laplace et al., 2020; Zhang, Dugbartey, Juriasingani, & Sener, 2021). In addition, supplementation with the probiotics *Lactobacillus reuteri* and *Clostridium butyricum* was demonstrated to alleviate cisplatin nephrotoxicity by restoring gut microbiome dysbiosis, thus further reducing uremic toxin production in rats (Hsiao et al., 2021). Other compounds, such as the epoxyeicosatrienoic acid analog (EET-F01) by kidney-targeted delivery, and Pevonedistat, a NEDD8-activating enzyme inhibitor, were recently found to downregulate the expression of inflammatory mediators to relieve cisplatin-induced nephrotoxicity (El-Far & El-Mesery, 2021; Imig et al., 2021). Tempol is a nitroxide that has an antioxidant property and is shown to reverse cisplatin AKI in mice through decreasing kidney injury markers and restoration of aquaporins (AQP2) (Afjal et al., 2020). Currently, Tempol is undergoing clinical trial in head and neck cancer patients with cisplatin and radiation treatment (ClinicalTrials.gov Identifier: NCT03480971).

Overall, the exploration of new targets and mechanism of mitigation by clinically available medications should provide a pathway for the discovery of novel and readily translatable treatments for kidney protection in cancer therapy. Also, targeting the prospective signaling pathways, such as those discussed above, the transcription factor of xenobiotic detoxification, microbiome, supplementation of sodium thiosulfate and kidney-targeted drug delivery, are promising strategies to attenuate kidney injury caused by cisplatin.

3.2 Alkylating agents

Ifosfamide and cyclophosphamide are two of the most widely used alkylating chemotherapeutic drugs that are also currently undergoing clinical trials in combination treatment with different novel cancer therapies (Fox et al., 2021; Zsiros et al., 2021). This class of drugs is quite frequently associated with kidney dysfunction (Bhat, Kalthur, Padmashali, & Monappa, 2018; Ensergueix et al., 2020). Either the prodrug forms, their metabolites or intermediates may mediate nephrotoxic effects to renal tubular cells (Dobrek, Nalik-Iwaniak, Fic, & Arent, 2020). A retrospective study revealed that in patients who received ifosfamide treatment, kidney biopsies showed tubular necrosis, vacuolation and nuclear atypias in renal epithelial cells, as well as interstitial inflammation and renal fibrosis (Ensergueix et al., 2020). Management of ifosfamide-induced nephrotoxicity has generally been limited to supportive care. However, a growing body of reports have identified several new targets mediating alkylating agents-induced kidney injury through e.g. increased oxidative stress, mitochondrial dysfunction, apoptosis or activation of the arginine vasopressin V2 receptor that leads to anticancer drug-induced nephrogenic syndrome of inappropriate antidiuresis (NSIAD) (Kim, Jo, & Kim, 2021).

A recent study found that the antioxidant, Carnosine, alleviated ifosfamide-induced oxidative stress as well as mitochondrial impairment, and consequently attenuated renal injury and electrolyte imbalance (Ommati et al., 2020). Similarly, *Annona* species ethanolic extracts reduced the renal toxicity of ifosfamide by suppressing oxidative stress, inflammation and apoptosis in a rat kidney model (Abd-Elrazek, Shapana, Shukry, & Galilah, 2021). Mesna is the clinically approved drug to prevent ifosfamide- and cyclophosphamide-induced hemorrhagic cystitis; however, its preventive action towards tubular toxicity of ifosfamide and cyclophosphamide remains undetermined (Reddy & Winston, 2021). In the rat model of cyclophosphamide- and ifosfamide-induced cystitis, the antioxidant, acetylcysteine, which acts as a donor of —SH groups similar to the action of mesna, was reported to produce both uro- and nephron-protective effect against cyclophosphamide and ifosfamide treatment (Dobrek et al., 2020). Alogliptin, a selective inhibitor of the enzyme dipeptidyl peptidase-4 (DPP-4), is an FDA-approved drug for type 2 diabetes. Interestingly, Alogliptin was recently shown to produce beneficial effects in cyclophosphamide-induced nephrotoxicity as indicated by inhibiting the MAP3K/JNK/SMAD3 signaling cascade, which can initiate oxidative stress and production of inflammatory and fibrotic mediators (Salama, Nasr, Abdelhakeem, Roshdy, & ElGamal, 2020). The dual beneficial effects of Alogliptin could be a novel therapeutic approach for cancer patients with diabetes.

A pentadecapeptide derived from *Cyclina sinensis* was shown to mitigate cyclophosphamide-induced kidney injury by activating the antioxidative enzymes superoxide dismutase, glutathione peroxidase and catalase (Jiang et al., 2020). A more recent study reported that cyclophosphamide-induced inflammation and oxidative stress could be ameliorated by pretreatment with gallic acid, a nature phenolic compound, in mice (Baharmi et al., 2021). Cyclophosphamide is known to be one of the major causes of the syndrome of inappropriate antidiuretic hormone secretion (SIADH) in cancer patients (Khan, Waguespack, & Ahmed, 2019; Kitchlu & Rosner, 2019). Tolvaptan, a selective vasopressin V2 receptor antagonist, is used for the treatment of hyponatremia in patients with heart failure or SIADH in cancer patients. The protective effect of tolvaptan against cyclophosphamide-induced nephrotoxicity was recently demonstrated through the attenuation of apoptosis markers in rat models (El-Shabrawy et al., 2020). While this growing body of evidence does identify new potential strategies for mitigation of alkylating agents-induce nephrotoxicity, it should be noted that those findings need to be confirmed in clinical trials.

3.3 Antitumor antibiotics

Patients receiving antitumor antibiotics such as mitomycin C, actinomycin and doxorubicin can develop nephrotoxicity (Groff, Kozak, Boehmer, Demko, & Diamond, 1997; Sun et al., 2013). While mitomycin C was described to be involved in end-stage renal disease, a retrospective report showed that the combination of cisplatin and mitomycin C in laparoscopic hyperthermic intraperitoneal chemotherapy (HIPEC) did not, in fact, increase the incidence of AKI in gastric cancer (Kapoor et al., 2019).

Doxorubicin (adriamycin) has been reported to cause AKI and nephrotic syndrome (Carron, Padilla, & Maurizi Balzan, 2014; Yemm, Alwan, Malik, & Salazar, 2019). More recent studies identified several new compounds/agents to protect against doxorubicin-induced nephropathy, such as omega-3 fatty acids, flavonoids, *Acacia hydaspica* tannin-rich ethyl acetate fraction and naringenin, through mechanisms of reducing oxidative stress, inflammation, apoptosis as well as renal podocyte detachment (Afsar, Razak, Almajwal, & Al-Disi, 2020; Khan et al., 2020; Navarro-Hortal et al., 2020; Saleh et al., 2020). A novel compound, YH0618, was reported to reduce renal toxicity of adriamycin through a FOXO4-mediated Bcl-2 Bax/Bcl-2 mechanism (You et al., 2019). Curcumin demonstrated improvement of doxorubicin-induced toxicity in renal podocytes through the activation of Nrf2, inhibition of NF-κB activity, and the upregulation of podocin (Fan et al., 2020). Loss of podocytes is one of the mechanisms involved in adriamycin-induced glomerulosclerosis; however, the mechanistic pathway is unclear. Deficiency of angiopoietin-like-3 (ANGPTL3) was found to protect against adriamycin-induced glomerulosclerosis and podocyte loss in mice (Dai et al., 2019). Though inhibition of ANGPTL3 seems a promising approach against adriamycin nephrotoxicity, it might block the beneficial effect of ANGPTL3 in renal cell carcinoma metastasis (Zhang, Zhang, Feng, & Cao, 2021; Zhao et al., 2019). Therefore, the prospective treatment using the ANGPTL3 inhibitor, evinacumab, which is currently tested in patients with homozygous familial hypercholesterolemia (Raal et al., 2020), in adriamycin-induced nephropathy should be carefully evaluated in patients with renal cell carcinoma metastases (Wilson et al., 2021).

3.4 Antimetabolites cancer drugs

Antimetabolites including methotrexate, 5-fluorouracil, clofarabine, and gemcitabine have been shown to cause renal insufficiency and tubular injury. Methotrexate (MTX) is a classical antifolate, which interferes with folate metabolism. MTX is used to treat a broad range of cancers such as acute lymphoblastic leukemia, head and neck cancer, and breast cancer, among others (Koźmiński, Halik, Chesori, & Gniazdowska, 2020). In addition to plasma MTX concentration monitoring, hydration, and alkalinization to prevent/ameliorate the MTX-induced nephrotoxicity (Howard, McCormick, Pui, Buddington, & Harvey, 2016), recent consensus guidelines and case reports have suggested that glucarpidase (Carboxypeptidase G2) can provide nonrenal elimination of MTX and should be used with folic acid rescue to attenuate MTX-induced nephrotoxicity (Misra, Santagostino, Dine, & Bonhomme Faivre, 2019; Ramsey et al., 2018; Young et al., 2019).

A number of studies have identified natural compounds and other potential approaches to mitigate nephrotoxicity of MTX. Natural compounds such as Rosmarinic Acid and Apigenin that exhibit antioxidant and anti-inflammatory properties have been reported to attenuate MTX nephrotoxicity in rodent models (Jafaripour et al., 2021; Sahindokuyucu-Kocasari, Akyol, Ozmen, Erdemli-Kose, & Garli, 2021). Another natural product,

Dioscin, can relieve MTX-induced kidney damage via inhibiting miRNA-145-5p-mediated oxidative stress (Li et al., 2021). A MTX metabolite, 7-OH MTX, has been considered as a primary toxic metabolite responsible for nephrotoxicity due to its lower water solubility than MTX (Holmboe, Andersen, Mørkrid, Slørdal, & Hall, 2012). A study found that Nobiletin, a flavonoid isolated from *Citrus aurantium* L, could reduce 7-OH MTX nephrotoxicity via endoplasmic reticulum stress-dependent PERK/CHOP signaling and protect tubular cell survival (Song et al., 2021). A compound modified from Paeoniflorin, paeoniflorin-6′-O-benzene sulfonate (CP-25), showed protection against MTX-renal toxicity by preventing tubular cells apoptosis and facilitating MTX excretion through recovering OAT3 expression (Wei et al., 2021).

Rebamipide is a gastroprotective drug for the treatment of gastric ulcers and gastritis. It was found that rebamipide could potentially mitigate nephrotoxicity of MTX through activation of NRF-2/SIRT-1/FOXO-3 and mTOR/PI3K/AKT signaling while inhibiting NF-κB-p65/TLR-4 (Elmansy, Seleem, Mahmoud, Hassanein, & Ali, 2021). Moreover, using HA-230 adsorber in hemadsorption procedure was found to be a new therapeutic approach to reduce MTX toxicity in pediatric patients with acute lymphocytic leukemia who had delayed MTX clearance after high-dose MTX treatment (Sazonov et al., 2021).

5-Fluorouracil (5-FU), another antimetabolite cancer drug, has multiple mechanisms of action through inhibition of DNA synthesis and misincorporation into DNA. Although patients with renal impairment do not require dose adjustment of 5-FU (Lexicomp, 2022), there is increasing *in vivo* and *in vitro* evidence that 5-FU can cause renal dysfunction as indicated by the promotion of apoptosis, induction of oxidative stress, and tubular injury. Recent studies have demonstrated potential approaches to mitigate 5-FU nephrotoxicity. Inhibition of miR-181a has been found to attenuate 5-FU-induced mesangial cell apoptosis, inflammation and kidney injury (Liu et al., 2018). Camel milk is shown to have renoprotection against 5-FU, which could be mediated by suppressing MAPKs, NF-kappaB and PI3K/Akt/eNOS (Arab, Salama, & Maghrabi, 2018). In addition, hesperidin and curcumin play beneficial roles in 5-FU-induced nephrotoxicity by inhibiting oxidative stress, lipid peroxidation, apoptosis, and renal dysfunction (Gelen et al., 2021). Herein, we have summarized some recently discovered targets and pathways, such as nonrenal metabolizing enzyme, transporter, the utilization of HA-230 adsorber and natural compounds, which could be used as potential therapeutic strategies in MTX and 5-FU nephrotoxicity in the future.

3.5 Epidermal growth factor receptor pathway inhibitors (EGFR inhibitors)

EGFR inhibitors have been used in the treatment of various malignancies including colon, head, neck, and non-small cell lung cancers. Monoclonal antibodies such as cetuximab and panitumumab are EGFR inhibitors associated with tubular toxicity, glomerulopathies and electrolyte disorders. (Izzedine, Boostandoost, & Mathian, 2017; Jhaveri, Wanchoo, Sakhiya, Ross, & Fishbane, 2017). Panitumumab-induced immune complex glomerulonephritis could be mitigated by drug discontinuation and glucocorticoids treatment (Izzedine et al., 2017). Hypomagnesaemia of EGFR inhibitors is the most common side effect that could lead to renal toxicity (Jhaveri et al., 2017). Therefore, electrolyte disturbances should be carefully monitored and managed with subsequent fluid and sodium restriction, diuretics treatment, and magnesium supplementation or discontinuation of EGFR inhibitors.

3.6 Vascular endothelial growth factor pathway inhibitors (VEGF inhibitors)

The vascular endothelial growth factor (VEGF) plays an essential role in angiogenesis, which provides oxygen and nutrients to support tumor growth and metastasis. The VEGF inhibitors bevacizumab, sorafenib and sunitinib are clinically used for metastatic colon cancer, rectal cancer, non-small cell lung cancer, and breast cancer, among others (Qin et al., 2019). Common renal toxicities of VEGF inhibitors include proteinuria, glomerular disease and thrombotic microangiopathy (TMA) (Hanna et al., 2019; Shye et al., 2020). Currently, there are no molecule-specific targeted therapies for VEGF inhibitor-induced renal adverse effects, except for supportive care, dose reduction, medication discontinuation or treatment with ACEi or ARBs for proteinuria. It should be noted that ACEi and ARBs also contribute to an increased risk of AKI under certain conditions. Hence, it would be appropriate to monitor renal function and individualize treatment in patients with VEGF inhibitor if using ACEi or ARBs for renal side effects (Porta et al., 2020).

Results from a rat experimental model demonstrated that the phosphodiesterase type 5 (PDE5) inhibitor, sildenafil, reduced sunitinib-induced proteinuria (Lankhorst et al., 2014). Unfortunately, the clinical application of sildenafil in sunitinib-induced nephrotoxicity has not yet been verified. A better understanding of the underlying mechanism(s) of VEGF inhibitor-induced nephrotoxicity could lead to the development of a selective compound or adjuvant drug to minimize the off-target effects of VEGF inhibitors. The side effects of VEGF inhibitors are partially attributed to the off-target effect from VEGF inhibitors-induced circulating endothelin-1 levels (Lankhorst et al., 2015). A recent study showed that sunitinib-induced albuminuria can be improved by a selective endothelin (ET_A) receptor antagonist (sitaxentan) but not $ET_{A/B}$ receptor antagonist (macitentan) in rats. Sitaxentan and ambrisentan are FDA-approved selective ET_A receptor antagonists for pulmonary hypertension treatment. Despite the withdrawal of Sitaxentan, ambrisentan seems a promising therapeutic approach against VEGF inhibitor-induced nephrotoxicity. Moreover, it is possible that targeting downstream of the ET-1 signaling pathway could also provide the renoprotection against VEGF inhibitors (Mirabito Colafella et al., 2020). Therefore, it is clear that additional research is needed to develop new strategies for mitigation of VEGF inhibitor-induced nephrotoxicity.

3.7 Immune checkpoint inhibitors

Immune checkpoint inhibitors (CPIs) directly target inhibitory receptors on immune and tumor cells, allowing for the activation of T cells and antitumor action. These receptors include programmed cell death 1 ligand (PD-L1) and cytotoxic T-lymphocyte-associated antigen 4 (CTLA-4). CPIs can reactivate the immune cells to attack the cancer cells (Robert, 2020; Weber, 2010). Immune checkpoint inhibitors such as CTLA-4 antagonists (ipilimumab) and PD-L1 antagonists (nivolumab and pembrolizumab) are approved by the FDA for treatment of renal cell cancer, non-small lung cancer, and melanoma (Wanchoo et al., 2017). Extensive evidence has shown that CPIs are often associated with the incidence of AKI and acute interstitial nephritis (AIN) (Belliere et al., 2016; Cortazar et al., 2020; Gupta et al., 2021; Oleas et al., 2020; Patel et al., 2020). Current management for CPIs-induced AIN involves treatment with glucocorticoids (GCs) or discontinuation of the medications (Manohar et al., 2019; Oleas et al., 2020; Qu et al., 2021). A recent multicenter study recommends that patient can return to CPIs treatment after renal injury has been relieved (Cortazar et al., 2020; Koks et al., 2021); nonetheless, fewer than 50% of patients receiving GCs can fully recover from kidney injury. A retrospective study reported that infliximab, an TNF-α (tumor necrosis factor) blocker, would be another treatment option for relapsed CPI-AIN in patients who do not respond to or tolerate the side effects of GCs (Lin et al., 2021). However, a larger sample size and standardization in the timing of the infliximab treatment in CPI-AIN are required for future clinical studies.

A study suggested that the checkpoint inhibition was not a major mechanism for efficacy of anti-CTLA-4 antibodies. The investigators identified a weak checkpoint blockade by the anti-CTLA-4 antibody, GIGA-564, which had reduced kidney damage compared with the CTLA-4 antagonist, ipilimumab, in murine models without affecting antitumor effects. This finding suggests a novel strategy for optimizing anti-CTLA-4 drugs based on regulatory T cells instead of checkpoint inhibition (Stone et al., 2021). Considering the very limited strategies available for mitigation of CPIs-induced kidney adverse effects, further research should emphasize the structure or affinity-guided design of antagonist/inhibitors for immune cells or tumor cells.

3.8 Proteasome inhibitors

Carfilzomib is a proteasome inhibitor used to treat patients with relapsed and refractory myeloma. Carfilzomib-induced renal manifestations include thrombotic microangiopathy (TMA), proteinuria and AKI (Fotiou, Roussou, Gakiopoulou, Psimenou, & Gavriatopoulou, 2020). The pathophysiology behind carfilzomib-induced nephrotoxicity is associated with tumor lysis syndrome, endothelial injury, and podocyte injury (Fotiou et al., 2020). Current management for acute renal function impairment is achieved by hydration and drug discontinuation (Bringhen et al., 2019). A case report showed that acetylcysteine could prevent carfilzomib-induced vasoconstriction-related renal injury (Wanchoo, Khan, Kolitz, & Jhaveri, 2015). Rutin, a bioflavonoid, ameliorates carfilzomib-induced oxidative stress and inflammation in nephrotoxicity through the inhibition of the iNOS-mediated NF-κB signaling pathway (Al-Harbi et al., 2019).

Another proteasome inhibitor, Bortezomib, is effective in multiple myeloma and mantle cell lymphoma. Common renal side effects of bortezomib include TMA and acute interstitial nephritis (AIN) (Chiruvella et al., 2020). A case reported showed that the bortezomib renal impairment was improved after glucocorticoid therapy and discontinuation, but AKI reoccurred following reinitiating bortezomib (Cheungpasitporn et al., 2015). Therefore, the utilization of glucocorticoid therapy for bortezomib-induced AIN remains undetermined.

3.9 mTOR protein kinase inhibitors

The mammalian target of rapamycin (mTOR) is a serine/threonine kinase, which regulates many biological processes, including cell survival, proliferation, autophagy, and metabolism (Hua et al., 2019). Inhibitors of mTOR, sirolimus and everolimus, are used for the treatment of breast cancer, renal cell carcinoma and pancreatic neuroendocrine tumors (Roskoski, 2019). It has been shown that both everolimus and sirolimus induce proteinuria and AKI. Although the chances of mTOR inhibitors leading to life-threatening AKI is minimal, they may cause development of CKD and other long-term complications (Chandra, Rao, Malhotra, Rastogi, & Khurana, 2017; Paluri et al., 2019; Wanchoo et al., 2017). Novel therapeutics and strategies are not currently available. Current strategies to mitigate mTOR inhibitors-induced nephrotoxicity depend primarily on monitoring of renal injury and proteinuria, early use of ACEi and ARBs and drug discontinuation.

3.10 Biologic agents

Biologic agents such as interleukin-2 (IL-2) and interferon alpha (IFN-α) have utility in cancer therapy. High-dose interleukin-2 (IL-2) is effective in metastatic renal cancer, and metastatic melanoma (Marabondo & Kaufman, 2017; Perazella & Shirali, 2018). However, high-dose IL-2 causes severe hypotension and cytokine-mediated inflammation, leading to AKI (Guleria et al., 1994; Marabondo & Kaufman, 2017). Current managements for IL-2 nephrotoxicity are primarily reliant on urine output monitoring and fluid management, which can improve mild oliguria or increased serum creatinine caused by IL-2 (Marabondo & Kaufman, 2017). Low and intermediate-dose of vasopressors such as dopamine can also be used in the treatment of hypotension and oliguria (Perazella & Shirali, 2018). However, advanced therapeutic approaches are limited.

A novel mechanism underlying IL-2-induced vascular leak syndrome is associated with increased circulating angiopoietin-2 levels. Endothelial damage is one of the major causes of AKI. A study revealed that diabetes mellitus can increase angiopoietin-2. Angiopoietin-2 is the competitive antagonist for angiopoietin-1 and a partial agonist/antagonist of the receptor tyrosine kinase TIE2 in endothelial cells (Yuan, Khankin, Karumanchi, & Parikh, 2009). Increased circulating angiopoietin-2 leads to vascular wall destabilization and promotes neovascularization (Fiedler & Augustin, 2006). Vascular endothelial protein tyrosine phosphatase (VE-PTP) is involved in balancing TIE2 signaling to stabilize the vasculature. Inhibition of VE-PTP by AKB-9778 can reset TIE2 signaling and therefore decrease serum angiopoietin-2 level and rescue vascular stability (Li, Sachdev,

Peters, Liang, & Lotze, 2019). In addition, targeting VE-PTP phosphatase was demonstrated to protect against diabetic kidney injury (Carota et al., 2019). Therefore, VE-PTP inhibitor AKB-9778 could be a promising candidate to attenuate IL-2-induced vascular leak and endothelial dysfunction which could lead to AKI.

3.11 BRAF inhibitors

BRAF is a member of rapidly accelerated fibrosarcoma kinase family proteins, and also a proto-oncogene. BRAF inhibitors (dabrafenib and vemurafenib) are approved for the treatment of BRAF V600E mutation-positive melanoma (Holderfield, Deuker, McCormick, & McMahon, 2014). BRAF inhibitors can cause tubular interstitial nephritis, acute tubular necrosis, increases of serum creatinine, proteinuria, and electrolyte disorders, including hypophosphatemia, hyponatremia and hypokalemia (Wanchoo, Jhaveri, Deray, & Launay-Vacher, 2016). Overall, compared with vemurafenib, dabrafenib is associated with a lower incidence of kidney disease (Wanchoo et al., 2016). A vemurafenib-induced increase in serum creatinine is usually immediate and reversible (Hurabielle et al., 2016). Routine monitoring of electrolytes and serum creatinine are recommended during treatment with BRAF inhibitors (Hurabielle et al., 2016). A recent study showed that combined therapy of MEK inhibitor, cobimetinib, and BRAF inhibitor, vemurafenib, in the treatment of BRAF V600-mutated metastatic melanoma exhibited a 60% reduction of AKI compared with BRAF inhibitor monotherapy (Teuma et al., 2017). Therefore, it is important to note that patients with different combined therapies of BRAF and MEK inhibitors could result in different responses in renal disorders (Meirson, Asher, Bomze, & Markel, 2020). In this regard, future investigations should also highlight the optimal timing of drug switching and the best combination therapy to reduce renal toxicities.

3.12 Anaplastic lymphoma kinase inhibitor

Anaplastic lymphoma kinase (ALK) is a receptor tyrosine kinase, which regulates diverse cellular process, including cellular proliferation, survival, and proliferation. Mutation of ALK is associated with the development of neuroblastoma, diffuse large B-cell lymphoma, anaplastic large-cell lymphoma, myofibroblastic tumor, esophageal and squamous cell carcinoma, as well as other malignancies (Webb et al., 2009). Crizotinib is an ALK inhibitor for the treatment of non-small cell lung cancer with the ALK mutation.

To this point in time, only a few cases of crizotinib-associated nephrotoxicity have been described. Electrolyte disorder and renal cyst are common adverse events in crizotinib treatment, which can be reversed after crizotinib discontinuation (Izzedine, El-Fekih, & Perazella, 2016; Lin et al., 2014). Crizotinib was reported to increase serum creatinine and reduce the eGFR, which could be recovered after cessation of therapy (Brosnan et al., 2014; Camidge, Brosnan, DeSilva, Koo, & Chonchol, 2014; Martín Martorell, Huerta Alvaro, Solís Salguero, & Insa Molla, 2014). Recently, kidney biopsy data on cirzotinib-induced renal injury showed that crizotinib induced acute tubular necrosis and glomerular mesangiolysis after drug rechallenge (Gastaud et al., 2013). Another case study described renal arteriolar myocyte vacuolization following crizotinib treatment (Izzedine, Brocheriou, Amoura, & Mathian, 2021). However, the mechanism of crizotinib-induced kidney damage remains unknown. In a rodent study, it was demonstrated that long-term crizotinib treatment could induce fibrosis and renal dysfunction via activating the TNF-α/NF-κB signaling pathway (Yasuma et al., 2018). It is also speculated that the crizotinib-induced kidney damage can be mediated by the inhibition of the mesenchymal epithelial transition growth factor (c-Met) pathway (Izzedine et al., 2021). Due to limited data and unclear pathophysiological explanations for the kidney damage induced by crizotinib, careful monitoring of renal function and examination of kidney biopsies during crizotinib therapy are recommended.

4. Renoprotective effects of some novel anticancer therapeutics

Because emerging preclinical and clinical studies are devoted to developing and testing promising novel anticancer drugs, efforts should be made to prevent/reduce the potential nephrotoxicity. Developing and discovering an anticancer drug that exhibits dual action of anti-tumor and renoprotection would be very beneficial for cancer patients. Therefore, identifying the common targets and mechanisms shared by both carcinogenesis and nephrotoxicity could serve to develop a novel and effective drug that simultaneously provides anticancer as well as renoprotective activities. Herein, we discuss some strategies associated with novel anticancer therapies in mitigating renal toxicities in the context of the immune system, sphingolipid signaling, DNA repair, histone modifications, and non-coding RNAs.

4.1 The immune system

The innate immune system contributes to the destruction of tumor cells that present tumor antigens through various pattern recognition receptors (PRRs), including toll-like receptors (TLRs) and macrophage-inducible C-type lectin (Mincle). TLRs can regulate immune responses both positively and negatively (Bai et al., 2020). A specific TLR2 agonist designed by Feng et al can generate macrophages that have strong anti-tumor properties in mice (Feng et al., 2019). Furthermore, the combination of TLR9 agonists and immune checkpoint inhibitors can maximize TLR9-incduced T-cell activity and provoke the antitumor immune response (Buss & Bhatia, 2020; Chuang et al., 2020). TLRs have been recognized to play multifaceted roles in AKI (Habib, 2021). TLR-2 and TLR-9 were shown to protect against cisplatin-caused nephrotoxicity in mice (Alikhan et al., 2016; Andrade-Silva et al., 2018). However, recent studies showed that depletion of TLR9 reduced renal ischemia-reperfusion injury and that nanoparticle-mediated selective targeting of renal tubular TLR9 decreased renal tubular inflammation, apoptosis and necrosis after ischemia reperfusion (Han et al., 2020; Zheng et al., 2021). These apparent discrepancies could be due to disease/cells-dependent effects of TLR9. In addition to the anticancer effects, TLR agonists may at the same time exert renal protective effects. However, care should be taken in the interpretation of these preclinical data because either overaction of TLRs by agonist or deletion/blockade of TLRs can dysregulate tissue repair during different phases of AKI in human.

Mincle (Macrophage-inducible C-type lectin), a pattern recognition receptor, was recently demonstrated as a novel target for cancer treatment (Li et al., 2020). Furthermore, it was found to be associated with M1 macrophage activation during cisplatin-induced AKI. Interestingly, adoptive transfer of Mincle-knockdown macrophages reduced the nephrotoxicity caused by cisplatin (Inoue, 2017). Additionally, the natural anti-inflammatory compound curcumin elicits inhibitory effect on the Mincle-maintained M1 macrophage phenotype, providing a promising therapeutic strategy for cisplatin-induced nephrotoxicity (Tan et al., 2019).

Cytokines released from innate immune cells are responsible for regulating the host innate immune response toward tumor cells and promoting their apoptosis. Cytokine therapy has drawn attention to treat cancers in recent years (Xue, Hsu, Fu, & Peng, 2021). Interestingly, an IL-2 and IL-33 hybrid cytokine, IL233, was shown to protect mice from cisplatin- and doxorubicin-induced nephrotoxic injury in mice (Sabapathy, Cheru,

Corey, Mohammad, & Sharma, 2019; Stremska et al., 2017). Thus, modulation of the immune system could be a promising strategy for mitigation of nephrotoxicity in cancer treatment.

4.2 Sphingolipid signaling

Sphingolipids, including ceramide, sphingosine and sphingosine-1-phosphate (S1P) have been proposed to participate in important cellular functions, such as cell growth, cell senescence, differentiation, cellular inflammation and cell cycle regulation (Hannun & Obeid, 2018). It has been increasingly recognized that inflammatory mediators, growth factors, and cellular stress could disrupt the balance between ceramide-induced cell death and S1P-induced cell growth, which are responsible for the cell fate. Sphingosine kinases (SPHKs) represent two isoforms, SPHK1 and SPHK2, that catalyze the conversion of sphingosine to sphingosine-1-phosphate (S1P). SPHKs are shown to be involved in tumorigenesis in many types of cancers (Pitman, Oehler, & Pitson, 2021; Pyne & Pyne, 2020). Recently, SPHK1 and SPHK2 inhibitors have been under development to treat breast cancer, ovarian cancer and cholangiocarcinoma (Alshaker, Thrower, & Pchejetski, 2020; Ding et al., 2016; Khan, Lai, Anwer, Azim, & Khan, 2020). ABC294640, a specific SPHK2 inhibitor, enhanced the antitumor effects of TNF-related apoptosis-inducing ligand (TRAIL) by inducing apoptosis in non-small-cell lung cancer (Yang et al., 2015). Moreover, ABC294640 (Yeliva ®) is currently being tested in a phase IIa clinical trial with hydroxychloroquine sulfate in treatment of patients with advanced cholangiocarcinoma (ClinicalTrials.gov Identifier: NCT03377179). It has also been reported that the antitumor action of ABC294640 can be enhanced by sorafenib, a multiple tyrosine kinase inhibitor in human cholangiocarcinoma cells (Ding et al., 2016; Evangelisti et al., 2016).

Despite the role of sphingolipids in cancers, they are increasingly being viewed as bioactive factors in the regulation of renal physiology. It has been shown that the activation of the sphingosine kinase/S1P/S1P receptor (SphK/S1P/S1PR) pathway contributes to different kidney diseases (Lyu, Wang, Ji, Ritter, & Li, 2020; Yokota, Bhunu, Toba, & Intapad, 2021; Zhang, Ritter, & Li, 2018; Zhang, Wang, Ji, Ritter, & Li, 2019). A more recent study demonstrated that oral administration of ABC294640 could attenuate cisplatin-induced nephrotoxicity in mice (Xie et al., 2020).

Therefore, sphingolipid signaling is a novel potential target for anticancer agent-induced nephrotoxicity in combination with the antitumor effect.

4.3 DNA repair pathway

Many antitumor therapies, including antimetabolites (5-fluorouracil, Methotrexate), DNA cross linking and alkylating agents (cisplatin, temozolomide, ifosfamide), topoisomerase II inhibitors (etoposide, doxorubicin), and radiotherapy, target proliferating cells. Several inhibitors of DNA damage repair have been tested in clinical trials; these include poly (ADP-ribose) polymerase (PARP) inhibitors, ataxia-telangiectasia mutated (ATM) inhibitors, checkpoint kinases (CHK1/2) inhibitor and DNA methyltransferases (DNMT1) inhibitor (Chan, Tan, & Cornelissen, 2021; Isono, Okubo, Asano, & Sato, 2021; Smith, Southgate, Tweddle, & Curtin, 2020; Wong, 2020; Yap et al., 2021). Generally, these agents could impair DNA repair ability or cause excessive DNA damage resulting in cell death following DNA replication (Huang & Zhou, 2021). Herein, we discuss the current inhibitors targeting DNA damage repair and their role in kidney injury.

4.3.1 Poly(ADP-ribose) polymerase (PARP), PARP-1

Poly (ADP-ribose) polymerase (PARP) is responsible for single-stranded DNA break (SSB) or double-strand DNA break (DSB) repair. The occurrence of SSB or DSB will increase the activity of PARP-1, enhancing poly (ADP-ribose) (PAR) activity, which in turn helps synthesize long branched PAR chains in order to recruit base excision repair (BER) enzymes to the damage site (Cerrato, Morra, & Celetti, 2016; Huang et al., 2018). Several PARP inhibitors have been approved for the treatment of breast and ovarian cancers. However, three PARP inhibitors (olaparib, rucaparib, and velaparib) also increase serum creatinine by inhibiting proximal tubular transporter channels such as multidrug and toxin extruder 1 and 2 (MATE1 and MATE 2) and the organic cationic transporters 1and 2 (OCT 1 and OCT 2) (LaFargue, Dal Molin, Sood, & Coleman, 2019). A more recent study showed that creatinine-derived eGFR may not actually reflect the renal function affected by olaparib. An alternative renal marker, cystatin C should be considered to more accurately measure eGFR in patients taking olaparib (Bruin et al., 2021), which would be more helpful in the evaluation of kidney damage by this class of drugs. A clinical trial is active to test whether cystatin C could be an early renal function marker for children with nephrotoxic chemotherapy (ClinicalTrials.gov Identifier: NCT02822404).

Recently, niraparib (ZEJULA), a PARP inhibitor, was approved in 2020 for patients with advanced ovarian cancer following front-line platinum treatment (González-Martín et al., 2019). Based on previous studies, it is not necessary for the renal adjustment of niraparib dosage in patients with mild or moderate declines in kidney function (Deshpande, Perazella, & Jhaveri, 2021; Zibetti Dal Molin et al., 2020). However, it remains undetermined to what extent dosing modification for PARP inhibitors might be indicated in cancer patients with advanced CKD and end-stage kidney disease (ESKD). It is still unclear about the long-term effects of PARP inhibitors on kidney function and proteinuria (Deshpande et al., 2021). Therefore, it should be noted that patients could develop nephrotoxicity due to preexistent changes in intrarenal homeostasis (Lazareth et al., 2020). While PARP-1 inhibition was demonstrated to attenuate ischemic AKI (Jang et al., 2020), one study revealed that PARP-1 deficiency promoted an alkylating agent methyl methanesulfonate (MMS)-induced nephrotoxicity in alkyladenine DNA glycosylase-transgenic mice, which was sex dependent, i.e. MMS-induced nephrotoxicity was observed in male, but not female mice (Calvo et al., 2016). This discrepancy of PARP inhibition impact in kidney damage could be due to different phases of AKI caused by different insults. Therefore, more studies are required to determine the role of PARP on anticancer agents-induced nephrotoxicity.

4.4 Histone modifications

4.4.1 Histone acetyltransferases (HATs)

Histone acetyltransferases (HATs) are the enzymes responsible for transferring an acetyl group to the lysine residue of cellular proteins, including transcription factors and histones, to promote gene expression. p300, a member of HATs, plays a definitive role in cell proliferation, differentiation, and apoptosis. p300 has been implicated in tumorigenesis and epithelial-mesenchymal transition in non-small cell lung cancer cells (Hou et al., 2018; Iyer, Özdag, & Caldas, 2004). Cisplatin was reported to induce kidney injury through activating one member of HATs, p300 activity, thereby increasing acetylation of histone H3 and further enhancing oxidative stress, inflammation, and apoptosis. It was shown that a potent p300 inhibitor, Garcinol, reversed cisplatin-induced kidney injury (Kim, Jo, Leem, & Park, 2020). It is speculated that targeting HAT could be a potential strategy to treat anticancer drug-induced kidney damage. Interestingly, p300 inhibition has demonstrated anticancer effects (Liu et al., 2020; Wang et al., 2017).

Targeting p300 may provide dual action of protection against renal chemotoxicity and an enhanced antitumor effect.

4.4.2 Histone deacetylases (HDACs)

Histone deacetylases (HDACs) are enzymes involved in the removal of acetyl group from lysine residue of histones or non-histones proteins. HDACs are grouped into four classes, which have distinct functions. HDACs participate in tumorigenesis in several cancers (Pant, Peixoto, Richard, & Gradilone, 2020; Wang, Wang, & Liu, 2020). HDAC inhibitors have shown synergistic antitumor effects in combination with other cancer drugs in preclinical and clinical tests (Hontecillas-Prieto et al., 2020; Jenke, Reßing, Hansen, Aigner, & Büch, 2021). Emerging studies suggest that HDAC inhibitors are a potential strategy to decrease the AKI induced by cisplatin. For example, HDAC inhibitors such as trichostatin A (TSA) and suberoylanilide hydroxamic acid (SAHA) are effective in reducing cisplatin-induced AKI through increasing autophagy in renal tubular cells (Liu et al., 2018; Tang et al., 2018). More recently, a highly selective HDAC6 inhibitor, 2-methylquinazoline derivative 23BB, was demonstrated to improve cisplatin-induced AKI (Hao et al., 2020). Another selective HDAC6 inhibitor, Ricolinostat (ACY-1215), elicited suppressive effects on TGF-β and EGFR signaling pathways to mitigate kidney damage in obstructive nephropathy (Chen et al., 2020). Given their antitumor potential, HDAC inhibitors could represent an additional candidate to protect against renal chemotoxicity and simultaneously enhance the antitumor action of cisplatin.

4.4.3 Sirtuins (SIRT)

Sirtuins (SIRT) are NAD+-dependent histone deacetylases belonging to class III HDACs. Unlike other HDACs, the SIRT are not affected by the inhibitors of HDACs. Different SIRTs have distinct roles in cancers as some SIRTs act as oncoproteins or as tumor suppressors. Accumulating evidence have indicated diverse roles of SIRT in kidney injuries (Peasley, Chiba, Goetzman, & Sims-Lucas, 2021). Activation of SIRT1, SIRT3 or SIRT6 has demonstrated attenuation of cisplatin-induced kidney damage through various mechanisms such as repressing apoptosis, inflammation and oxidative stress (Kim et al., 2019; Li et al., 2018; Yoon & Kim, 2016). Additionally, SIRT5 mediates the balance between mitochondrial and peroxisomal fatty acid oxidation in proximal tubular epithelial cells, which protect against injury in AKI (Chiba et al., 2019). Downregulation of SIRT3 by cisplatin can be restored through PARP-1 inhibition, which mediates protection against cisplatin-induced oxidative stress in tubular cells (Yoon & Kim, 2016). Consistent with this finding, deletion of SIRT3 exacerbates cisplatin nephrotoxicity via increasing apoptosis and inflammatory response (Kim et al., 2018). Moreover, upregulation of SIRT6 inhibits ERK1/2 expression and thereby mitigates renal dysfunction, inflammation and apoptosis caused by cisplatin (Li et al., 2018). Recently, SIRT2 overexpression was shown to reverse cisplatin-downregulated mitogen-activated protein kinase phosphatase-1 (MKP-1) and further ameliorate renal injury (Jung, Park, Kang, & Kim, 2020). Hence, manipulation of SIRT could be a potential approach to simultaneously mitigate cancer and anticancer drugs-induced nephrotoxicity by targeting various pathways. However, it should be noted that either overactivation or deactivation of HDAC or SIRT could lead to opposing effect in cancers and kidneys. Further testing in animal studies and human are needed due to distinct functions of HDACs and SIRTs.

4.5 Non-coding RNAs

Non-coding RNAs, such as miRNA, lncRNAs, and circRNAs, are believed to account for a variety of physiological and pathological functions at both transcriptional and posttranscriptional regulations. In addition to numerous studies in cancers, they have gained increasing attentions from researchers in the kidney area in the recent decade. Their roles in cisplatin-induced renal injuries have also been extensively studied (Du et al., 2017; Guo et al., 2018; Loren et al., 2021; Pavkovic et al., 2016). A natural steroid saponin, Dioscin, exerts antioxidant effect against cisplatin-induced nephrotoxicity through a mechanism of increased SIRT1 expression regulated by microRNA-34a (Zhang et al., 2017). Urolithin A (UA) is a gut metabolite of dietary polymeric polyphenols ellagitannins. Oral gavage of biocompatible nanoparticle urolithin A attenuates the reduction of miRNA (miR-192-5p and miR-140-5p) by cisplatin and reduces renal oxidative stress (Zou et al., 2019). Liposomes carrying microRNA-500a-3P elicit inhibitory effect on necroptosis-related protein expression and inflammatory responses by cisplatin (Zhang, Sun, Kong, & Zhang, 2020). Several potential agents such as scutellarin, puerarin, curcumin and pentoxifylline also show renoprotective effect involving miRNAs regulation (El Magdoub, Schaalan, Rahmo, Farag, & Khedr, 2020; Huang et al., 2020; Sun et al., 2019; Wu, Li, Li, Li, & Lu, 2020).

Ginkgo Biloba extract mediates renoprotection against methotrexate-induced renal injury by interrupting the PI3K/Akt/mTOR signaling and the expression of long non-coding RNA (lncRNA), metastasis-associated lung adenocarcinoma transcript-1 (MALAT1) (Sherif, Al-Shaalan, & Sabry, 2019). Overexpression of lncRNA XLOC_032768 is revealed to repress the gene expression of TNF-α thus attenuating cisplatin-induced apoptosis and inflammatory effects of renal tubular cells (Zhou et al., 2020). Interestingly, cisplatin-induced IL-1β upregulates lncRNA9884 expression via the NF-κB pathway; lncRNA9884 subsequently activates transcription of macrophage migration inhibitory factor (MIF) gene via binding to MIF promoter region, which in turn promotes IL-1β/NF-κB signaling and triggers a cytokine storm (Zhang et al., 2020). Furthermore, lncRNA PRNCR1 overexpression is shown to reduce cisplatin-induced renal epithelial cell apoptosis through interference with the miR-182-5p/EZH1 axis (Li, Fan, Wang, Gong, & Guo, 2021).

Besides microRNAs and lncRNAs, circular RNAs (circRNAs) have recently attracted increasing attention as promising targets in both cancers and kidney diseases (van Zonneveld, Kölling, Bijkerk, & Lorenzen, 2021). One study uncovered that 368 circRNAs were expressed differentially in cisplatin treatment mice (Li et al., 2019). Another study demonstrated that the elevated circRNA, circ-0114427, was found in the early stage of a cisplatin-AKI model, which was implicated in the binding of miR-494 and the increase of ATF3 expression; as a result, circ-0114427 led to a decrease in the production of inflammatory cytokines. Therefore, circ-0114427 could be a novel target for early treatment strategies for cisplatin-induced AKI (Cao et al., 2020). Overall, studies have demonstrated that the non-coding RNAs, including miRNA, lncRNAs, and circRNAs, are involved in both cancer therapy and kidney chemotoxicity, indicating that non-coding RNAs could be promising targets for developing novel drugs in the management of anticancer agent-induced nephrotoxicity, and intriguingly, bear potential dual action of renoprotection and anticancer.

5. Conclusions

There are currently no effective drugs targeting specific biochemical or molecular pathways associated with anticancer therapy-induced nephrotoxicity in clinical use or clinical trials, although several clinically used drugs, such as gemigliptin (DPP-4 inhibitor), pantoprazole (proton pump inhibitor) and Tempol, are under clinical trials for renal chemotoxicity. This chapter reviews and summarizes the kidney- and drug-associated risk factors of anticancer drug-induced nephrotoxicity that will allow clinicians and researchers to carefully assess the risk and benefits of treatments. This review also proposes prospective strategies to diagnose/mitigate nephrotoxicity arising from cancer therapeutics. In addition to the assessment of route of drug administration, timing of drug treatments, and pharmacodynamics, pharmacokinetics and pharmacogenetics among patients with cancer, future studies should also focus on identifying novel mechanisms and targets that may lead to the development of new approaches in the management of chemotherapy-induced nephrotoxicity.

References

Abd-Elrazek, A., Shapana, H., Shukry, W., & Galilah, D. (2021). Comparison between Annona squamosa, Annona cherimolia and Annona atemoya ethanolic extracts extenuative impact against oxidative stress, inflammation and apoptosis in rat kidney induced by Ifosfamid. *Toxicology Research, 10*(4), 947–958. https://doi.org/10.1093/toxres/tfab078.

Afjal, M. A., Goswami, P., Ahmad, S., Dabeer, S., Akhter, J., Salman, M., et al. (2020). Tempol (4-hydroxy tempo) protects mice from cisplatin-induced acute kidney injury via modulation of expression of aquaporins and kidney injury molecule-1. *Drug and Chemical Toxicology, 1–9*. https://doi.org/10.1080/01480545.2020.1831011.

Afsar, T., Razak, S., Almajwal, A., & Al-Disi, D. (2020). Doxorubicin-induced alterations in kidney functioning, oxidative stress, DNA damage, and renal tissue morphology; Improvement by Acacia hydaspica tannin-rich ethyl acetate fraction. *Saudi Journal of Biological Sciences, 27*(9), 2251–2260. https://doi.org/10.1016/j.sjbs.2020.07.011.

Al-Harbi, N. O., Imam, F., Al-Harbi, M. M., Al-Shabanah, O. A., Alotaibi, M. R., As Sobeai, H. M., et al. (2019). Rutin inhibits carfilzomib-induced oxidative stress and inflammation via the NOS-mediated NF-κB signaling pathway. *Inflammopharmacology, 27*(4), 817–827. https://doi.org/10.1007/s10787-018-0550-5.

Alikhan, M. A., Summers, S. A., Gan, P. Y., Chan, A. J., Khouri, M. B., Ooi, J. D., et al. (2016). Endogenous toll-like receptor 9 regulates AKI by promoting regulatory T cell recruitment. *Journal of the American Society of Nephrology, 27*(3), 706–714. https://doi.org/10.1681/asn.2014090927.

Alshaker, H., Thrower, H., & Pchejetski, D. (2020). Sphingosine kinase 1 in breast cancer—A new molecular marker and a therapy target. *Frontiers in Oncology, 10*(289). https://doi.org/10.3389/fonc.2020.00289.

Andrade-Silva, M., Cenedeze, M. A., Perandini, L. A., Felizardo, R. J. F., Watanabe, I. K. M., Agudelo, J. S. H., et al. (2018). TLR2 and TLR4 play opposite role in autophagy associated with cisplatin-induced acute kidney injury. *Clinical Science, 132*(16), 1725–1739. https://doi.org/10.1042/cs20170262.

Andreev, E., Koopman, M., & Arisz, L. (1999). A rise in plasma creatinine that is not a sign of renal failure: which drugs can be responsible? *Journal of Internal Medicine, 246*(3), 247–252. https://doi.org/10.1046/j.1365-2796.1999.00515.x.

Arab, H. H., Salama, S. A., & Maghrabi, I. A. (2018). Camel milk ameliorates 5-fluorouracil-induced renal injury in rats: Targeting MAPKs, NF-κB and PI3K/Akt/eNOS pathways. *Cellular Physiology and Biochemistry, 46*(4), 1628–1642. https://doi.org/10.1159/000489210.

Azouz, A. A., Abdel-Nassir Abdel-Razek, E., & Abo-Youssef, A. M. (2020). Amlodipine alleviates cisplatin-induced nephrotoxicity in rats through gamma-glutamyl transpeptidase (GGT) enzyme inhibition, associated with regulation of Nrf2/HO-1, MAPK/NF-κB, and Bax/Bcl-2 signaling. *Saudi Pharmaceutical Journal, 28*(11), 1317–1325. https://doi.org/10.1016/j.jsps.2020.08.022.

Baharmi, S., Kalantari, H., Kalantar, M., Goudarzi, M., Mansouri, E., & Kalantar, H. (2021). Pretreatment with gallic acid mitigates cyclophosphamide induced inflammation and oxidative stress in mice. *Current Molecular Pharmacology*. https://doi.org/10.2174/1874467214666210531162741.

Bai, L., Li, W., Zheng, W., Xu, D., Chen, N., & Cui, J. (2020). Promising targets based on pattern recognition receptors for cancer immunotherapy. *Pharmacological Research, 159*, 105017. https://doi.org/10.1016/j.phrs.2020.105017.

Basit, A., Radi, Z., Vaidya, V. S., Karasu, M., & Prasad, B. (2019). Kidney cortical transporter expression across species using quantitative proteomics. *Drug Metabolism and Disposition*. https://doi.org/10.1124/dmd.119.086579. dmd.119.086579.

Bauchet, A.-L., Masson, R., Guffroy, M., & Slaoui, M. (2011). Immunohistochemical identification of kidney nephron segments in the dog, rat, mouse, and cynomolgus monkey. *Toxicologic Pathology, 39*(7), 1115–1128. https://doi.org/10.1177/0192623311425060.

Belliere, J., Meyer, N., Mazieres, J., Ollier, S., Boulinguez, S., Delas, A., et al. (2016). Acute interstitial nephritis related to immune checkpoint inhibitors. *British Journal of Cancer, 115*(12), 1457–1461. https://doi.org/10.1038/bjc.2016.358.

Bhat, N., Kalthur, S. G., Padmashali, S., & Monappa, V. (2018). Toxic effects of different doses of cyclophosphamide on liver and kidney tissue in swiss albino mice: A histopathological study. *Ethiopian Journal of Health Sciences, 28*(6), 711–716. https://doi.org/10.4314/ejhs.v28i6.5.

Bozina, N., Bradamante, V., & Lovrić, M. (2009). Genetic polymorphism of metabolic enzymes P450 (CYP) as a susceptibility factor for drug response, toxicity, and cancer risk. *Arhiv za higijenu rada i toksikologiju, 60*(2), 217–242. https://doi.org/10.2478/10004-1254-60-2009-1885.

Bringhen, S., Milan, A., D'Agostino, M., Ferri, C., Wäsch, R., Gay, F., et al. (2019). Prevention, monitoring and treatment of cardiovascular adverse events in myeloma patients receiving carfilzomib A consensus paper by the European Myeloma Network and the Italian Society of Arterial Hypertension. *Journal of Internal Medicine, 286*(1), 63–74. https://doi.org/10.1111/joim.12882.

Brosnan, E. M., Weickhardt, A. J., Lu, X., Maxon, D. A., Barón, A. E., Chonchol, M., et al. (2014). Drug-induced reduction in estimated glomerular filtration rate in patients with ALK-positive non-small cell lung cancer treated with the ALK inhibitor crizotinib. *Cancer, 120*(5), 664–674. https://doi.org/10.1002/cncr.28478.

Brown, A., Kumar, S., & Tchounwou, P. B. (2019). Cisplatin-based chemotherapy of human cancers. *Journal of Cancer Science & Therapy, 11*(4).

Bruin, M. A. C., Korse, C. M., van Wijnen, B., de Jong, V. M. T., Linn, S. C., van Triest, B., et al. (2021). A real or apparent decrease in glomerular filtration rate in patients using olaparib? *European Journal of Clinical Pharmacology*, 77(2), 179–188. https://doi.org/10.1007/s00228-020-03070-0.

Buss, C. G., & Bhatia, S. N. (2020). Nanoparticle delivery of immunostimulatory oligonucleotides enhances response to checkpoint inhibitor therapeutics. *Proceedings of the National Academy of Sciences of the United States of America, 117*(24), 13428–13436. https://doi.org/10.1073/pnas.2001569117.

Caglar, K., Kinalp, C., Arpaci, F., Turan, M., Saglam, K., Ozturk, B., et al. (2002). Cumulative prior dose of cisplatin as a cause of the nephrotoxicity of high-dose chemotherapy followed by autologous stem-cell transplantation. *Nephrology, Dialysis, Transplantation, 17*(11), 1931–1935. https://doi.org/10.1093/ndt/17.11.1931.

Calvo, J. A., Allocca, M., Fake, K. R., Muthupalani, S., Corrigan, J. J., Bronson, R. T., et al. (2016). Parp1 protects against Aag-dependent alkylation-induced nephrotoxicity in a sex-dependent manner. *Oncotarget*, 7(29). Retrieved from https://www.oncotarget.com/article/10440/text/.

Camidge, D. R., Brosnan, E. M., DeSilva, C., Koo, P. J., & Chonchol, M. (2014). Crizotinib effects on creatinine and non-creatinine-based measures of glomerular filtration rate. *Journal of Thoracic Oncology*, *9*(11), 1634–1637. https://doi.org/10.1097/jto.0000000000000321.

Cao, Y., Mi, X., Zhang, D., Wang, Z., Zuo, Y., & Tang, W. (2020). Transcriptome sequencing of circular RNA reveals a novel circular RNA-has_circ_0114427 in the regulation of inflammation in acute kidney injury. *Clinical Science*, *134*(2), 139–154. https://doi.org/10.1042/cs20190990.

Carota, I. A., Kenig-Kozlovsky, Y., Onay, T., Scott, R., Thomson, B. R., Souma, T., et al. (2019). Targeting VE-PTP phosphatase protects the kidney from diabetic injury. *Journal of Experimental Medicine*, *216*(4), 936–949. https://doi.org/10.1084/jem.20180009.

Carron, P. L., Padilla, M., & Maurizi Balzan, J. (2014). Nephrotic syndrome and acute renal failure during pegylated liposomal doxorubicin treatment. *Hemodialysis International*, *18*(4), 846–847. https://doi.org/10.1111/hdi.12196.

Cerrato, A., Morra, F., & Celetti, A. (2016). Use of poly ADP-ribose polymerase [PARP] inhibitors in cancer cells bearing DDR defects: the rationale for their inclusion in the clinic. *Journal of Experimental & Clinical Cancer Research*, *35*(1), 179. https://doi.org/10.1186/s13046-016-0456-2.

Chan, C. Y., Tan, K. V., & Cornelissen, B. (2021). PARP inhibitors in cancer diagnosis and therapy. *Clinical Cancer Research*, *27*(6), 1585–1594. https://doi.org/10.1158/1078-0432.Ccr-20-2766.

Chandra, A., Rao, N. S., Malhotra, K. P., Rastogi, M., & Khurana, R. (2017). Everolimus-associated acute kidney injury in patients with metastatic breast cancer. *Indian Journal of Nephrology*, *27*(5), 406–409. https://doi.org/10.4103/ijn.IJN_304_16.

Chávez-Iñiguez, J. S., Medina-Gonzalez, R., Aguilar-Parra, L., Torres-Vázquez, E. J., Maggiani-Aguilera, P., Cervantes-Pérez, E., et al. (2018). Oral acyclovir induced hypokalemia and acute tubular necrosis a case report. *BMC Nephrology*, *19*(1), 324. https://doi.org/10.1186/s12882-018-1121-0.

Chen, X., Yu, C., Hou, X., Li, J., Li, T., Qiu, A., et al. (2020). Histone deacetylase 6 inhibition mitigates renal fibrosis by suppressing TGF-β and EGFR signaling pathways in obstructive nephropathy. *American Journal of Physiology-Renal Physiology*, *319*(6), F1003–F1014. https://doi.org/10.1152/ajprenal.00261.2020.

Cheungpasitporn, W., Leung, N., Rajkumar, S. V., Cornell, L. D., Sethi, S., Angioi, A., et al. (2015). Bortezomib-induced acute interstitial nephritis. *Nephrology, Dialysis, Transplantation*, *30*(7), 1225–1229. https://doi.org/10.1093/ndt/gfv222.

Chiba, T., Peasley, K. D., Cargill, K. R., Maringer, K. V., Bharathi, S. S., Mukherjee, E., et al. (2019). Sirtuin 5 regulates proximal tubule fatty acid oxidation to protect against AKI. *Journal of the American Society of Nephrology*, *30*(12), 2384–2398. https://doi.org/10.1681/asn.2019020163.

Chiruvella, V., Annamaraju, P., & Guddati, A. K. (2020). Management of nephrotoxicity of chemotherapy and targeted agents: 2020. *American Journal of Cancer Research*, *10*(12), 4151–4164.

Choi, S. H., Leem, J., & Lee, I. K. (2017). Protective effects of gemigliptin, a dipeptidyl peptidase-4 inhibitor, against cisplatin-induced nephrotoxicity in mice. *Mediators of Inflammation*, *2017*, 4139439. https://doi.org/10.1155/2017/4139439.

Chuang, Y. C., Tseng, J. C., Huang, L. R., Huang, C. M., Huang, C. F., & Chuang, T. H. (2020). Adjuvant effect of toll-like receptor 9 activation on cancer immunotherapy using checkpoint blockade. *Frontiers in Immunology*, *11*, 1075. https://doi.org/10.3389/fimmu.2020.01075.

Cortazar, F. B., Kibbelaar, Z. A., Glezerman, I. G., Abudayyeh, A., Mamlouk, O., Motwani, S. S., et al. (2020). Clinical features and outcomes of immune checkpoint inhibitor-associated AKI: A multicenter study. *Journal of the American Society of Nephrology*, *31*(2), 435–446. https://doi.org/10.1681/asn.2019070676.

Dai, R., Liu, H., Han, X., Liu, J., Zhai, Y., Rao, J., et al. (2019). Angiopoietin-like-3 knockout protects against glomerulosclerosis in murine adriamycin-induced nephropathy by attenuating podocyte loss. *BMC Nephrology*, *20*(1), 185. https://doi.org/10.1186/s12882-019-1383-1.

Deshpande, P. P., Perazella, M. A., & Jhaveri, K. D. (2021). PARP inhibitors and the kidney. *Journal of Onco-Nephrology*, *5*(1), 42–47. https://doi.org/10.1177/2399369320987090.

Ding, X., Chaiteerakij, R., Moser, C. D., Shaleh, H., Boakye, J., Chen, G., et al. (2016). Antitumor effect of the novel sphingosine kinase 2 inhibitor ABC294640 is enhanced by inhibition of autophagy and by sorafenib in human cholangiocarcinoma cells. *Oncotarget*, 7(15). Retrieved from https://www.oncotarget.com/article/7914/text/.

Dobrek, L., Nalik-Iwaniak, K., Fic, K., & Arent, Z. (2020). The effect of acetylcysteine on renal function in experimental models of cyclophosphamide-and ifosfamide-induced cystitis. *Current Urology*, *14*(3), 150–162. https://doi.org/10.1159/000499245.

Drozdzik, M., Drozdzik, M., & Oswald, S. (2021). Membrane carriers and transporters in kidney physiology and disease. *Biomedicine*, *9*(4). https://doi.org/10.3390/biomedicines9040426.

Du, B., Dai, X.-M., Li, S., Qi, G.-L., Cao, G.-X., Zhong, Y., et al. (2017). MiR-30c regulates cisplatin-induced apoptosis of renal tubular epithelial cells by targeting Bnip3L and Hspa5. *Cell Death & Disease*, *8*(8), e2987. https://doi.org/10.1038/cddis.2017.377.

Duan, Z., Cai, G., Li, J., & Chen, X. (2020). Cisplatin-induced renal toxicity in elderly people. *Therapeutic Advances in Medical Oncology*, *12*, 1758835920923430. https://doi.org/10.1177/1758835920923430.

El Magdoub, H. M., Schaalan, M. F., Rahmo, R. M., Farag, D. B., & Khedr, L. H. (2020). Implications of miRNAs on TGF-β/TAK1/mTOR pathway in mediating the renoprotective effects of pentoxifylline against cisplatin-induced nephrotoxicity in rats. *Toxicology and Applied Pharmacology*, *404*, 115184. https://doi.org/10.1016/j.taap.2020.115184.

El-Far, Y. M., & El-Mesery, M. (2021). Pevonedistat attenuates cisplatin-induced nephrotoxicity in mice by downregulating the release of inflammatory mediators. *Journal of Biochemical and Molecular Toxicology*, e22908. https://doi.org/10.1002/jbt.22908.

Elmansy, R. A., Seleem, H. S., Mahmoud, A. R., Hassanein, E. H. M., & Ali, F. E. M. (2021). Rebamipide potentially mitigates methotrexate-induced nephrotoxicity via inhibition of oxidative stress and inflammation: A molecular and histochemical study. *Anatomical Record (Hoboken)*, *304*(3), 647–661. https://doi.org/10.1002/ar.24482.

El-Shabrawy, M., Mishriki, A., Attia, H., Emad Aboulhoda, B., Emam, M., & Wanas, H. (2020). Protective effect of tolvaptan against cyclophosphamide-induced nephrotoxicity in rat models. *Pharmacology Research & Perspectives*, *8*(5), e00659. https://doi.org/10.1002/prp2.659.

Enomoto, A., & Endou, H. (2005). Roles of organic anion transporters (OATs) and a urate transporter (URAT1) in the pathophysiology of human disease. *Clinical and Experimental Nephrology*, *9*(3), 195–205. https://doi.org/10.1007/s10157-005-0368-5.

Ensergueix, G., Pallet, N., Joly, D., Levi, C., Chauvet, S., Trivin, C., et al. (2020). Ifosfamide nephrotoxicity in adult patients. *Clinical Kidney Journal*, *13*(4), 660–665. https://doi.org/10.1093/ckj/sfz183.

Esmaeeli, A., Keshavarz, Z., Dehdar, F., Assadi, M., & Seyedabadi, M. (2020). The effects of carvedilol, metoprolol and propranolol on cisplatin-induced kidney injury. *Drug and Chemical Toxicology*, *1*–7. https://doi.org/10.1080/01480545.2020.1846551.

Estrela, G. R., Wasinski, F., Gregnani, M. F., Freitas-Lima, L. C., Arruda, A. C., Morais, R. L., et al. (2020). Angiotensin-converting enzyme inhibitor protects against cisplatin nephrotoxicity by modulating kinin B1 receptor expression and aminopeptidase P activity in mice. *Frontiers in Molecular Biosciences*, 7(96). https://doi.org/10.3389/fmolb.2020.00096.

Evangelisti, C., Evangelisti, C., Buontempo, F., Lonetti, A., Orsini, E., Chiarini, F., et al. (2016). Therapeutic potential of targeting sphingosine kinases and sphingosine 1-phosphate in hematological malignancies. *Leukemia, 30*(11), 2142–2151. https://doi.org/10.1038/leu.2016.208.

Ewees, M. G. E., Abdel-Bakky, M. S., Bayoumi, A. M. A., Abo-Saif, A. A., Altowayan, W. M., Alharbi, K. S., et al. (2021). Dabigatran mitigates cisplatin-mediated nephrotoxicity through down regulation of thrombin pathway. *Journal of Advanced Research, 31*, 127–136. https://doi.org/10.1016/j.jare.2020.12.014.

Fan, H. Y., Wang, X. K., Li, X., Ji, K., Du, S. H., Liu, Y., et al. (2020). Curcumin, as a pleiotropic agent, improves doxorubicin-induced nephrotic syndrome in rats. *Journal of Ethnopharmacology, 250*, 112502. https://doi.org/10.1016/j.jep.2019.112502.

Feidakis, A., Panagiotou, M.-R., Tsoukakis, E., Bacharaki, D., Gounari, P., Nikolopoulos, P., et al. (2021). Impact of angiotensin-converting enzyme inhibitors or angiotensin receptor blockers on acute kidney injury in emergency medical admissions. *Journal of Clinical Medicine, 10*(3), 412. Retrieved from https://www.mdpi.com/2077-0383/10/3/412.

Feng, Y., Mu, R., Wang, Z., Xing, P., Zhang, J., Dong, L., et al. (2019). A toll-like receptor agonist mimicking microbial signal to generate tumor-suppressive macrophages. *Nature Communications, 10*(1), 2272. https://doi.org/10.1038/s41467-019-10354-2.

Fiedler, U., & Augustin, H. G. (2006). Angiopoietins: A link between angiogenesis and inflammation. *Trends in Immunology, 27*(12), 552–558. https://doi.org/10.1016/j.it.2006.10.004.

Fotiou, D., Roussou, M., Gakiopoulou, C., Psimenou, E., & Gavriatopoulou, M. (2020). Carfilzomib-associated renal toxicity is common and unpredictable: A comprehensive analysis of 114 multiple myeloma patients. *Blood Cancer Journal, 10*(11), 109. https://doi.org/10.1038/s41408-020-00381-4.

Fox, C. P., Ali, A. S., McIlroy, G. W., Thust, S. C., Martinez-Calle, N., Jackson, A. E., et al. (2021). A phase I/II study of thiotepa-based immunochemotherapy in relapsed/refractory primary CNS lymphoma: The TIER trial. *Blood Advances*. https://doi.org/10.1182/bloodadvances.2021004779.

Fujita, K., Niioka, T., Motoyama, S., Masatomo Miura, N. T., Motoyama, S., & Miura, M. (2016). Influence of renal ABC and SLC transporter polymorphisms on cisplatin-induced nephrotoxicity in patients with esophageal cancer. *Clinical Pharmacology & Biopharmaceutics, 5*(158). https://doi.org/10.4172/2167-065X.1000158.

Gao, H., Wang, X., Qu, X., Zhai, J., Tao, L., Zhang, Y., et al. (2020). Omeprazole attenuates cisplatin-induced kidney injury through suppression of the TLR4/NF-κB/NLRP3 signaling pathway. *Toxicology, 440*, 152487. https://doi.org/10.1016/j.tox.2020.152487.

Garnier, A. S., Dellamaggiore, J., Brilland, B., Lagarce, L., Abgueguen, P., Furber, A., et al. (2020). High incidence of amoxicillin-induced crystal nephropathy in patients receiving high dose of intravenous amoxicillin. *Journal of Clinical Medicine, 9*(7). https://doi.org/10.3390/jcm9072022.

Gastaud, L., Ambrosetti, D., Otto, J., Marquette, C. H., Coutts, M., Hofman, P., et al. (2013). Acute kidney injury following crizotinib administration for non-small-cell lung carcinoma. *Lung Cancer, 82*(2), 362–364. https://doi.org/10.1016/j.lungcan.2013.08.007.

Gelen, V., Şengül, E., Yıldırım, S., Senturk, E., Tekin, S., & Kükürt, A. (2021). The protective effects of hesperidin and curcumin on 5-fluorouracil-induced nephrotoxicity in mice. *Environmental Science and Pollution Research International, 28*(34), 47046–47055. https://doi.org/10.1007/s11356-021-13969-5.

Ghane Shahrbaf, F., & Assadi, F. (2015). Drug-induced renal disorders. *Journal of Renal Injury Prevention, 4*(3), 57–60. https://doi.org/10.12861/jrip.2015.12.

Ghonaim, E., El-Haggar, S., & Gohar, S. (2021). Possible protective effect of pantoprazole against cisplatin-induced nephrotoxicity in head and neck cancer patients: A randomized controlled trial. *Medical Oncology, 38*(9), 108. https://doi.org/10.1007/s12032-021-01558-y.

González, E., Gutiérrez, E., Galeano, C., Chevia, C., de Sequera, P., Bernis, C., et al. (2008). Early steroid treatment improves the recovery of renal function in patients with drug-induced acute interstitial nephritis. *Kidney International, 73*(8), 940–946. https://doi.org/10.1038/sj.ki.5002776.

González-Martín, A., Pothuri, B., Vergote, I., DePont Christensen, R., Graybill, W., Mirza, M. R., et al. (2019). Niraparib in patients with newly diagnosed advanced ovarian cancer. *New England Journal of Medicine, 381*(25), 2391–2402. https://doi.org/10.1056/NEJMoa1910962.

Groff, J. A., Kozak, M., Boehmer, J. P., Demko, T. M., & Diamond, J. R. (1997). Endotheliopathy: A continuum of hemolytic uremic syndrome due to mitomycin therapy. *American Journal of Kidney Diseases, 29*(2), 280–284. https://doi.org/10.1016/S0272-6386(97)90042-1.

Guleria, A. S., Yang, J. C., Topalian, S. L., Weber, J. S., Parkinson, D. R., MacFarlane, M. P., et al. (1994). Renal dysfunction associated with the administration of high-dose interleukin-2 in 199 consecutive patients with metastatic melanoma or renal carcinoma. *Journal of Clinical Oncology, 12*(12), 2714–2722. https://doi.org/10.1200/jco.1994.12.12.2714.

Güntürk, I., Yazici, C., Köse, S. K., Dağli, F., Yücel, B., & Yay, A. H. (2019). The effect of N-acetylcysteine on inflammation and oxidative stress in cisplatin-induced nephrotoxicity: A rat model. *Turkish Journal of Medical Sciences, 49*(6), 1789–1799. https://doi.org/10.3906/sag-1903-225.

Guo, Y., Ni, J., Chen, S., Bai, M., Lin, J., Ding, G., et al. (2018). MicroRNA-709 mediates acute tubular injury through effects on mitochondrial function. *Journal of the American Society of Nephrology, 29*(2), 449–461. https://doi.org/10.1681/asn.2017040381.

Gupta, S., Short, S. A. P., Sise, M. E., Prosek, J. M., Madhavan, S. M., Soler, M. J., et al. (2021). Acute kidney injury in patients treated with immune checkpoint inhibitors. *Journal for Immunotherapy of Cancer, 9*(10). https://doi.org/10.1136/jitc-2021-003467.

Habib, R. (2021). Multifaceted roles of Toll-like receptors in acute kidney injury. *Heliyon,* 7(3), e06441. https://doi.org/10.1016/j.heliyon.2021.e06441.

Han, S. J., Williams, R. M., D'Agati, V., Jaimes, E. A., Heller, D. A., & Lee, H. T. (2020). Selective nanoparticle-mediated targeting of renal tubular Toll-like receptor 9 attenuates ischemic acute kidney injury. *Kidney International, 98*(1), 76–87. https://doi.org/10.1016/j.kint.2020.01.036.

Hanna, R. M., Barsoum, M., Arman, F., Selamet, U., Hasnain, H., & Kurtz, I. (2019). Nephrotoxicity induced by intravitreal vascular endothelial growth factor inhibitors: Emerging evidence. *Kidney International, 96*(3), 572–580. https://doi.org/10.1016/j.kint.2019.02.042.

Hannun, Y. A., & Obeid, L. M. (2018). Sphingolipids and their metabolism in physiology and disease. *Nature Reviews Molecular Cell Biology, 19*(3), 175–191. https://doi.org/10.1038/nrm.2017.107.

Hao, Y., Guo, F., Huang, Z., Feng, Y., Xia, Z., Liu, J., et al. (2020). 2-Methylquinazoline derivative 23BB as a highly selective histone deacetylase 6 inhibitor alleviated cisplatin-induced acute kidney injury. *Bioscience Reports, 40*(1). https://doi.org/10.1042/bsr20191538.

Harrison, D. J., Kharbanda, R., Cunningham, D. S., McLellan, L. I., & Hayes, J. D. (1989). Distribution of glutathione S-transferase isoenzymes in human kidney: basis for possible markers of renal injury. *Journal of Clinical Pathology, 42*(6), 624–628. https://doi.org/10.1136/jcp.42.6.624.

Hartmann, J. T., Kollmannsberger, C., Kanz, L., & Bokemeyer, C. (1999). Platinum organ toxicity and possible prevention in patients with testicular cancer. *International Journal of Cancer, 83*(6), 866–869. https://doi.org/10.1002/(sici)1097-0215(19991210)83:6<866::aid-ijc34>3.0.co;2-9.

Holderfield, M., Deuker, M. M., McCormick, F., & McMahon, M. (2014). Targeting RAF kinases for cancer therapy: BRAF-mutated melanoma and beyond. *Nature Reviews. Cancer, 14*(7), 455–467. https://doi.org/10.1038/nrc3760.

Holmboe, L., Andersen, A. M., Mørkrid, L., Slørdal, L., & Hall, K. S. (2012). High dose methotrexate chemotherapy: pharmacokinetics, folate and toxicity in osteosarcoma patients. *British Journal of Clinical Pharmacology, 73*(1), 106–114. https://doi.org/10.1111/j.1365-2125.2011.04054.x.

Hong, S. I., Ahn, S., Lee, Y. S., Kim, W. Y., Lim, K. S., Lee, J. H., et al. (2016). Contrast-induced nephropathy in patients with active cancer undergoing contrast-enhanced computed tomography. *Support Care Cancer, 24*(3), 1011–1017. https://doi.org/10.1007/s00520-015-2875-6.

Hontecillas-Prieto, L., Flores-Campos, R., Silver, A., de Álava, E., Hajji, N., & García-Domínguez, D. J. (2020). Synergistic enhancement of cancer therapy using HDAC inhibitors: Opportunity for clinical trials. *Frontiers in Genetics, 11*(1113). https://doi.org/10.3389/fgene.2020.578011.

Hou, X., Gong, R., Zhan, J., Zhou, T., Ma, Y., Zhao, Y., et al. (2018). p300 promotes proliferation, migration, and invasion via inducing epithelial-mesenchymal transition in non-small cell lung cancer cells. *BMC Cancer, 18*(1), 641. https://doi.org/10.1186/s12885-018-4559-3.

Howard, S. C., McCormick, J., Pui, C. H., Buddington, R. K., & Harvey, R. D. (2016). Preventing and managing toxicities of high-dose methotrexate. *The Oncologist, 21*(12), 1471–1482. https://doi.org/10.1634/theoncologist.2015-0164.

Hsiao, Y. P., Chen, H. L., Tsai, J. N., Lin, M. Y., Liao, J. W., Wei, M. S., et al. (2021). Administration of Lactobacillus reuteri combined with Clostridium butyricum attenuates cisplatin-induced renal damage by gut microbiota reconstitution, increasing butyric acid production, and suppressing renal inflammation. *Nutrients, 13*(8). https://doi.org/10.3390/nu13082792.

Hua, H., Kong, Q., Zhang, H., Wang, J., Luo, T., & Jiang, Y. (2019). Targeting mTOR for cancer therapy. *Journal of Hematology & Oncology, 12*(1), 71. https://doi.org/10.1186/s13045-019-0754-1.

Huang, S. J., Huang, J., Yan, Y. B., Qiu, J., Tan, R. Q., Liu, Y., et al. (2020). The renoprotective effect of curcumin against cisplatin-induced acute kidney injury in mice: Involvement of miR-181a/PTEN axis. *Renal Failure, 42*(1), 350–357. https://doi.org/10.1080/0886022x.2020.1751658.

Huang, Y., Shao, Q., Luo, X., Yang, D., Zeng, B., Xiang, T., et al. (2018). Poly(ADP-ribose) polymerase-1 promotes recruitment of meiotic recombination-11 to chromatin and DNA double-strand break repair in Ku70-deficient breast cancer cells. *The FASEB Journal*. https://doi.org/10.1096/fj.201800092R. fj201800092R.

Huang, K. M., Uddin, M. E., DiGiacomo, D., Lustberg, M. B., Hu, S., & Sparreboom, A. (2020). Role of SLC transporters in toxicity induced by anticancer drugs. *Expert Opinion on Drug Metabolism & Toxicology, 16*(6), 493–506. https://doi.org/10.1080/17425255.2020.1755253.

Huang, S., You, J., Wang, K., Li, Y., Zhang, Y., Wei, H., et al. (2019). N-acetylcysteine attenuates cisplatin-induced acute kidney injury by inhibiting the C5a receptor. *BioMed Research International, 2019*, 4805853. https://doi.org/10.1155/2019/4805853.

Huang, R., & Zhou, P.-K. (2021). DNA damage repair: historical perspectives, mechanistic pathways and clinical translation for targeted cancer therapy. *Signal Transduction and Targeted Therapy, 6*(1), 254. https://doi.org/10.1038/s41392-021-00648-7.

Hucke, A., & Ciarimboli, G. (2016). The role of transporters in the toxicity of chemotherapeutic drugs: Focus on transporters for organic cations. *The Journal of Clinical Pharmacology, 56*(S7), S157–S172. https://doi.org/10.1002/jcph.706.

Hurabielle, C., Pillebout, E., Stehlé, T., Pagès, C., Roux, J., Schneider, P., et al. (2016). Mechanisms underpinning increased plasma creatinine levels in patients receiving vemurafenib for advanced melanoma. *PLoS One, 11*(3), e0149873. https://doi.org/10.1371/journal.pone.0149873.
Imig, J. D., Hye Khan, M. A., Burkhan, A., Chen, G., Adebesin, A. M., & Falck, J. R. (2021). Kidney-targeted epoxyeicosatrienoic acid analog, EET-F01, reduces inflammation, oxidative stress, and cisplatin-induced nephrotoxicity. *International Journal of Molecular Sciences, 22*(6). https://doi.org/10.3390/ijms22062793.
Inoue, T. (2017). M1 macrophage triggered by Mincle leads to a deterioration of acute kidney injury. *Kidney International, 91*(3), 526–529. https://doi.org/10.1016/j.kint.2016.11.026.
Ismail, R. S., El-Awady, M. S., & Hassan, M. H. (2020). Pantoprazole abrogated cisplatin-induced nephrotoxicity in mice via suppression of inflammation, apoptosis, and oxidative stress. *Naunyn-Schmiedeberg's Archives of Pharmacology, 393*(7), 1161–1171. https://doi.org/10.1007/s00210-020-01823-3.
Isono, M., Okubo, K., Asano, T., & Sato, A. (2021). Inhibition of checkpoint kinase 1 potentiates anticancer activity of gemcitabine in bladder cancer cells. *Scientific Reports, 11*(1), 10181. https://doi.org/10.1038/s41598-021-89684-5.
Iwakura, T., Fukasawa, H., Kitamura, A., Ishibuchi, K., Yasuda, H., & Furuya, R. (2020). Effect of dipeptidyl peptidase-4 inhibitors on cisplatin-induced acute nephrotoxicity in cancer patients with diabetes mellitus: A retrospective study. *PLoS One, 15*(2), e0229377. https://doi.org/10.1371/journal.pone.0229377.
Iwakura, T., Zhao, Z., Marschner, J. A., Devarapu, S. K., Yasuda, H., & Anders, H. J. (2019). Dipeptidyl peptidase-4 inhibitor teneligliptin accelerates recovery from cisplatin-induced acute kidney injury by attenuating inflammation and promoting tubular regeneration. *Nephrology, Dialysis, Transplantation, 34*(10), 1669–1680. https://doi.org/10.1093/ndt/gfy397.
Iyer, N. G., Özdag, H., & Caldas, C. (2004). p300/CBP and cancer. *Oncogene, 23*(24), 4225–4231. https://doi.org/10.1038/sj.onc.1207118.
Izzedine, H., Boostandoost, H., & Mathian, A. (2017). Panitumumab-induced immune complex glomerulonephritis. *American Journal of Kidney Diseases, 69*(2), 320–321. https://doi.org/10.1053/j.ajkd.2016.09.011.
Izzedine, H., Brocheriou, I., Amoura, Z., & Mathian, A. (2021). Acute tubular injury and renal arterial myocyte vacuolization following crizotinib administration. *Kidney International Reports, 6*(2), 526–528. https://doi.org/10.1016/j.ekir.2020.11.029.
Izzedine, H., El-Fekih, R. K., & Perazella, M. A. (2016). The renal effects of ALK inhibitors. *Investigational New Drugs, 34*(5), 643–649. https://doi.org/10.1007/s10637-016-0379-y.
Izzedine, H., & Perazella, M. A. (2017). Anticancer drug-induced acute kidney injury. *Kidney International Reports, 2*(4), 504–514. https://doi.org/10.1016/j.ekir.2017.02.008.
Jafaripour, L., Naserzadeh, R., Alizamani, E., Javad Mashhadi, S. M., Moghadam, E. R., Nouryazdan, N., et al. (2021). Effects of rosmarinic acid on methotrexate-induced nephrotoxicity and hepatotoxicity in wistar rats. *Indian Journal of Nephrology, 31*(3), 218–224. https://doi.org/10.4103/ijn.IJN_14_20.
Jang, H. R., Lee, K., Jeon, J., Kim, J.-R., Lee, J. E., Kwon, G. Y., et al. (2020). Poly (ADP-ribose) polymerase inhibitor treatment as a novel therapy attenuating renal ischemia-reperfusion injury. *Frontiers in Immunology, 11*(2574). https://doi.org/10.3389/fimmu.2020.564288.
Jenke, R., Reßing, N., Hansen, F. K., Aigner, A., & Büch, T. (2021). Anticancer therapy with HDAC inhibitors: Mechanism-based combination strategies and future perspectives. *Cancers (Basel), 13*(4). https://doi.org/10.3390/cancers13040634.
Jhaveri, K. D., Wanchoo, R., Sakhiya, V., Ross, D. W., & Fishbane, S. (2017). Adverse renal effects of novel molecular oncologic targeted therapies: A narrative review. *Kidney International Reports, 2*(1), 108–123. https://doi.org/10.1016/j.ekir.2016.09.055.

Jiang, X., Ren, Z., Zhao, B., Zhou, S., Ying, X., & Tang, Y. (2020). Ameliorating effect of pentadecapeptide derived from Cyclina sinensis on cyclophosphamide-induced nephrotoxicity. *Marine Drugs, 18*(9). https://doi.org/10.3390/md18090462.

Jung, Y. J., Park, W., Kang, K. P., & Kim, W. (2020). SIRT2 is involved in cisplatin-induced acute kidney injury through regulation of mitogen-activated protein kinase phosphatase-1. *Nephrology, Dialysis, Transplantation, 35*(7), 1145–1156. https://doi.org/10.1093/ndt/gfaa042.

Kapoor, R., Robinson, K. A., Cata, J. P., Owusu-Agyemang, P., Soliz, J. M., Hernandez, M., et al. (2019). Assessment of nephrotoxicity associated with combined cisplatin and mitomycin C usage in laparoscopic hyperthermic intraperitoneal chemotherapy. *International Journal of Hyperthermia, 36*(1), 492–497. https://doi.org/10.1080/02656736.2019.1597175.

Khan, T. H., Ganaie, M. A., Alharthy, K. M., Madkhali, H., Jan, B. L., & Sheikh, I. A. (2020). Naringenin prevents doxorubicin-induced toxicity in kidney tissues by regulating the oxidative and inflammatory insult in Wistar rats. *Archives of Physiology and Biochemistry, 126*(4), 300–307. https://doi.org/10.1080/13813455.2018.1529799.

Khan, F. I., Lai, D., Anwer, R., Azim, I., & Khan, M. K. A. (2020). Identifying novel sphingosine kinase 1 inhibitors as therapeutics against breast cancer. *Journal of Enzyme Inhibition and Medicinal Chemistry, 35*(1), 172–186. https://doi.org/10.1080/14756366.2019.1692828.

Khan, M. I., Waguespack, S. G., & Ahmed, I. (2019). Recent advances in the management of hyponatremia in cancer patients. *Journal of Cancer Metastasis and Treatment, 5*, 71. https://doi.org/10.20517/2394-4722.2019.017.

Kim, S., Jo, C. H., & Kim, G.-H. (2021). The role of vasopressin V2 receptor in drug-induced hyponatremia. *Frontiers in Physiology, 12*(2272). https://doi.org/10.3389/fphys.2021.797039.

Kim, J. Y., Jo, J., Kim, K., An, H. J., Gwon, M. G., Gu, H., et al. (2019). Pharmacological activation of sirt1 ameliorates cisplatin-induced acute kidney injury by suppressing apoptosis, oxidative stress, and inflammation in mice. *Antioxidants (Basel), 8*(8). https://doi.org/10.3390/antiox8080322.

Kim, J.-Y., Jo, J., Leem, J., & Park, K.-K. (2020). Inhibition of p300 by garcinol protects against cisplatin-induced acute kidney injury through suppression of oxidative stress, inflammation, and tubular cell death in mice. *Antioxidants, 9*(12), 1271. Retrieved from https://www.mdpi.com/2076-3921/9/12/1271.

Kim, D., Park, W., Lee, S., Kim, W., Park, S. K., & Kang, K. P. (2018). Absence of Sirt3 aggravates cisplatin nephrotoxicity via enhanced renal tubular apoptosis and inflammation. *Molecular Medicine Reports, 18*(4), 3665–3672. https://doi.org/10.3892/mmr.2018.9350.

Kitchlu, A., McArthur, E., Amir, E., Booth, C. M., Sutradhar, R., Majeed, H., et al. (2019). Acute kidney injury in patients receiving systemic treatment for cancer: A population-based cohort study. *Journal of the National Cancer Institute, 111*(7), 727–736. https://doi.org/10.1093/jnci/djy167.

Kitchlu, A., & Rosner, M. H. (2019). Hyponatremia in patients with cancer. *Current Opinion in Nephrology and Hypertension, 28*(5), 433–440. https://doi.org/10.1097/mnh.0000000000000525.

Klaus, R., Niyazi, M., & Lange-Sperandio, B. (2021). Radiation-induced kidney toxicity: Molecular and cellular pathogenesis. *Radiation Oncology, 16*(1), 43. https://doi.org/10.1186/s13014-021-01764-y.

Knights, K. M., Rowland, A., & Miners, J. O. (2013). Renal drug metabolism in humans: The potential for drug–endobiotic interactions involving cytochrome P450 (CYP) and UDP-glucuronosyltransferase (UGT). *British Journal of Clinical Pharmacology, 76*(4), 587–602. https://doi.org/10.1111/bcp.12086.

Kobayashi, R., Suzuki, A., Matsuura, K., Yamada, N., Nakano, M., Deguchi, T., et al. (2016). Risk analysis for cisplatin-induced nephrotoxicity during first cycle of chemotherapy. *International Journal of Clinical and Experimental Medicine, 9*(2), 3635–3641.

Koks, M. S., Ocak, G., Suelmann, B. B. M., Hulsbergen-Veelken, C. A. R., Haitjema, S., Vianen, M. E., et al. (2021). Immune checkpoint inhibitor-associated acute kidney injury and mortality: An observational study. *PLoS One, 16*(6), e0252978. https://doi.org/10.1371/journal.pone.0252978.

Koźmiński, P., Halik, P. K., Chesori, R., & Gniazdowska, E. (2020). Overview of dual-acting drug methotrexate in different neurological diseases, autoimmune pathologies and cancers. *International Journal of Molecular Sciences, 21*(10). https://doi.org/10.3390/ijms21103483.

Kwon, Y. (2021). Possible beneficial effects of N-acetylcysteine for treatment of triple-negative breast cancer. *Antioxidants (Basel), 10*(2). https://doi.org/10.3390/antiox10020169.

LaFargue, C. J., Dal Molin, G. Z., Sood, A. K., & Coleman, R. L. (2019). Exploring and comparing adverse events between PARP inhibitors. *The Lancet Oncology, 20*(1), e15–e28. https://doi.org/10.1016/S1470-2045(18)30786-1.

Lameire, N. H., Flombaum, C. D., Moreau, D., & Ronco, C. (2005). Acute renal failure in cancer patients. *Annals of Medicine, 37*(1), 13–25. https://doi.org/10.1080/07853890510007205.

Lankhorst, S., Baelde, H. J., Kappers, M. H. W., Smedts, F. M. M., Hansen, A., Groningen, M. C. C.-V., et al. (2015). Greater sensitivity of blood pressure than renal toxicity to tyrosine kinase receptor inhibition with sunitinib. *Hypertension, 66*(3), 543–549. https://doi.org/10.1161/hypertensionaha.115.05435.

Lankhorst, S., Kappers, M. H., van Esch, J. H., Smedts, F. M., Sleijfer, S., Mathijssen, R. H., et al. (2014). Treatment of hypertension and renal injury induced by the angiogenesis inhibitor sunitinib: Preclinical study. *Hypertension, 64*(6), 1282–1289. https://doi.org/10.1161/hypertensionaha.114.04187.

Laplace, N., Kepenekian, V., Friggeri, A., Vassal, O., Ranchon, F., Rioufol, C., et al. (2020). Sodium thiosulfate protects from renal impairement following hyperthermic intraperitoneal chemotherapy (HIPEC) with Cisplatin. *International Journal of Hyperthermia, 37*(1), 897–902. https://doi.org/10.1080/02656736.2020.1795277.

Launay-Vacher, V., Oudard, S., Janus, N., Gligorov, J., Pourrat, X., Rixe, O., et al. (2007). Prevalence of renal insufficiency in cancer patients and implications for anticancer drug management: The renal insufficiency and anticancer medications (IRMA) study. *Cancer, 110*(6), 1376–1384. https://doi.org/10.1002/cncr.22904.

Lazareth, H., Delanoy, N., Cohen, R., Boissier, E., Ayari, H., Combe, P., et al. (2020). Nephrotoxicity associated with niraparib. *American Journal of Kidney Diseases, 76*(6), 898–900. https://doi.org/10.1053/j.ajkd.2020.05.021.

Lexicomp. (2022). *Fluorouracil (systemic): Drug information.* Retrieved from https://www-uptodate-com.proxy.library.vcu.edu/contents/fluorouracil-systemic-drug-information?search=Fluorouracil&source=panel_search_result&selectedTitle=1~147&usage_type=panel&display_rank=1&showDrugLabel=true.

Li, J., Fan, X., Wang, Q., Gong, Y., & Guo, L. (2021). Long noncoding RNA PRNCR1 reduces renal epithelial cell apoptosis in cisplatin-induced AKI by regulating miR-182-5p/EZH1. *Kidney and Blood Pressure Research, 46*(2), 162–172. https://doi.org/10.1159/000510157.

Li, Y., Gao, M., Yin, L. H., Xu, L. N., Qi, Y., Sun, P., et al. (2021). Dioscin ameliorates methotrexate-induced liver and kidney damages via adjusting miRNA-145-5p-mediated oxidative stress. *Free Radical Biology & Medicine, 169*, 99–109. https://doi.org/10.1016/j.freeradbiomed.2021.03.035.

Li, C.-M., Li, M., Ye, Z.-C., Huang, J.-Y., Li, Y., Yao, Z.-Y., et al. (2019). Circular RNA expression profiles in cisplatin-induced acute kidney injury in mice. *Epigenomics, 11*(10), 1191–1207. https://doi.org/10.2217/epi-2018-0167.

Li, G., Sachdev, U., Peters, K., Liang, X., & Lotze, M. T. (2019). The VE-PTP inhibitor AKB-9778 improves antitumor activity and diminishes the toxicity of interleukin 2 (IL-2) administration. *Journal of Immunotherapy*, *42*(7), 237–243. https://doi.org/10.1097/cji.0000000000000290.

Li, Z., Xu, K., Zhang, N., Amador, G., Wang, Y., Zhao, S., et al. (2018). Overexpressed SIRT6 attenuates cisplatin-induced acute kidney injury by inhibiting ERK1/2 signaling. *Kidney International*, *93*(4), 881–892. https://doi.org/10.1016/j.kint.2017.10.021.

Li, C., Xue, V. W., Wang, Q.-M., Lian, G.-Y., Huang, X.-R., Lee, T.-L., et al. (2020). The mincle/Syk/NF-κB signaling circuit is essential for maintaining the protumoral activities of tumor-associated macrophages. *Cancer Immunology Research*, *8*(8), 1004–1017. https://doi.org/10.1158/2326-6066.Cir-19-0782.

Lin, J. S., Mamlouk, O., Selamet, U., Tchakarov, A., Glass, W. F., Sheth, R. A., et al. (2021). Infliximab for the treatment of patients with checkpoint inhibitor associated acute tubular interstitial nephritis. *OncoImmunology*, *10*(1), 1877415. https://doi.org/10.1080/2162402X.2021.1877415.

Lin, Y. T., Wang, Y. F., Yang, J. C., Yu, C. J., Wu, S. G., Shih, J. Y., et al. (2014). Development of renal cysts after crizotinib treatment in advanced ALK-positive non-small-cell lung cancer. *Journal of Thoracic Oncology*, *9*(11), 1720–1725. https://doi.org/10.1097/jto.0000000000000326.

Liu, J., Livingston, M. J., Dong, G., Tang, C., Su, Y., Wu, G., et al. (2018). Histone deacetylase inhibitors protect against cisplatin-induced acute kidney injury by activating autophagy in proximal tubular cells. *Cell Death & Disease*, *9*(3), 322. https://doi.org/10.1038/s41419-018-0374-7.

Liu, Y., Yang, E. J., Shi, C., Mou, P. K., Zhang, B., Wu, C., et al. (2020). Histone acetyltransferase (HAT) P300/CBP inhibitors induce synthetic lethality in PTEN-deficient colorectal cancer cells through destabilizing AKT. *International Journal of Biological Sciences*, *16*(11), 1774–1784. https://doi.org/10.7150/ijbs.42197.

Liu, X.-Y., Zhang, F.-R., Shang, J.-Y., Liu, Y.-Y., Lv, X.-F., Yuan, J.-N., et al. (2018). Renal inhibition of miR-181a ameliorates 5-fluorouracil-induced mesangial cell apoptosis and nephrotoxicity. *Cell Death & Disease*, *9*(6), 610. https://doi.org/10.1038/s41419-018-0677-8.

Loren, P., Saavedra, N., Saavedra, K., Zambrano, T., Moriel, P., & Salazar, L. A. (2021). Epigenetic mechanisms involved in cisplatin-induced nephrotoxicity: An update. *Pharmaceuticals (Basel)*, *14*(6). https://doi.org/10.3390/ph14060491.

Luan, Z., Wei, Y., Huo, X., Sun, X., Zhang, C., Ming, W., et al. (2021). Pregnane X receptor (PXR) protects against cisplatin-induced acute kidney injury in mice. *Biochimica et Biophysica Acta - Molecular Basis of Disease*, *1867*(3), 165996. https://doi.org/10.1016/j.bbadis.2020.165996.

Lyu, B., Wang, W., Ji, X. Y., Ritter, J. K., & Li, N. (2020). Detrimental role of sphingosine kinase 1 in kidney damage in DOCA-salt hypertensive model: Evidence from knockout mice. *BMC Nephrology*, *21*(1), 173. https://doi.org/10.1186/s12882-020-01815-8.

Manohar, S., Kompotiatis, P., Thongprayoon, C., Cheungpasitporn, W., Herrmann, J., & Herrmann, S. M. (2019). Programmed cell death protein 1 inhibitor treatment is associated with acute kidney injury and hypocalcemia: Meta-analysis. *Nephrology, Dialysis, Transplantation*, *34*(1), 108–117. https://doi.org/10.1093/ndt/gfy105.

Manohar, S., & Leung, N. (2018). Cisplatin nephrotoxicity: A review of the literature. *Journal of Nephrology*, *31*(1), 15–25. https://doi.org/10.1007/s40620-017-0392-z.

Marabondo, S., & Kaufman, H. L. (2017). High-dose interleukin-2 (IL-2) for the treatment of melanoma: Safety considerations and future directions. *Expert Opinion on Drug Safety*, *16*(12), 1347–1357. https://doi.org/10.1080/14740338.2017.1382472.

Markowitz, G. S., & Perazella, M. A. (2005). Drug-induced renal failure: A focus on tubulointerstitial disease. *Clinica Chimica Acta*, *351*(1-2), 31–47. https://doi.org/10.1016/j.cccn.2004.09.005.

Martín Martorell, P., Huerta Alvaro, M., Solís Salguero, M. A., & Insa Molla, A. (2014). Crizotinib and renal insufficiency: A case report and review of the literature. *Lung Cancer, 84*(3), 310–313. https://doi.org/10.1016/j.lungcan.2014.03.001.

Martínez-Valenzuela, L., Draibe, J., Fulladosa, X., Gomà, M., Gómez, F., Antón, P., et al. (2021). Acute tubulointerstitial nephritis in clinical oncology: A comprehensive review. *International Journal of Molecular Sciences, 22*(5), 2326. Retrieved from https://www.mdpi.com/1422-0067/22/5/2326.

McKibbin, T., Cheng, L. L., Kim, S., Steuer, C. E., Owonikoko, T. K., Khuri, F. R., et al. (2016). Mannitol to prevent cisplatin-induced nephrotoxicity in patients with squamous cell cancer of the head and neck (SCCHN) receiving concurrent therapy. *Support Care Cancer, 24*(4), 1789–1793. https://doi.org/10.1007/s00520-015-2978-0.

McSweeney, K. R., Gadanec, L. K., Qaradakhi, T., Ali, B. A., Zulli, A., & Apostolopoulos, V. (2021). Mechanisms of cisplatin-induced acute kidney injury: Pathological mechanisms, pharmacological interventions, and genetic mitigations. *Cancers (Basel), 13*(7). https://doi.org/10.3390/cancers13071572.

Meirson, T., Asher, N., Bomze, D., & Markel, G. (2020). Safety of BRAF + MEK inhibitor combinations: Severe adverse event evaluation. *Cancers, 12*(6), 1650. Retrieved from https://www.mdpi.com/2072-6694/12/6/1650.

Mercantepe, T., Topcu, A., Rakici, S., Tumkaya, L., Yilmaz, A., & Mercantepe, F. (2019). The radioprotective effect of N-acetylcysteine against x-radiation-induced renal injury in rats. *Environmental Science and Pollution Research International, 26*(28), 29085–29094. https://doi.org/10.1007/s11356-019-06110-0.

Mérida, E., & Praga, M. (2019). NSAIDs and nephrotic syndrome. *Clinical Journal of the American Society of Nephrology, 14*(9), 1280–1282. https://doi.org/10.2215/cjn.08090719.

Mirabito Colafella, K. M., Neves, K. B., Montezano, A. C., Garrelds, I. M., van Veghel, R., de Vries, R., et al. (2020). Selective ETA vs. dual ETA/B receptor blockade for the prevention of sunitinib-induced hypertension and albuminuria in WKY rats. *Cardiovascular Research, 116*(10), 1779–1790. https://doi.org/10.1093/cvr/cvz260.

Misra, S. C., Santagostino, A., Dine, G., & Bonhomme Faivre, L. (2019). Acute kidney injury following high-dose methotrexate administration in a day care hospital. *Drug Safety - Case Reports, 6*(1), 12. https://doi.org/10.1007/s40800-019-0106-7.

Miteva-Marcheva, N. N., Ivanov, H. Y., Dimitrov, D. K., & Stoyanova, V. K. (2020). Application of pharmacogenetics in oncology. *Biomarker Research, 8*(1), 32. https://doi.org/10.1186/s40364-020-00213-4.

Mlejnek, P., Dolezel, P., Kriegova, E., & Pastvova, N. (2021). N-acetylcysteine can induce massive oxidative stress, resulting in cell death with apoptotic features in human leukemia cells. *International Journal of Molecular Sciences, 22*(23). https://doi.org/10.3390/ijms222312635.

Moledina, D. G., & Perazella, M. A. (2016). Proton pump inhibitors and CKD. *Journal of the American Society of Nephrology, 27*(10), 2926–2928. https://doi.org/10.1681/asn.2016020192.

Nagata, M., Ohji, G., & Iwata, K. (2019). Acute tubulointerstitial nephritis caused by rifampicin: An increasing and often overlooked side effect in elderly patients. *International Journal of Clinical Pharmacology and Therapeutics, 57*(5), 264–269. https://doi.org/10.5414/cp203287.

Naughton, C. A. (2008). Drug-induced nephrotoxicity. *American Family Physician, 78*(6), 743–750.

Navarro-Hortal, M. D., Varela-López, A., Romero-Márquez, J. M., Rivas-García, L., Speranza, L., Battino, M., et al. (2020). Role of flavonoids against adriamycin toxicity. *Food and Chemical Toxicology, 146*, 111820. https://doi.org/10.1016/j.fct.2020.111820.

Oleas, D., Bolufer, M., Agraz, I., Felip, E., Muñoz, E., Gabaldón, A., et al. (2020). Acute interstitial nephritis associated with immune checkpoint inhibitors: a single-centre experience. *Clinical Kidney Journal, 14*(5), 1364–1370. https://doi.org/10.1093/ckj/sfaa008.

Ommati, M. M., Farshad, O., Ghanbarinejad, V., Mohammadi, H. R., Khadijeh, M., Negar, A., et al. (2020). The nephroprotective role of carnosine against ifosfamide-induced renal injury and electrolytes imbalance is mediated via the regulation of mitochondrial function and alleviation of oxidative stress. *Drug Research, 70*(1), 49–56. https://doi.org/10.1055/a-1017-5085.

Ozkok, S., & Ozkok, A. (2017). Contrast-induced acute kidney injury: A review of practical points. *World Journal of Nephrology, 6*(3), 86–99. https://doi.org/10.5527/wjn.v6.i3.86.

Paluri, R. K., Sonpavde, G., Morgan, C., Rojymon, J., Mar, A. H., & Gangaraju, R. (2019). Renal toxicity with mammalian target of rapamycin inhibitors: A meta-analysis of randomized clinical trials. *Oncology Reviews, 13*(2), 455. https://doi.org/10.4081/oncol.2019.455.

Pant, K., Peixoto, E., Richard, S., & Gradilone, S. A. (2020). Role of histone deacetylases in carcinogenesis: Potential role in cholangiocarcinoma. *Cells, 9*(3). https://doi.org/10.3390/cells9030780.

Patel, V., Elias, R., Formella, J., Schwartzman, W., Christie, A., Cai, Q., et al. (2020). Acute interstitial nephritis, a potential predictor of response to immune checkpoint inhibitors in renal cell carcinoma. *Journal for Immunotherapy of Cancer, 8*(2). https://doi.org/10.1136/jitc-2020-001198.

Patel, J. B., & Sapra, A. (2021). *Nephrotoxic medications*. Retrieved from https://www.ncbi.nlm.nih.gov/books/NBK553144/.

Pavkovic, M., Robinson-Cohen, C., Chua, A. S., Nicoara, O., Cárdenas-González, M., Bijol, V., et al. (2016). Detection of drug-induced acute kidney injury in humans using urinary KIM-1, miR-21, -200c, and -423. *Toxicological Sciences, 152*(1), 205–213. https://doi.org/10.1093/toxsci/kfw077.

Peasley, K., Chiba, T., Goetzman, E., & Sims-Lucas, S. (2021). Sirtuins play critical and diverse roles in acute kidney injury. *Pediatric Nephrology*. https://doi.org/10.1007/s00467-020-04866-z.

Perazella, M. A. (2005). Drug-induced nephropathy: An update. *Expert Opinion on Drug Safety, 4*(4), 689–706. https://doi.org/10.1517/14740338.4.4.689.

Perazella, M. A. (2009). Renal vulnerability to drug toxicity. *Clinical Journal of the American Society of Nephrology, 4*(7), 1275–1283. https://doi.org/10.2215/cjn.02050309.

Perazella, M. A. (2010). Toxic nephropathies: Core curriculum 2010. *American Journal of Kidney Diseases, 55*(2), 399–409. https://doi.org/10.1053/j.ajkd.2009.10.046.

Perazella, M. A. (2012). Onco-nephrology: Renal toxicities of chemotherapeutic agents. *Clinical Journal of the American Society of Nephrology*, 7(10), 1713–1721. https://doi.org/10.2215/cjn.02780312.

Perazella, M. A. (2018). Pharmacology behind common drug nephrotoxicities. *Clinical Journal of the American Society of Nephrology, 13*(12), 1897–1908. https://doi.org/10.2215/cjn.00150118.

Perazella, M. A., & Markowitz, G. S. (2010). Drug-induced acute interstitial nephritis. *Nature Reviews Nephrology, 6*(8), 461–470. https://doi.org/10.1038/nrneph.2010.71.

Perazella, M. A., & Moeckel, G. W. (2010). Nephrotoxicity from chemotherapeutic agents: Clinical manifestations, pathobiology, and prevention/therapy. *Seminars in Nephrology, 30*(6), 570–581. https://doi.org/10.1016/j.semnephrol.2010.09.005.

Perazella, M. A., & Shirali, A. C. (2018). Nephrotoxicity of cancer immunotherapies: Past, present and future. *Journal of the American Society of Nephrology, 29*(8), 2039–2052. https://doi.org/10.1681/asn.2018050488.

Pitman, M., Oehler, M. K., & Pitson, S. M. (2021). Sphingolipids as multifaceted mediators in ovarian cancer. *Cellular Signalling, 81*, 109949. https://doi.org/10.1016/j.cellsig.2021.109949.

Porta, C., Bamias, A., Danesh, F. R., Dębska-Ślizień, A., Gallieni, M., Gertz, M. A., et al. (2020). KDIGO Controversies Conference on onco-nephrology: Understanding kidney impairment and solid-organ malignancies, and managing kidney cancer. *Kidney International*, *98*(5), 1108–1119. https://doi.org/10.1016/j.kint.2020.06.046.

Pyne, N. J., & Pyne, S. (2020). Recent advances in the role of sphingosine 1-phosphate in cancer. *FEBS Letters*, *594*(22), 3583–3601. https://doi.org/10.1002/1873-3468.13933.

Qin, S., Li, A., Yi, M., Yu, S., Zhang, M., & Wu, K. (2019). Recent advances on anti-angiogenesis receptor tyrosine kinase inhibitors in cancer therapy. *Journal of Hematology & Oncology*, *12*(1), 27. https://doi.org/10.1186/s13045-019-0718-5.

Qu, J., Ding, Y., Jiang, K., Hao, J., Li, Y., Zhang, A., et al. (2021). Nephrotoxicity of immune checkpoint inhibitors: A disproportionality analysis from 2013 to 2020. *The Tohoku Journal of Experimental Medicine*, *254*(4), 275–282. https://doi.org/10.1620/tjem.254.275.

Raal, F. J., Rosenson, R. S., Reeskamp, L. F., Hovingh, G. K., Kastelein, J. J. P., Rubba, P., et al. (2020). Evinacumab for homozygous familial hypercholesterolemia. *The New England Journal of Medicine*, *383*(8), 711–720. https://doi.org/10.1056/NEJMoa2004215.

Ramsey, L. B., Balis, F. M., O'Brien, M. M., Schmiegelow, K., Pauley, J. L., Bleyer, A., et al. (2018). Consensus guideline for use of glucarpidase in patients with high-dose methotrexate induced acute kidney injury and delayed methotrexate clearance. *The Oncologist*, *23*(1), 52–61. https://doi.org/10.1634/theoncologist.2017-0243.

Reddy, V., & Winston, N. R. (2021). *Mesna*. Retrieved from https://www.ncbi.nlm.nih.gov/books/NBK556021/.

Richter, S. K., & Crannage, A. J. (2015). Evaluation of N-acetylcysteine for the prevention of contrast-induced nephropathy. *Journal of Community Hospital Internal Medicine Perspectives*, *5*(3), 27297. https://doi.org/10.3402/jchimp.v5.27297.

Robert, C. (2020). A decade of immune-checkpoint inhibitors in cancer therapy. *Nature Communications*, *11*(1), 3801. https://doi.org/10.1038/s41467-020-17670-y.

Roskoski, R., Jr. (2019). Properties of FDA-approved small molecule protein kinase inhibitors. *Pharmacological Research*, *144*, 19–50. https://doi.org/10.1016/j.phrs.2019.03.006.

Sabapathy, V., Cheru, N. T., Corey, R., Mohammad, S., & Sharma, R. (2019). A novel hybrid cytokine IL233 mediates regeneration following doxorubicin-induced nephrotoxic injury. *Scientific Reports*, *9*(1), 3215. https://doi.org/10.1038/s41598-019-39886-9.

Sahindokuyucu-Kocasari, F., Akyol, Y., Ozmen, O., Erdemli-Kose, S. B., & Garli, S. (2021). Apigenin alleviates methotrexate-induced liver and kidney injury in mice. *Human & Experimental Toxicology*. https://doi.org/10.1177/09603271211009964. 9603271211009964.

Salama, R. M., Nasr, M. M., Abdelhakeem, J. I., Roshdy, O. K., & ElGamal, M. A. (2020). Alogliptin attenuates cyclophosphamide-induced nephrotoxicity: A novel therapeutic approach through modulating MAP3K/JNK/SMAD3 signaling cascade. *Drug and Chemical Toxicology*, *1–10*. https://doi.org/10.1080/01480545.2020.1814319.

Saleh, D., Abdelbaset, M., Hassan, A., Sharaf, O., Mahmoud, S., & Hegazy, R. (2020). Omega-3 fatty acids ameliorate doxorubicin-induced cardiorenal toxicity: In-vivo regulation of oxidative stress, apoptosis and renal Nox4, and in-vitro preservation of the cytotoxic efficacy. *PLoS One*, *15*(11), e0242175. https://doi.org/10.1371/journal.pone.0242175.

Sazonov, V., Tobylbayeva, Z., Saparov, A., Jubaniyazov, B., Issakov, S., & Gaipov, A. (2021). New therapeutic approach to reduce methotrexate toxicity after high-dose chemotherapy in a child with acute lymphocytic leukemia: Efficacy and safety of hemoadsorption with HA-230 adsorber. *Blood Purification*, *1–5*. https://doi.org/10.1159/000514135.

Schiffl, H. (2020). Gender differences in the susceptibility of hospital-acquired acute kidney injury: More questions than answers. *International Urology and Nephrology*, *52*(10), 1911–1914. https://doi.org/10.1007/s11255-020-02526-7.

Sharp, C. N., & Siskind, L. J. (2017). Developing better mouse models to study cisplatin-induced kidney injury. *American Journal of Physiology-Renal Physiology*, *313*(4), F835–F841. https://doi.org/10.1152/ajprenal.00285.2017.

Sherif, I. O., Al-Shaalan, N. H., & Sabry, D. (2019). Ginkgo biloba extract alleviates methotrexate-induced renal injury: New impact on PI3K/Akt/mTOR signaling and MALAT1 expression. *Biomolecules*, *9*(11), 691. Retrieved from https://www.mdpi.com/2218-273X/9/11/691.

Shye, M., Hanna, R. M., Patel, S. S., Tram-Tran, N., Hou, J., McCannel, C., et al. (2020). Worsening proteinuria and renal function after intravitreal vascular endothelial growth factor blockade for diabetic proliferative retinopathy. *Clinical Kidney Journal*, *13*(6), 969–980. https://doi.org/10.1093/ckj/sfaa049.

Smith, H. L., Southgate, H., Tweddle, D. A., & Curtin, N. J. (2020). DNA damage checkpoint kinases in cancer. *Expert Reviews in Molecular Medicine*, *22*, e2. https://doi.org/10.1017/erm.2020.3.

Song, Y., Liu, L., Liu, B., Liu, R., Chen, Y., Li, C., et al. (2021). Interaction of nobiletin with methotrexate ameliorates 7-OH methotrexate-induced nephrotoxicity through endoplasmic reticulum stress-dependent PERK/CHOP signaling pathway. *Pharmacological Research*, *165*, 105371. https://doi.org/10.1016/j.phrs.2020.105371.

Stone, E. L., Carter, K. P., Wagner, E. K., Asensio, M. A., Benzie, E., Chiang, Y. Y., et al. (2021). Lack of blocking activity in anti-CTLA-4 antibodies reduces toxicity, but not anti-tumor efficacy. *bioRxiv*. https://doi.org/10.1101/2021.07.12.452090. 2021.2007.2012.452090.

Stremska, M. E., Jose, S., Sabapathy, V., Huang, L., Bajwa, A., Kinsey, G. R., et al. (2017). IL233, a novel IL-2 and IL-33 hybrid cytokine, ameliorates renal injury. *Journal of the American Society of Nephrology*, *28*(9), 2681–2693. https://doi.org/10.1681/asn.2016121272.

Sun, C. Y., Nie, J., Zheng, Z. L., Zhao, J., Wu, L. M., Zhu, Y., et al. (2019). Renoprotective effect of scutellarin on cisplatin-induced renal injury in mice: Impact on inflammation, apoptosis, and autophagy. *Biomedicine & Pharmacotherapy*, *112*, 108647. https://doi.org/10.1016/j.biopha.2019.108647.

Sun, Y. B., Qu, X., Zhang, X., Caruana, G., Bertram, J. F., & Li, J. (2013). Glomerular endothelial cell injury and damage precedes that of podocytes in adriamycin-induced nephropathy. *PLoS One*, *8*(1), e55027. https://doi.org/10.1371/journal.pone.0055027.

Tan, R. Z., Liu, J., Zhang, Y. Y., Wang, H. L., Li, J. C., Liu, Y. H., et al. (2019). Curcumin relieved cisplatin-induced kidney inflammation through inhibiting Mincle-maintained M1 macrophage phenotype. *Phytomedicine*, *52*, 284–294. https://doi.org/10.1016/j.phymed.2018.09.210.

Tang, J., Shi, Y., Liu, N., Xu, L., Zang, X., Li, P., et al. (2018). Blockade of histone deacetylase 6 protects against cisplatin-induced acute kidney injury. *Clinical Science (London, England)*, *132*(3), 339–359. https://doi.org/10.1042/cs20171417.

Tchounwou, P. B., Dasari, S., Noubissi, F. K., Ray, P., & Kumar, S. (2021). Advances in our understanding of the molecular mechanisms of action of cisplatin in cancer therapy. *Journal of Experimental Pharmacology*, *13*, 303–328. https://doi.org/10.2147/jep.S267383.

Teuma, C., Pelletier, S., Amini-Adl, M., Perier-Muzet, M., Maucort-Boulch, D., Thomas, L., et al. (2017). Adjunction of a MEK inhibitor to Vemurafenib in the treatment of metastatic melanoma results in a 60% reduction of acute kidney injury. *Cancer Chemotherapy and Pharmacology*, *79*(5), 1043–1049. https://doi.org/10.1007/s00280-017-3300-2.

Townsend, D. M., & Tew, K. D. (2003). The role of glutathione-S-transferase in anti-cancer drug resistance. *Oncogene*, *22*(47), 7369–7375. https://doi.org/10.1038/sj.onc.1206940.

Trendowski, M. R., El-Charif, O., Ratain, M. J., Monahan, P., Mu, Z., Wheeler, H. E., et al. (2019). Clinical and genome-wide analysis of serum platinum levels after cisplatin-based chemotherapy. *Clinical Cancer Research, 25*(19), 5913–5924. https://doi.org/10.1158/1078-0432.Ccr-19-0113.

Un, H., Ugan, R. A., Kose, D., Bayir, Y., Cadirci, E., Selli, J., et al. (2020). A novel effect of aprepitant: Protection for cisplatin-induced nephrotoxicity and hepatotoxicity. *European Journal of Pharmacology, 880*, 173168. https://doi.org/10.1016/j.ejphar.2020.173168.

van Zonneveld, A. J., Kölling, M., Bijkerk, R., & Lorenzen, J. M. (2021). Circular RNAs in kidney disease and cancer. *Nature Reviews. Nephrology*. https://doi.org/10.1038/s41581-021-00465-9.

Volarevic, V., Djokovic, B., Jankovic, M. G., Harrell, C. R., Fellabaum, C., Djonov, V., et al. (2019). Molecular mechanisms of cisplatin-induced nephrotoxicity: A balance on the knife edge between renoprotection and tumor toxicity. *Journal of Biomedical Science, 26*(1), 25. https://doi.org/10.1186/s12929-019-0518-9.

Wanchoo, R., Abudayyeh, A., Doshi, M., Edeani, A., Glezerman, I. G., Monga, D., et al. (2017). Renal toxicities of novel agents used for treatment of multiple myeloma. *Clinical Journal of the American Society of Nephrology, 12*(1), 176–189. https://doi.org/10.2215/cjn.06100616.

Wanchoo, R., Jhaveri, K. D., Deray, G., & Launay-Vacher, V. (2016). Renal effects of BRAF inhibitors: A systematic review by the cancer and the kidney international network. *Clinical Kidney Journal, 9*(2), 245–251. https://doi.org/10.1093/ckj/sfv149.

Wanchoo, R., Karam, S., Uppal, N. N., Barta, V. S., Deray, G., Devoe, C., et al. (2017). Adverse renal effects of immune checkpoint inhibitors: A narrative review. *American Journal of Nephrology, 45*(2), 160–169. https://doi.org/10.1159/000455014.

Wanchoo, R., Khan, S., Kolitz, J. E., & Jhaveri, K. D. (2015). Carfilzomib-related acute kidney injury may be prevented by N-acetyl-L-cysteine. *Journal of Oncology Pharmacy Practice, 21*(4), 313–316. https://doi.org/10.1177/1078155214531804.

Wang, Y. M., Gu, M. L., Meng, F. S., Jiao, W. R., Zhou, X. X., Yao, H. P., et al. (2017). Histone acetyltransferase p300/CBP inhibitor C646 blocks the survival and invasion pathways of gastric cancer cell lines. *International Journal of Oncology, 51*(6), 1860–1868. https://doi.org/10.3892/ijo.2017.4176.

Wang, P., Wang, Z., & Liu, J. (2020). Role of HDACs in normal and malignant hematopoiesis. *Molecular Cancer, 19*(1), 5. https://doi.org/10.1186/s12943-019-1127-7.

Webb, T. R., Slavish, J., George, R. E., Look, A. T., Xue, L., Jiang, Q., et al. (2009). Anaplastic lymphoma kinase: Role in cancer pathogenesis and small-molecule inhibitor development for therapy. *Expert Review of Anticancer Therapy, 9*(3), 331–356. https://doi.org/10.1586/14737140.9.3.331.

Weber, J. (2010). Immune checkpoint proteins: a new therapeutic paradigm for cancer—Preclinical background: CTLA-4 and PD-1 blockade. *Seminars in Oncology, 37*(5), 430–439. https://doi.org/10.1053/j.seminoncol.2010.09.005.

Wei, X., Wu, Y., Tang, H., Wang, B., Wang, Y., Sun, W., et al. (2021). CP-25 ameliorates methotrexate induced nephrotoxicity via improving renal apoptosis and methotrexate excretion. *Journal of Pharmacological Sciences, 146*(1), 21–28. https://doi.org/10.1016/j.jphs.2021.02.007.

Wilson, N., Wiele, A. J., Surasi, D. S., Rao, P., Sircar, K., Shah, A. Y., et al. (2021). Efficacy of gemcitabine plus doxorubicin (Gem + Dox) in patients with renal medullary carcinoma (RMC). *Journal of Clinical Oncology, 39*(6_Suppl), 324. https://doi.org/10.1200/JCO.2021.39.6_suppl.324.

Wong, K. K. (2020). DNMT1 as a therapeutic target in pancreatic cancer: Mechanisms and clinical implications. *Cellular Oncology, 43*(5), 779–792. https://doi.org/10.1007/s13402-020-00526-4.

Workeneh, B. T., Uppal, N. N., Jhaveri, K. D., & Rondon-Berrios, H. (2021). Hypomagnesemia in the Cancer Patient. *Kidney360, 2*(1), 154–166. https://doi.org/10.34067/kid.0005622020.

Wu, Z., Li, C., Li, Q., Li, J., & Lu, X. (2020). Puerarin alleviates cisplatin-induced acute renal damage and upregulates microRNA-31-related signaling. *Experimental and Therapeutic Medicine*, *20*(4), 3122–3129. https://doi.org/10.3892/etm.2020.9081.

Xie, D. L., Hu, G., Wang, W., Li, P.-L., Chen, C., & Li, N. (2020). Inhibition of sphingosine kinase 2 attenuated cisplatin-induced nephrotoxicity in mice. *The FASEB Journal*, *34*(S1), 1. https://doi.org/10.1096/fasebj.2020.34.s1.06411.

Xu, Y.-Y., Jiang, N., Liu, T.-S., Qu, H.-Q., & Wang, T. (2012). Evaluation of the effect of glutathione on cisplatin antitumor activity and kidney injury at different administration times. *Molecular Medicine Reports*, *6*(5), 1075–1080. https://doi.org/10.3892/mmr.2012.1033.

Xu, Y., Ma, H., Shao, J., Wu, J., Zhou, L., Zhang, Z., et al. (2015). A role for tubular necroptosis in cisplatin-induced AKI. *Journal of American Society of Nephrology*, *26*(11), 2647–2658. https://doi.org/10.1681/asn.2014080741.

Xue, D., Hsu, E., Fu, Y.-X., & Peng, H. (2021). Next-generation cytokines for cancer immunotherapy. *Antibody Therapeutics*, *4*(2), 123–133. https://doi.org/10.1093/abt/tbab014.

Yang, J., Yang, C., Zhang, S., Mei, Z., Shi, M., Sun, S., et al. (2015). ABC294640, a sphingosine kinase 2 inhibitor, enhances the antitumor effects of TRAIL in non-small cell lung cancer. *Cancer Biology & Therapy*, *16*(8), 1194–1204. https://doi.org/10.1080/15384047.2015.1056944.

Yap, T. A., Tan, D. S. P., Terbuch, A., Caldwell, R., Guo, C., Goh, B. C., et al. (2021). First-in-human trial of the oral Ataxia telangiectasia and RAD3-related (ATR) inhibitor BAY 1895344 in patients with advanced solid tumors. *Cancer Discovery*, *11*(1), 80–91. https://doi.org/10.1158/2159-8290.Cd-20-0868.

Yarlagadda, S. G., & Perazella, M. A. (2008). Drug-induced crystal nephropathy: An update. *Expert Opinion on Drug Safety*, *7*(2), 147–158. https://doi.org/10.1517/14740338.7.2.147.

Yasuma, T., Kobayashi, T., D'Alessandro-Gabazza, C. N., Fujimoto, H., Ito, K., Nishii, Y., et al. (2018). Renal injury during long-term crizotinib therapy. *International Journal of Molecular Sciences*, *19*(10). https://doi.org/10.3390/ijms19102902.

Yee, S. W., Brackman, D. J., Ennis, E. A., Sugiyama, Y., Kamdem, L. K., Blanchard, R., et al. (2018). Influence of transporter polymorphisms on drug disposition and response: A perspective from the international transporter consortium. *Clinical Pharmacology & Therapeutics*, *104*(5), 803–817. https://doi.org/10.1002/cpt.1098.

Yemm, K. E., Alwan, L. M., Malik, A. B., & Salazar, L. G. (2019). Renal toxicity with liposomal doxorubicin in metastatic breast cancer. *Journal of Oncology Pharmacy Practice*, *25*(7), 1738–1742. https://doi.org/10.1177/1078155218798157.

Yeung, E., Yong, E., & Wong, F. (2004). Renal dysfunction in cirrhosis: Diagnosis, treatment and prevention. *MedGenMed*, *6*(4), 9.

Yokota, R., Bhunu, B., Toba, H., & Intapad, S. (2021). Sphingolipids and kidney disease: Possible role of preeclampsia and intrauterine growth restriction (IUGR). *Kidney360*, *3*(2), 534–541. https://doi.org/10.34067/kid.0006322020.

Yoon, S. P., & Kim, J. (2016). Poly(ADP-ribose) polymerase 1 contributes to oxidative stress through downregulation of sirtuin 3 during cisplatin nephrotoxicity. *Anatomy & Cell Biology*, *49*(3), 165–176. https://doi.org/10.5115/acb.2016.49.3.165.

Yoshida, T., Niho, S., Toda, M., Goto, K., Yoh, K., Umemura, S., et al. (2014). Protective effect of magnesium preloading on cisplatin-induced nephrotoxicity: A retrospective study. *Japanese Journal of Clinical Oncology*, *44*(4), 346–354. https://doi.org/10.1093/jjco/hyu004.

You, J., Gao, F., Tang, H., Peng, F., Jia, L., Huang, K., et al. (2019). A medicinal and edible formula YH0618 ameliorates the toxicity induced by doxorubicin via regulating the expression of Bax/Bcl-2 and FOXO4. *Journal of Cancer*, *10*, 3665–3677.

Young, A., Beriault, D., Jung, B., Nikonova, A., Abosh, D., Lee, S., et al. (2019). DAMPAned methotrexate: A case report and review of the management of acute methotrexate toxicity. *Canadian Journal of Kidney Health and Disease*, *6*, 2054358119895078. https://doi.org/10.1177/2054358119895078.

Yuan, H. T., Khankin, E. V., Karumanchi, S. A., & Parikh, S. M. (2009). Angiopoietin 2 is a partial agonist/antagonist of Tie2 signaling in the endothelium. *Molecular and Cellular Biology*, *29*(8), 2011–2022. https://doi.org/10.1128/mcb.01472-08.

Zazuli, Z., Duin, N., Jansen, K., Vijverberg, S. J. H., Maitland-van der Zee, A. H., & Masereeuw, R. (2020). The impact of genetic polymorphisms in organic cation transporters on renal drug disposition. *International Journal of Molecular Sciences*, *21*(18). https://doi.org/10.3390/ijms21186627.

Zhang, M. Y., Dugbartey, G. J., Juriasingani, S., & Sener, A. (2021). Hydrogen sulfide metabolite, sodium thiosulfate: Clinical applications and underlying molecular mechanisms. *International Journal of Molecular Sciences*, *22*(12). https://doi.org/10.3390/ijms22126452.

Zhang, X. Y., & Elfarra, A. A. (2018). Toxicity mechanism-based prodrugs: Glutathione-dependent bioactivation as a strategy for anticancer prodrug design. *Expert Opinion on Drug Discovery*, *13*(9), 815–824. https://doi.org/10.1080/17460441.2018.1508207.

Zhang, X., Ritter, J. K., & Li, N. (2018). Sphingosine-1-phosphate pathway in renal fibrosis. *American Journal of Physiology-Renal Physiology*, *315*(4), F752–F756. https://doi.org/10.1152/ajprenal.00596.2017.

Zhang, S., Sun, H., Kong, W., & Zhang, B. (2020). Functional role of microRNA-500a-3P-loaded liposomes in the treatment of cisplatin-induced AKI. *IET Nanobiotechnology*, *14*(6), 465–469. https://doi.org/10.1049/iet-nbt.2019.0247.

Zhang, Y., Tang, P. M.-K., Niu, Y., García Córdoba, C. A., Huang, X.-R., Yu, C., et al. (2020). Long non-coding RNA LRNA9884 promotes acute kidney injury via regulating NF-kB-mediated transcriptional activation of MIF. *Frontiers in Physiology*, *11*(1399). https://doi.org/10.3389/fphys.2020.590027.

Zhang, Y., Tao, X., Yin, L., Xu, L., Xu, Y., Qi, Y., et al. (2017). Protective effects of dioscin against cisplatin-induced nephrotoxicity via the microRNA-34a/sirtuin 1 signalling pathway. *British Journal of Pharmacology*, *174*(15), 2512–2527. https://doi.org/10.1111/bph.13862.

Zhang, D. Y., Tu, T., Younis, M. R., Zhu, K. S., Liu, H., Lei, S., et al. (2021). Clinically translatable gold nanozymes with broad spectrum antioxidant and anti-inflammatory activity for alleviating acute kidney injury. *Theranostics*, *11*(20), 9904–9917. https://doi.org/10.7150/thno.66518.

Zhang, X., Wang, W., Ji, X. Y., Ritter, J. K., & Li, N. (2019). Knockout of sphingosine kinase 1 attenuates renal fibrosis in unilateral ureteral obstruction model. *American Journal of Nephrology*, *50*(3), 196–203. https://doi.org/10.1159/000502448.

Zhang, J., Ye, Z.-W., Tew, K. D., & Townsend, D. M. (2021). Chapter eight—Cisplatin chemotherapy and renal function. In K. D. Tew, & P. B. Fisher (Eds.), *Vol. 152. Advances in cancer research* (pp. 305–327). Academic Press.

Zhang, Y.-J., Zhang, L., Feng, F., & Cao, Q.-F. (2021). ANGPTL3 overexpression suppresses the development of oncogenic properties in renal cell carcinoma via the Wnt-catenin signaling pathway and predicts good prognosis. *Disease Markers*, *2021*, 2863856. https://doi.org/10.1155/2021/2863856.

Zhang, J., Zou, Y., Cheng-Jing, Y., Xiang-Heng, L., Wang, X. P., Yu, X. J., et al. (2020). Pioglitazone alleviates cisplatin nephrotoxicity by suppressing mitochondria-mediated apoptosis via SIRT1/p53 signalling. *Journal of Cellular and Molecular Medicine*, *24*(20), 11718–11728. https://doi.org/10.1111/jcmm.15782.

Zhao, T., Liang, X., Chen, J., Bao, Y., Wang, A., Gan, X., et al. (2019). ANGPTL3 inhibits renal cell carcinoma metastasis by inhibiting VASP phosphorylation. *Biochemical and Biophysical Research Communications*, *516*(3), 880–887. https://doi.org/10.1016/j.bbrc.2019.06.120.

Zheng, H., Zhang, Y., Li, L., Zhang, R., Luo, Z., Yang, Z., et al. (2021). Depletion of toll-like receptor-9 attenuates renal tubulointerstitial fibrosis after ischemia-reperfusion injury. *Frontiers in Cell and Development Biology*, *9*(236). https://doi.org/10.3389/fcell.2021.641527.

Zhou, X., Jiang, K., Luo, H., Wu, C., Yu, W., & Cheng, F. (2020). Novel lncRNA XLOC_032768 alleviates cisplatin-induced apoptosis and inflammatory response of renal tubular epithelial cells through TNF-α. *International Immunopharmacology*, *83*, 106472. https://doi.org/10.1016/j.intimp.2020.106472.

Zhou, S., Zeng, S., & Shu, Y. (2021). Drug-drug interactions at organic cation transporter 1. *Frontiers in Pharmacology*, *12*, 628705. https://doi.org/10.3389/fphar.2021.628705.

Zibetti Dal Molin, G., Westin, S. N., Msaouel, P., Gomes, L. M., Dickens, A., & Coleman, R. L. (2020). Discrepancy in calculated and measured glomerular filtration rates in patients treated with PARP inhibitors. *International Journal of Gynecological Cancer*, *30*(1), 89–93. https://doi.org/10.1136/ijgc-2019-000714.

Zsiros, E., Lynam, S., Attwood, K. M., Wang, C., Chilakapati, S., Gomez, E. C., et al. (2021). Efficacy and safety of pembrolizumab in combination with bevacizumab and oral metronomic cyclophosphamide in the treatment of recurrent ovarian cancer: A phase 2 nonrandomized clinical trial. *JAMA Oncology*, 7(1), 78–85. https://doi.org/10.1001/jamaoncol.2020.5945.

Zunino, F., Pratesi, G., Micheloni, A., Cavalletti, E., Sala, F., & Tofanetti, O. (1989). Protective effect of reduced glutathione against cisplatin-induced renal and systemic toxicity and its influence on the therapeutic activity of the antitumor drug. *Chemico-Biological Interactions*, *70*(1), 89–101. https://doi.org/10.1016/0009-2797(89)90065-3.

Zou D., Ganugula R., Arora M., Nabity M.B., Sheikh-Hamad D., Ravi Kumar M.N.V., 2019. Oral delivery of nanoparticle urolithin A normalizes cellular stress and improves survival in mouse model of cisplatin-induced AKI, American Journal of Physiology. Renal Physiology 317 (5), F1255-F1264. doi: 10.1152/ajprenal.00346.2019. Epub 2019 Sep 18.

CHAPTER FOUR

Chemotherapy induced gastrointestinal toxicities

Hamid I. Akbarali*, Karan H. Muchhala, Donald K. Jessup, and Stanley Cheatham
Department of Pharmacology and Toxicology, School of Medicine, Virginia Commonwealth University, Richmond, VA, United States
*Corresponding author: e-mail address: hamid.akbarali@vcuhealth.org

Contents

Abstract

Chemotherapy—induced gastrointestinal dysfunction is a common occurrence associated with many different classes of chemotherapeutic agents. Gastrointestinal toxicity includes mucositis, diarrhea, and constipation, and can often be a dose-limiting complication, induce cessation of treatment and could be life threatening. The gastrointestinal epithelium is rich in rapidly dividing cells and hence is a prime target for chemotherapeutic drugs. The incidence of gastrointestinal toxicity, including diarrhea and mucositis, is extremely high for a wide array of chemotherapeutic and radiation regimens. In fact, 60%–100% of patients on high-dose chemotherapy suffer from gastrointestinal side effects. Unfortunately, treatment options are limited, and therapy is often restricted to palliative care. Therefore, there is a great unmet therapeutic need for preventing and treating chemotherapy-induced gastrointestinal toxicities in the clinic. In this review, we discuss our current understanding of the mechanisms underlying chemotherapy-induced diarrhea and mucositis, and emerging mechanisms

Advances in Cancer Research, Volume 155
ISSN 0065-230X
https://doi.org/10.1016/bs.acr.2022.02.007

involving the enteric nervous system, smooth muscle cells and enteric immune cells. Recent evidence has also implicated gut dysbiosis in the pathogenesis of not only chemotherapy-induced mucositis and diarrhea, but also chemotherapy-induced peripheral neuropathy. Oxidative stress induced by chemotherapeutic agents results in post-translational modification of ion channels altering neuronal excitability. Thus, investigating how chemotherapy-induced changes in the gut- microbiome axis may lead to gut-related toxicities will be critical in the discovery of new drug targets for mitigating adverse gastrointestinal effects associated with chemotherapy treatment.

1. Introduction

Current chemotherapeutic regimens do not differentiate between cancer cells and normal cells such as those lining the epithelium of the gastrointestinal tract. This non-specific targeting of rapidly dividing cells results in many gastrointestinal related side effects. Chemotherapy-induced gastrointestinal (GI) toxicities are prevalent among a wide array of chemotherapeutic and radiation regimens. It is estimated that 40% of patients receiving standard dose chemotherapy will develop GI-related toxicities, while significantly greater incidence rates have been reported in patients receiving higher drug doses. In fact, it is estimated that approximately 60%–100% of patients on high dose chemotherapy will experience GI toxicities (Cinausero et al., 2017; Dahlgren et al., 2021; Sonis, 2004). Clinical symptoms typically manifest as nausea, constipation, vomiting, diarrhea, abdominal pain, weight loss and ulcerations within the mucosa (Cinausero et al., 2017; Kwon, 2016; Sonis, 2004). The prevalence and severity of these GI toxicities is dependent on the type of chemotherapy and the dose regimen. Treatment of patients is often limited to symptom management and palliative care as currently no preventative treatments exist. Clinicians are frequently left with the difficult choice of interrupting or altering the chemotherapy regimen or even prematurely discontinuing treatment, ultimately reducing desirable outcomes, increasing hospital stays, impairing patients' quality of life and increasing their economic burden (Cinausero et al., 2017; Elting et al., 2003). Given the pervasiveness and challenges in treating chemotherapy-induced gastrointestinal toxicities, it is of great interest to review the mechanisms contributing to development of these morbidities. A large number of chemotherapeutic agents used for the treatment of different cancers affect the epithelial barrier integrity. The gut epithelium contains rapidly dividing cells and therefore presents a significant target for the chemotherapeutic agents acting either directly or indirectly to initiate the disruption of the epithelial barrier. The current understanding of underlying mechanisms that contribute to these toxicities involve enterocytes, smooth muscle, enteric neurons, and immune cells.

2. Chemotherapy-induced diarrhea

Diarrhea is a common side effect of chemotherapy, especially in patients suffering from advanced cancers. It is estimated that about 50%–80% of cancer patients suffer from chemotherapy-induced diarrhea (CID) (Stein et al., 2010). CID is associated with a failure to retain fluid and electrolytes resulting in severe dehydration and electrolyte imbalances, malnutrition, or renal and cardiac dysfunction, all of which can lead to hospitalization and in severe cases, death. Diarrhea is characterized by increased stool frequency, liquidity, and decreased consistency of stools (Table 1). Diarrhea can be classified as secretory, osmotic, or infectious. Secretory diarrhea occurs due to increased chloride secretion from the epithelium to the lumen. The increase in intraluminal fluid can stimulate gastrointestinal motility and can be initiated by drugs acting on intestinal epithelial cells (Said, 2018). Osmotic diarrhea occurs as a result of increased water being drawn into the lumen. This can be a product of both malabsorption and maldigestion pathways. This diarrhea is often associated with high stool osmolarity gap (Said, 2018). Infectious diarrhea due to viruses such as Rotavirus or by bacteria including Escherichia coli and Salmonella, increase chloride secretion by crypt cells to produce watery diarrhea.

The pathophysiology of CID is complex and involves multiple factors. It has been suggested that acute damage to the intestinal epithelium, including to the architecture of the crypts and villi, plays a significant role in disrupting

Table 1 Diarrhea scoring criteria adapted from Common Terminology Criteria for Adverse Events (CTCAE) Version 5.0 (National Cancer Institute, 2017).

Grade	Description
1	Increase of <4 stools per day over baseline
2	4–6 stools per day
3	<7 stools per day, incontinence, hospitalization
4	Life threatening, urgent intervention
5	Death

the secretory and absorptive functions of the intestinal wall. Preclinical studies have shown that chemotherapeutics, such as irinotecan, induce significant apoptosis in intestinal crypt cells within the colon and jejunum. The need to compensate for the enhanced rate of apoptosis can result in metaplasia of goblet cells and increased production of immature secretory cells, thus contributing to the exacerbated mucus production associated with CID (Gibson et al., 2003; Ikuno et al., 1995; McQuade, Stojanovska, et al., 2016). Impaired absorption across the intestinal epithelium due to atrophy of the villi can result in greater retention of contents within the lumen. This can produce an osmotic shift which drives water into the lumen, further contributing to the onset of diarrhea (Gibson & Keefe, 2006; McQuade, Stojanovska, et al., 2016; Stringer et al., 2007). CID is largely believed to be a form, or by-product, of intestinal mucositis characterized by mucosal injury presenting as inflammation and ulceration, resulting in alterations of intestinal microflora and GI secretion. Additionally, emerging evidence also implicates the enteric nervous system, which is known to regulate the motor and secretomotor functions of the GI tract. The mechanisms underlying the development of CID are discussed in detail in the subsequent sections. Table 2 provides a list of drugs that are known to cause CID and specific drug types commonly associated with CID have been described in more detail below.

3. Chemotherapy-induced mucositis (CIM)

Damage to the gastrointestinal mucosa is a common consequence of both radiation therapy and chemotherapy. Chemotherapy-induced mucositis (CIM) is a major dose limiting side effect that can affect many parts of the gastrointestinal tract, although predominantly in the small intestine, oral and oropharyngeal mucosal linings (Cinausero et al., 2017; Dahlgren et al., 2021). As such CIM has been considered a great unmet clinical challenge to improving patient's cancer outcomes. To date a majority of the research has been focused on oral mucositis rather than gastrointestinal mucositis. This is largely due to accessibility of the tissue area. In this section we first describe the structure of the intestinal epithelium and the function of constituent cell types. We then cover the specific pathobiological mechanisms as well as current and emerging treatment targets, with a focus on the gastrointestinal changes brought on by chemotherapy. For a more thorough review of chemotherapy induced oral mucositis and radiation therapy

Table 2 Drugs associated with high incidence of chemotherapy-induced diarrhea.

Drug class	Drug examples	Severity (diarrhea grade 1–4)
Topoisomerase inhibitors	Irinotecan	16%–22% grade 3–4 (Stein et al., 2010)
	Topotecan	
	Etoposide	
	Teniposide	
Alkylating agents	Cyclophosphamide	All grades 20% (Boussios et al., 2012; Fraiser et al., 1991; McQuade, Stojanovska, et al., 2016; Sang et al., 1997)
	Ifosfamide	
	Melphalan	
	Busulfan	
	Carmustine and lomustine	
	Dacarbazine	
	Chlorambucil	
	Melphalan	
	Mechlorethamine	
Small molecules	Epidermal growth factor inhibitors	All grades 34%–96% (Secombe et al., 2020; Ustaris et al., 2015)
	Human epidermal growth factor receptor 2 inhibitors (neratinib, Lapatinib)	
	Tyrosine kinase inhibitors	
	Mitogen-activated protein kinase kinase inhibitors	
	BCR-ABL Kinase inhibitors	
Antimetabolites	5-Fluorouracil	80% all grades (Maroun et al., 2007)
	Capecitabine	
	Methotrexate	
Mitotic inhibitors	Paclitaxel	50% grade 1–2 (Boussios et al., 2012)
	Docetaxel	
	Vinblastine	
	vincristine	
Platinum based agents	Cisplatin	70%–80% grade 1–2 (McQuade, Stojanovska, et al., 2016)
	Carboplatin	
	Oxaliplatin	
m-TOR inhibitors	Everolimus	34% grade 1–2 (Kamp et al., 2013); 1%–4% grade 3 (Stein et al., 2010)
	Temsirolimus	

induced mucositis, readers are encouraged to consult the following reviews (Cinausero et al., 2017; Dahlgren et al., 2021; Kwon, 2016; Sonis, 2004).

3.1 Intestinal epithelium

The intestinal epithelium is a continuous monolayer of cells that forms a physical and biochemical barrier for host defense against luminal contents through a multitude of neuro-immune-epithelial interactions with the gut microbiota. These interactions are important for maintaining a symbiotic relationship with commensal gut bacteria. Perturbation of this steady state can result in microbial dysbiosis and invasion by pathogenic organisms, and lead to mucositis, dysregulation of intestinal activity and systemic inflammation.

The structure of the epithelium, and prevalence and function of the cell types constituting the epithelium vary between the small and the large intestines. In the small intestine, the epithelium extends into the lumen to form structures called villi, which greatly enhance the surface area for absorption of nutrients. Villi are not observed in the colon. In between the villi are invaginations called the crypts of Lieberkühn. Intestinal epithelial stem cells are present at the base of the crypts and give rise to transit-amplifying cells, which differentiate and mature into various cell types as they migrate towards the tip of the villi. These terminally differentiated cell-types, with the exception of Paneth cells, are subsequently shed into the lumen and as a result under homeostatic conditions the entire intestinal epithelium is renewed every 4–5 days (Allaire et al., 2018; van der Flier & Clevers, 2009). An unregulated increase in the intestinal epithelial cell death affects tissue restitution underlying many of the gastrointestinal toxicities.

The formation and maintenance of the crypt-villus organization of the intestinal epithelium rely on interactions between the epithelium and the underlying mesenchyme. Hedgehog signals from intestinal epithelial cells in the crypts induce bone morphogenetic protein (BMP) in the underlying mesenchyme, which in turn promotes the formation of villi, while inhibiting ectopic crypt formation. Studies have shown that inhibition of hedgehog signaling, or attenuation of BMP signal transduction prevents villi formation and overactivation of crypt-like domains (Haramis et al., 2004; He et al., 2004; Madison et al., 2005). Alternately, endogenous inhibitors of BMP such as noggin and gremlin produced in the inter-villus domains protect the crypts from BMP-mediated inhibition and sustains crypt proliferation. Furthermore, WNT-β-catenin and Notch signaling in intestinal epithelial cells in the crypts co-operate to maintain cell proliferation and determine cell fate i.e., differentiation into the various secretory and absorptive cell types (Crosnier et al., 2006; van der Flier et al., 2009).

The different species of epithelial cells include enterocytes (also referred to as colonocytes in the large intestine), which mediate absorption of nutrients and water, but can also secrete antimicrobial peptides; goblet cells, which secrete mucins and antimicrobial peptides, and play an important role in antigen sensing and priming of the immune system; enteroendocrine cells; and tuft cells that provide protection against helminths. Other cell types such as Paneth cells and microfold (M) cells are exclusively found in the small intestine. Paneth cells are intercalated with stem cells at the base of the crypts and secrete granules containing antimicrobial peptides, such as defensins, C-type lectin regenerating islet-derived III proteins, lysozyme C, cathelicidins, and phospholipases (Bevins & Salzman, 2011; Gallo & Hooper, 2012), to protect nearby stem cells, whereas M cells are specialized antigen presenting epithelial cells localized to Peyer's patches. In the Peyer's patches M cells present luminal antigens and live bacteria to the underlying mucosal immune system in order to prime the immune cells to the luminal microbial environment. The lymphoid follicle-associated epithelium of the Peyer's patches can produce low levels of antimicrobial peptides, mucus, and IgA. Additionally, the intestinal epithelium is surveilled by subepithelial dendritic cells and resident macrophages, which sample the luminal microbial environment through transepithelial dendritic processes (Allaire et al., 2018; Peterson & Artis, 2014).

3.2 Pathobiological mechanisms of CIM

The pathological mechanisms that drive the specific changes in CIM are generally tissue (oral vs intestinal) and insult (chemotherapy vs radiation therapy) independent (Sonis, 2004). Chemotherapy-induced gastrointestinal mucositis is histologically characterized by crypt loss, villus atrophy, loss of renewal capacity, and impairment of the gut absorptive and barrier function (Bajic et al., 2018; Dahlgren et al., 2021). Originally thought to be an insult purely to the epithelium of the GI tract, we now understand CIM to be the result of a complex biological pathway that begins before patients often experience symptoms (Cinausero et al., 2017; Sonis, 2004). A well-established five-step model was initially proposed for the development of mucositis by a panelist established by The Multinational Association of Supportive Care in Cancer and the International Society for oral oncology. The initial clinical practice guidelines have been periodically modified over time (Bowen et al., 2019).

The first stage of mucositis begins with an *initiation* phase and the formation of reactive oxygen species (ROS) with direct damage to the DNA of cells throughout the tissue (submucosa, basal layers, endothelium) initiating both the innate immune response and the subsequent secondary messenger cascade likely within short order after DNA strand damage (Cinausero et al., 2017; Dahlgren et al., 2021). These initial events are followed by an *upregulation and secondary messenger generation* phase. ROS and the local immune cells further disrupt the tissue by stimulating macrophages and activating a host of secondary messengers such as Wnt/β-catenin, p53, caspase-1/3, Bcl-2 (Cinausero et al., 2017; Dahlgren et al., 2021; Sonis, 2004). NF-κB is likewise activated and is one of the most studied secondary messengers in the pathogenesis of CIM mainly because of its role in activating pro-inflammatory cytokines such as tumor necrosis factor-α (TNF-α), IL-6, IL-1, IL-18, and IL-33 (Dahlgren et al., 2021; Guabiraba et al., 2014). This is then followed by a *signal amplification* phase, leading to increasing inflammation by the collective effects of many of the pathways that are activated by the aforementioned cytokines and secondary messengers. For example, the activation of TNF-α by NF-kB in turn activates yet more NF-kB which then activates MAPK and the JNK pathway and a breakdown of fibronectin and further macrophage activation. Ultimately the downstream effect of the NF-kB dependent and independent pathways is an increase in mucosal inflammation and increased apoptosis, resulting in an *ulceration* phase characterized by damage of the epithelial barrier promoting bacterial translocation and wound colonization. In this phase, loss of stem cells from the crypts due to apoptosis leads to a diminished renewal capacity and the inability to replenish the intestinal epithelial cells (IEC), and results in an ulcer formation allowing for increased translocation of luminal contents to the submucosa including bacterial colonization of the ulcer, ultimately promoting even more inflammatory signal cascades and activating pro apoptotic genes (Cinausero et al., 2017; Dahlgren et al., 2021; Sonis, 2004). Clinically, symptoms become evident at this stage, approximately 1–2 days after treatment in the case of CIM of the small intestines (enteritis) which is 10 times more likely to experience spontaneous apoptosis than the large intestines (Bowen et al., 2006; Dahlgren et al., 2021). Finally, a post ulcerative *healing phase* with cell proliferation and barrier recovery occurs once chemotherapy has ceased. This phase is characterized by the restoration of physiologically normal proliferation, maturation and differentiation of both progenitor stem cells within the crypts and subsequent epithelial layers (Cinausero et al., 2017; Dahlgren et al., 2021; Sonis, 2004).

Beyond the canonical DNA damage/ROS initiated pathway established by Bowen et al (Bowen et al., 2019), activation of toll-like receptors (TLRs), which are a group of pattern recognition receptors that can recognize pathogen associated molecular patterns such as lipoteichoic acid (LTA) and lipopolysaccharide (LPS) from various gram positive and gram-negative microbes is another important mechanism by which CIM may occur. TLRs are expressed throughout the gastrointestinal tract by epithelial and immune cells, and TLRs have been a focus of research due to their role in mediating the balance between the gut microbial flora and activated immunity. With specific regards to TLR-2, pre-clinical evidence has implicated TLR-2 in a mucosal protective role by downregulating the excessive activation of inflammatory cells and promoting the expression of the anti-inflammatory cytokine IL-10. However, some reports have also implicated TLRs in the induction of intestinal inflammation via production of NF-kB through MyD88-dependent and -independent pathways (Lupi et al., 2020; Wu et al., 2020). Studies examining the role different chemotherapeutics might play on the expression of TLRs have been recently examined by Wei et al. (Wei, Wen, & Xian, 2021), where the authors discussed the body of literature around TLRs with particular interest in TLR-2, TLR-4 and TLR-9. Interestingly, the authors found that while generally 5-FU and methotrexate (MTX) showed an increase in TLR-2, 4 and 9, conflicting evidence exists between animal models and species. Overall, findings from the clinical studies and animal model experiments suggested that the effects of chemotherapeutic agents on the expression of TLRs vary significantly with the types of drugs, dosages, species and routes of administration across models (Wei et al., 2021).

Drug metabolizing enzymes (DMEs) are an understudied but potentially important component of chemotherapy induced GI toxicity. In their recent review Tao et al. (2020) reported that altered activity of DMEs play an important role in chemotherapy-induced GI toxicity. The predominant DME in the gut are suggested to be the cytochrome P450 (CYP) and UDP-glucuronosyltransferase (UGT) families. These enzymes (particularly the CYP super family) are responsible for reactive metabolism with alkylating agents, antimetabolites, topoisomerase inhibitors and tyrosine kinase inhibitors. A noted example are the metabolizing effects on irinotecan (detailed below). Lapatinib has been demonstrated to inhibit CYP3A4, while paclitaxel, erlotinib, tamoxifen, ifosfamide, flutamide and docetaxel all activated the pregnane X receptor (PXR) and altered CYP3A4 expression (Tao et al., 2020).

3.3 Chemotherapeutics implicated in CIM and CID

Severity of CIM can often depend on the specific chemotherapeutic used. For example, the antimetabolites, 5-FU (Abdel-Rahman et al., 2016; Schwab et al., 2008) methotrexate, and irinotecan (Mayo et al., 2017; Stein et al., 2010), the alkylating agents such as cyclophosphamide, and cisplatin (Villa & Sonis, 2015), anthracyclines and lastly taxanes (Kwon, 2016) all tend to be more consistently associated with mucosal toxicities than bleomycin, hydroxyurea, or etoposide (Cinausero et al., 2017).

One of the most studied chemotherapeutics related to CIM/CID is irinotecan. Irinotecan is a prodrug that is metabolized by hepatic and systemic carboxylesterases to its active form, SN-38, which is 100–1000 times more potent than its parent compound (Stein et al., 2010). SN-38 inhibits topoisomerase I leading to single stranded DNA breaks. Activation of death signals induces pro-apoptotic proteins (caspase-9 and-3 and p53) via up-regulation of MAPK signaling and downregulates the anti-apoptotic PI3K/Akt and BcL-xL mediated pathways (Bowen et al., 2007; Koizumi et al., 2005; Takeba et al., 2007). SN-38 is inactivated through glucuronidation by hepatic uridine diphosphate glucuronosyltransferase to the inactive metabolite SN-38G and excreted into the gastrointestinal tract through the bile duct (Sun et al., 2020). In the GI tract β-glucuronidases produced by the gut bacteria cleave the glucuronide moiety, thus transforming the inactive SN-38G back to the active SN-38 (Bhatt et al., 2020). Consequently, accumulation of active SN-38 within the intestinal lumen leads to enhanced local toxicity and as a result causes significant damage to the intestinal wall (Stringer et al., 2008). As such, in animal models, irinotecan produces a dose and time dependent induction of villous atrophy and crypt ablation in the small intestine and severe colonic mucosal damage with crypt hypoplasia (Stringer, Gibson, Bowen, et al., 2009; Stringer, Gibson, Logan, et al., 2009). The damage to the mucosa contributes ultimately to the severe diarrhea, a hallmark of this chemotherapeutic. Diarrhea occurs within minutes of administration and can last up to 24 h. The diarrhea produced immediately post drug administration is believed to be mediated through inhibition of acetylcholinesterase, which leads to increased cholinergic transmission (Stein et al., 2010). Acetylcholine is the major excitatory neurotransmitter in the gut and enhanced cholinergic signaling can lead to increased muscle contractions and fluid secretion, the basis for enhanced motility and secretion in diarrhea (Furness, 2007). Thus, atropine, the muscarinic receptor antagonist, is prescribed as a first-line therapy for mitigating the initial diarrhea (Schiller et al., 2017). However, late-onset diarrhea is chronic, atropine-insensitive, and is associated with significant intestinal mucositis (Bowen et al., 2007; Stein et al., 2010). Interestingly, CIM occurs more frequently with a combination of irinotecan and fluoropyrimidines such as capecitabine. For example, patients treated with the capecitabine, and irinotecan (XELIRI) regimen reported higher gastrointestinal toxicity than with fluorouracil and irinotecan (FOLFIRI). For combination regimens with fluorouracil and oxaliplatin (FOLFOX) or FOLFIRI, CIM has been estimated at 50% and 89%, respectively (Basile et al., 2019).

5-Fluorouracil (5-FU) is used in the treatment of variety of malignancies. 5-FU is converted to fluorodeoxyuridine monophosphate (FdUMP), which complexes with thymidylate synthase and prevents DNA replication and repair. 5-FU can also serve as a pyrimidine analog incorporating in place of uracil or thymine. This inhibition of DNA repair machinery results in cell death. 5-FU induced apoptosis of enterocytes results in a reduction of crypt and villi length, fusion of villi, enterocyte hyperplasia and increased apoptosis of crypt cells (Cinausero et al., 2017; Dahlgren et al., 2021; Kwon, 2016; Lee, 2014; Soares et al., 2008). The most common adverse effect reported with the systemic administration of 5-FU is diarrhea due to mucositis. As described above, mucositis involves epithelial damage due to ROS which can have significant effects on secretory processes resulting in diarrhea. Additionally, use of leucovorin in combination with fluoropyrimidines results in increased prevalence of this 5-FU induced diarrhea. The specific mechanisms whereby 5-FU induces toxicity are varied across the preclinical literature. In rats, Logan et al. (Logan, Gibson, Sonis, & Keefe, 2007) found that 5-FU elevated levels of TNF-α and IL-1β but not of NF-κB and IL-6 while other studies have noted increases in transcription of NF-kB (Lee, 2014). In general, a specific role for pro-inflammatory cytokines and proteins involved in apoptosis regulation has been suggested by many studies involving 5-FU (Basile et al., 2019; Lee, 2014; Ribeiro et al., 2016).

Platinum-based chemotherapeutic drugs, including cisplatin, form interstrand and intra-strand crosslinks with nucleic acid bases creating DNA adducts which inhibit DNA synthesis and transcription (Boussios et al., 2012; Stojanovska et al., 2015). The oxidative stress resulting from mitochondrial damage induces pro-apoptotic proteins (Bak, Bax, p38 MAPK, and p53), the activation of NF-kB and TNF- α signaling pathways as well as downstream activation of inflammatory cytokines and chemokines in the intestinal epithelium to induce gastrointestinal toxicity (Qi et al., 2019). Platinum agents are also known to elicit immunogenic cell death through activation of damage associated molecular patterns (DAMPs), such as high mobility group protein B1(HMGB1) and calreticulin (Stojanovska et al., 2015, 2016). When administered as single drug therapies, oxaliplatin and carboplatin exhibit lower GI toxicity compared to cisplatin (Hartmann & Lipp, 2003). This could be because preclinical studies have reported that the ileal mucosa is more sensitive to cisplatin (Cinausero et al., 2017; Yamamoto et al., 2013). The GI toxicity of platinum salts, oxaliplatin and cisplatin are significantly increased when administered in combination with 5-FU (Lee, 2014).

3.4 CIM and the microbiome

The role of the gut microbiome in mucosal inflammation has been well recognized, including in chemotherapy—induced effects (Bajic et al., 2018). Bacteria of the GI tract are essential to the normal physiological regulation of intestinal barrier functions, maintenance of selective intestinal permeability, inflammation and innate immune response, repair mechanisms, cell apoptosis, and oxidative stress (Dahlgren et al., 2021; Prisciandaro et al., 2011). Both the direct and indirect toxic effects of chemotherapeutics on the gut microbiome impact the clinical manifestations of CIM, where they ultimately may lead to the development of bacteremia and diarrhea or sepsis. Hong-Li et al reported that 5-FU resulted in significantly changed profiles of inflammatory cytokines/chemokines in serum and colon (Li et al., 2017). 5-FU diminished bacterial diversity, leading to the relative lower abundance of Firmicutes and decreased Firmicutes/Bacteroidetes (F/B) ratio in feces and cecum contents. These investigators found that fecal transplant from healthy mice prevented body weight loss and colon shortening of 5-FU treated mice, although restoring the dysbiotic microbiota from 5-FU-treated animals reversed these effects. Thus, these findings suggested that 5-FU induced intestinal mucositis could be alleviated by altering the gut microbiome (Li et al., 2017). Similarly, Chun-Yan et al found that diarrhea developed in the 5-FU groups but was attenuated after oral probiotic suspension of *Lactobacillus casei* variety rhamnosus (Lcr35) or *Lactobacillus acidophilus* and *Bifidobacterium bifidum* (LaBi). Additionally, they found that pro-inflammatory cytokines (TNF- α, IL-1β and Il-6) were upregulated after 5-FU treatment but were attenuated by the probiotic mixture (Yeung et al., 2015). Despite the fact that many studies find a strong link between the microbiome and the development of mucositis (Chang et al., 2018, 2020; da Silva Ferreira et al., 2020; Stringer & Logan, 2015; Wu et al., 2019; Wu et al., 2021) therapies designed to ameliorate gut dysbiosis have so far proven unsuccessful in treating CIM in the clinical setting (Dahlgren et al., 2021; Touchefeu et al., 2014). This could be due to challenges associated with the use of antibiotics or delivery of pharmacologically relevant doses of compatible strains of microbiota. Nonetheless, modulating the gut microbiome to alleviate chemotherapy-induced gastrointestinal toxicities by combining antibiotics with patient-specific probiotics/fecal microbiota transplants to target pathogenic bacteria and replenish the commensal microbiome remains an exciting prospect.

4. Emerging mechanisms underlying chemotherapy-induced gastrointestinal toxicity

It has been reported that 13%–50% of patients exhibit diarrhea for up to 10 years after cessation of cancer treatment (Denlinger & Barsevick, 2009). However, the pathophysiology underlying chronic diarrhea is not well understood. The development of diarrhea often involves aberrations in the transport of sodium, chloride, and water across the epithelium, and/or enhanced contractility of gastrointestinal smooth muscle function regulated by the enteric nervous system (ENS). The ENS is a semi-autonomous nervous system intrinsic to the gastrointestinal tract (Kulkarni et al., 2018). It is organized into two interconnected ganglionated plexuses—myenteric plexus (Auerbach's) and submucosal (Meissner's) plexus. The myenteric plexus lies between the outer longitudinal and circular muscle layers and is responsible for coordinating motor patterns of the intestinal wall. The submucosal plexus is located in the submucosal layer between the circular muscle and muscularis mucosa and regulates secretomotor activity (fluid and electrolyte homeostasis), vascular tone and epithelial cell functions. Nerve fibers from both the myenteric and submucosal plexuses project into the mucosa (Brehmer, 2006; Furness, 2007). For a comprehensive review of the enteric nervous system and the pathways underlying the motor and secretomotor reflexes that mediate gastrointestinal peristalsis and regulate secretory flux across the intestinal epithelium, respectively, readers are encouraged to refer to 'The Enteric Nervous System' by Furness (Furness, 2007).

Several chemotherapeutic agents, including platinum compounds and taxanes, are known to induce neuropathy in peripheral afferents innervating the limbs and the mechanisms underlying neurotoxicity have been extensively investigated (Colvin, 2019). It has been reported that enteric neurons are replaced every 2 weeks in the myenteric plexus of adult mice (Kulkarni et al., 2017), thus rendering neurogenesis as a potential target for disruption by chemotherapy drugs. However, it is not entirely clear whether the regeneration occurs under insult and injury or is a homeostatic mechanism in health conditions (Pawolski & Schmidt, 2020). Interestingly, enteric neurogenesis is reported to be modulated by the gut microbiome (Vicentini et al., 2021). However, the toxic effects of chemotherapy on the ENS are not well characterized in the clinic. Moreover, the role of enteric neurotoxicity in the induction and prolongation of symptoms associated with chemotherapy-induced diarrhea is unclear. Emerging evidence from preclinical studies indicates that several chemotherapeutic agents, such as oxaliplatin, cisplatin, 5-FU, irinotecan and vincristine induce enteric neuron toxicity and/or alter the neurochemical composition of enteric ganglia within the myenteric and/or submucosal plexuses (McQuade et al., 2020).

Oxidative stress has been reported as a key mechanism underlying mitochondrial damage and apoptosis in intestinal epithelial cells and enteric neurons following chemotherapy treatment (Cinausero et al., 2017; McQuade, Carbone, et al., 2016). Reactive oxygen/nitrogen species are known to alter ion channel function by inducing post-translational modifications directly to its structure or indirectly by modifying proteins regulating channel activity or gene expression. Several ion channels have been extensively examined in neurons under neuropathic or inflammatory conditions and provide a strong body of evidence for shared mechanisms and pathways. For example, voltage-gated sodium channels are important drivers of he action potential upstroke in neurons (Akbarali, 2014). Reactive nitrogen species (RNS) have been shown to increase the activity of inactivation-resistant voltage-gated sodium channels in hippocampal neurons (Annunziato et al., 2002), whereas in baroreceptor neurons RNS decreased the activity of tetrodotoxin-resistant and tetrodotoxin-sensitive voltage-gated sodium channels (Li et al., 1998). ROS/RNS can also increase the activity of several members of the TRP channel superfamily, such as TRPM2, TRPM7, TRPC3/4 and TRPV1, many of which are expressed in the gut and thus seem a likely target for dysregulation by the inflammation induced by chemotherapy (Akbarali, 2014; Drokhlyansky et al., 2020; Holzer, 2011). Oxidative stress can evoke prolonged sensitization of TRPV1 channels (Schilling & Eder, 2009). Oxidation of TRPV1 channels can also render them resistant to capsaicin-induced desensitization and activity of the desensitized receptor can be rescued following significant oxidative stress (Chuang & Lin, 2009). TRPV1 channels are non-selective cation channels known to respond to noxious stimuli, such as acidic pH (<5.9), noxious heat ($>43\,^{\circ}C$) and capsaicin (Akbarali, 2014; Buckinx et al., 2013). They are highly expressed in nociceptive neurons, including those innervating the gut (Holzer, 2011), and have been implicated in chemotherapy-induced peripheral hypersensitivity (Aromolaran & Goldstein, 2017). Under physiological conditions, TRPV1-expressing gut-innervating neurons play an important role in maintaining homeostasis by responding to various changes in the gut (Chiu et al., 2013; Lai et al., 2020). Indeed, Lai et al. (2020) recently demonstrated that TRPV1-expressing primary afferent neurons protect against pathogenic bacteria invasion by regulating M cell density in ileum Peyer's patches via release of CGRP. Although, during pathophysiological conditions, such as colitis, TRPV1 activation in sensory neurons exacerbated inflammation (Engel et al., 2012). It is important to note that the presence of TRPV1 channels in the ENS is contested with some studies indicating expression exclusively in extrinsic afferents innervating the gut (including primary afferent dorsal root ganglia neurons), while others have reported expression in enteric neurons (Buckinx et al., 2013; Holzer, 2011). This discrepancy could be due to the presence of different isoforms that display differing immunoreactivity (Allais et al., 2017; Buckinx et al., 2013). TRPM2, TRPM7 and TRPC3/4 are expressed in enteric neurons (Drokhlyansky et al., 2020). The P2X2 receptor is another ion channel target for oxidative stress in the intestine. The P2X2 receptor can mediate synaptic transmission between neurons and the neuromuscular junction (Akbarali, 2014). Oxidative stress has been shown to modulate P2X2 receptor activity. For example, Coddou et al. reported increased P2X2 receptor activity due to oxidative stress (Coddou et al., 2009). In contrast, Mason et al. (Mason, Bourke, & Kemp, 2004) reported that mitochondrial ROS production decreased ATP-activated P2X2 currents.

Interspersed between neurons are enteric glia. In the myenteric and submucosal plexuses, enteric glia have been observed surrounding neuronal cell bodies and like neurons, they extend projections into the musculature and the mucosal epithelium (Seguella & Gulbransen, 2021). Within the gastrointestinal tract, enteric glia have been reported to play an integral role in supporting epithelial barrier function, peristalsis and inflammation (Bhave et al., 2017; Rao et al., 2017; Seguella & Gulbransen, 2021) through interactions with neurons, immune cells, enterocytes and enteroendocrine cells (Bohórquez et al., 2014; Ibiza et al., 2016; Liu et al., 2013; Neunlist et al., 2007). Recent studies have reported up-regulation of molecular markers associated with reactive gliosis (S100β and GFAP) in enteric glia after treatment with chemotherapeutic agents, such as oxaliplatin, 5-FU and irinotecan (Costa et al., 2019; Gibson et al., 2016; Grundmann et al., 2019; Nogueira et al., 2017; Robinson et al., 2016). Furthermore, S-100β inhibition mitigated chemotherapy-induced oxidative stress, enteric neuropathy, and insult to the crypt-villi architecture (Costa et al., 2019). In enteric glia S-100β is known to stimulate inducible nitric oxide synthase (iNOS) in models of ulcerative colitis and celiac disease (Sharkey, 2015). Increased nitric oxide (NO) due to iNOS overstimulation via S100β-mediated activation of the RAGE/NF-kB pathway has been shown to disrupt epithelial barrier function (Grundmann et al., 2019) and promote neuron cell death (Costa et al., 2019; Villarreal et al., 2011). NO produced by iNOS can potentiate connexin 43 (CX43) hemichannel opening in enteric glia, stimulate CX43-mediated ATP release and drive neuron cell death during intestinal inflammation (Brown et al., 2016). Similarly, colonic inflammation due to chronic morphine treatment was associated with enhanced ATP currents in enteric glia. CX43 inhibition downregulated pro-inflammatory cytokines in the colon, reduced ATP currents and reversed chronic morphine-induced decrease in neuron density in the myenteric ganglia (Bhave et al., 2017). Consequently, upstream mechanisms underlying gliosis or downstream mechanisms modulating ATP release, for example, CX43 hemichannels, may be important targets for alleviating chemotherapy-induced gastrointestinal toxicity.

While several chemotherapeutic agents are reported to produce toxicity in cardiac muscle, skeletal muscle, vascular smooth muscle and the detrusor smooth muscle, not much is known about the direct effects on intestinal smooth muscle (Hayward et al., 2013; Iguchi et al., 2019; Min et al., 2015). Anthracycline compounds such as doxorubicin are known to induce chronic constipation and ileus in patients. Millington, Pascasio, Halligan, and De Chadarévian (2012) recently reported moderate-to-severe fibrosis, loss of myofilaments and/or hyalinization in the muscularis externa and the muscularis mucosae in the gastrointestinal tract of doxorubicin-treated patients. Gastrointestinal motility is a highly integrated phenomena and requires coordination between myenteric plexus neurons, interstitial cells (Interstitial Cells of Cajal and PDGFRα-positive cells), and smooth muscle (Sanders et al., 2012). Therefore, myopathy in intestinal smooth muscle cells can also impair intestinal motility. However, the mechanisms underlying intestinal smooth muscle toxicity are unclear. Anthracycline compounds such as doxorubicin are known to induce cardiotoxicity and cardiomyopathy via mitochondrial production of ROS (Min et al., 2015). Thus, oxidative stress could be a potential mechanism underlying doxorubicin effects on intestinal smooth muscle. During intestinal inflammation, oxidative stress altered the gating kinetics of the L-type calcium channel, $Ca_V1.2b$, in intestinal smooth muscle cells by mutating residues that are necessary for interactions with c-src kinase. Reduced channel phosphorylation resulted in decreased calcium influx, which impaired excitation-contraction coupling (Akbarali, 2014; Kang et al., 2004, 2007; Ross et al., 2007; 2010). Reduced calcium influx can also alter gene transcription in smooth muscles. For example, during colonic inflammation, tyrosine nitration of L-type calcium channels prevented depolarization-mediated CREB phosphorylation (Kang et al., 2010). Interestingly, doxorubicin has also been reported to inhibit smooth muscle activity without inducing changes in oxidative stress markers (Iguchi et al., 2019). In this study by Iguchi et al. (2019), doxorubicin impaired detrusor smooth muscle contractility by downregulating the large-conductance Ca^{2+}-activated K^+ channel and altering the expression of myosin-light chain kinase isoforms. Thus, inflammation-mediated dysregulation of excitation-contraction and/or excitation/transcription coupling in intestinal smooth muscle cells might underlie anthracycline-induced chronic constipation.

Macrophages are critical mediators of intestinal inflammation during the initial phase of chemotherapy-induced gastrointestinal mucositis. Increased bacterial translocation as a consequence of the breakdown of the gut epithelial barrier results in the activation of macrophages and thus helps sustain or even exacerbate intestinal inflammation (Sougiannis et al., 2021). Recently, Luo et al. (2018) demonstrated that macrophages, specifically those localized in intestinal smooth muscles, can also directly promote chemotherapy-induced diarrhea. In their studies Luo et al. demonstrated that genetic ablation of TRPV4 receptors expressed on muscularis macrophages inhibited irinotecan-induced diarrhea by preventing the release of PGE2 from macrophages and downstream activation of E-prostanoid receptor 1 and 3 on intestinal smooth muscles (Luo et al., 2018). This pathway was independent of enteric neurons. Thus, targeting muscularis macrophage activation might be a viable option for preventing not only chemotherapy-induced diarrhea but also a host of inflammation-driven gastrointestinal motility disorders.

5. Treatment of CID

Despite our growing understanding of the mechanisms for chemotherapy-induced GI toxicities, pharmacological treatments are limited. Currently, treatment of chemotherapy-induced diarrhea is based on the severity of the diarrhea.

The first line defense in treatment of CID is loperamide. Loperamide is a peripherally restricted mu opioid receptor agonist that decreases secretion and reduces intestinal motility by inhibition of enteric neurons. Patients receive escalating doses of loperamide to control symptoms. High dose loperamide remains the primary treatment for the acute management of CID (grades 1–2) with the clinical guidance of a maximum daily dosing of up to 24 mg in man. However, in randomized clinical trial data, irinotecan-based therapy was found to progress to grades 3-4 diarrhea with a frequency ranging from 6% to 47% despite high doses of loperamide (Andreyev et al., 2014). Irinotecan-induced diarrhea therefore becomes refractory to loperamide. This may be due to the tolerance to opioids that develops in the myenteric neurons of the ileum (Muchhala et al., 2020).

The second line treatment in CID is often octreotide, a synthetic somatostatin analogue which decreases secretion and promotes absorption. Octreotide is generally recommended for patients refractory to loperamide (Zidan et al., 2001). The mechanism of action involves inhibition of serotonin, vasoactive intestinal peptide and gastrin and enhances intestinal motility (Brunton et al., 2017; Deng et al., 2017). However, prolonged use of octreotide may result in deleterious cardiovascular, CNS and endocrine effects.

Although other pharmacological interventions including corticosteroids may be used for inflammation, treatment for severe diarrhea are limited and can often result in cessation of therapy (Boussios et al., 2012).

5.1 Current and emerging pre-clinical strategies for treating CIM

Currently approved strategies for alleviating chemotherapy-induced GI toxicity are aimed towards managing symptoms associated with diarrhea and abdominal pain (Table 3). However, there is a major unmet clinical need to investigate and develop treatments for CIM, which is the basis of chemotherapy induced diarrhea (Cinausero et al., 2017; Dahlgren et al., 2021; Sonis, 2004). Novel strategies have aimed at targeting the pathway of interconnected pathobiological mechanisms including scavenging ROS with antioxidants, inhibiting pro-inflammatory cytokine (anti-inflammatory), and preventing runaway cell death particularly in the cryptic stem cells with apoptosis inhibitors (Dahlgren et al., 2021; Kwon, 2016).

For example, Amifostine, an aminothiol prodrug, which undergoes biotransformation to the active metabolite WR-1065, has been reported to reduce doxorubicin-induced CIM in rats and methotrexate-induced CIM

Table 3 Drugs used in the treatment of CID.

	Mechanism	**Side effects**
Loperamide	Loperamide is a peripherally restricted mu opioid receptor agonist that decreases secretion and intestinal motility by decreasing excitability of enteric neurons (Brunton et al., 2017; Wadler et al., 1998)	Constipation, nausea, vomiting, cardiovascular effects
Deodorized tincture of opium	Centrally active mu-opioid receptor agonist inhibits gastrointestinal motility (Gershon et al., 1994)	CNS effects, tolerance
Octreotide	Somatostatin analogue that decreases hormone secretion (e.g., vasoactive intestinal polypeptide), reduces motility and pancreatic secretions and promotes absorption (Brunton et al., 2017; Deng et al., 2017; Zidan et al., 2001)	Nausea, bloating, hypo or hyperglycemia, gallstone formation, endocrine, cardiovascular and CNS side effects
Budesonide	Topically active corticosteroid that might restore mucosal function and fluid absorption; a 90% first-pass effect in the liver results in low systemic availability (Brunton et al., 2017; de Man et al., 2018; Ribeiro et al., 2016)	Endocrine effects
Atropine	Inhibition of acetylcholine at muscarinic receptors located at neuromuscular junction to inhibit motility (Brunton et al., 2017; Stein et al., 2010)	Cognitive impairment, xerostomia, constipation, blurred vision, dyspepsia
Antibiotics	Broad spectrum antibiotic that targets small intestinal bacterial overgrowth (Boussios et al., 2012; Brunton et al., 2017)	Diarrhea, increased risk of *C. difficile* infection
Bile acid sequestrants	Prevent water secretion into the colon induced by non-sequestered bile acids (Brunton et al., 2017)	
Beta-glucuronidase inhibitors	Inhibits removal of glucuronic acid moiety by gut commensal bacteria (Bhatt et al., 2020)	

in mice (Chen et al., 2013; Jaćević et al., 2018). The therapeutic effect of WR-1065 has been attributed to scavenging free radicals produced by oxidative stress and mitochondrial damage (Bensadoun, Schubert, Lalla, & Keefe, 2006). Currently Amifostine is FDA-approved for prophylactic treatment of oral mucositis in post-operative head-neck-cancer patients treated with radiotherapy (Antonadou et al., 2002) However, significant side-effects such as nausea and hypotension have limited its broader clinical use.

The profound role that inflammatory cytokines, such as interleukin-1β (IL-1β), NF-kB, and COX-2, play in inducing and potentiating the development of mucositis has made them a prime potential therapeutic target. IL-1β, for example, expressed and released by intestinal epithelial cells due to DNA damage perturbs the epithelial barrier by disrupting tight junction proteins (Kanarek et al., 2014; Kwon, 2016). IL-1 receptor antagonists, such as the endogenously produced interleukin-1 receptor antagonist (IL-1RA)(Daig, 2000), has been shown to inhibit irinotecan-induced mucosal damage, inflammatory cell infiltration, ulceration, and decrease 5-FU-induced intestinal mucosal damage while also expediting post-ulcerative healing (Kanarek et al., 2014; Wu et al., 2010, 2011). NF-kB has been established as a key component in the development of mucositis, and thus directly inhibiting NF-kB or its downstream messengers has been an extensively investigated strategy for reducing mucositis in preclinical studies (Chen et al., 2011; Cinausero et al., 2017; Kwon, 2016). Cyclooxygenase 2 (COX-2), which is involved in the synthesis of prostanoids, plays an important role in the amplification of CIM (Dahlgren et al., 2021; Sonis, 2004). Interestingly, in preclinical studies, COX-2 inhibition was reported to reduce the histological changes induced by 5-FU (de Miranda et al., 2020) and irinotecan (Javle et al., 2007). However, the clinical efficacy of COX-2 inhibition in the treatment of intestinal mucositis caused by radiation or chemotherapy was limited (Dahlgren et al., 2021; Javle et al., 2007; Lalla et al., 2014).

The gastrointestinal tract mucosa and submucosa rely on rapidly dividing stem cells within the crypts for renewal and normal physiological function, and therefore preventing apoptosis induced by chemotherapeutics is another well-established pre-clinical strategy to manage the development of mucositis. For example, CXCL9 mitigated 5-FU-induced intestinal mucositis and mucosal damage by protecting stem/progenitor cells against unregulated cell death induced by chemotherapy. CXCL9 treatment not only decreased chemotherapy-induced apoptosis in intestinal crypts but also increased the proliferation rate of intestinal crypts after treatment. Furthermore, administration of CXCL9 raised the therapeutic window of 5-FU in treated animals

(Han et al., 2011). Likewise, GLP-1 and 2 and their metabolic enzyme (DPP-IV) have shown evidence for supporting mucosal regeneration after CIM, in part by reducing apoptosis (Boushey et al., 2001; Dahlgren et al., 2021).

Other potentially beneficial, but highly specific and cell restricted treatment targets outside of the traditional ROS—inflammatory cytokine—runaway apoptosis pathway have emerged in recent literature, such as the G-coupled estrogen receptor (GPER). GPER, which is expressed in intestinal crypts, has been reported to exhibit protective effects on intestinal architecture during colitis and ischemia by inhibiting inflammation-driven apoptosis of crypt cells, particularly during colitis and ischemia (Chai et al., 2019; Wang et al., 2021). Similarly, during treatment with 5-FU, Chen et al reported that the selective GPER agonist, G-1, significantly inhibited weight loss and inflammation-induced mucosal damage. Additionally, mucosal barrier function improved as G-1 treatment resulted in increased goblet cell density and reduced mucosal permeability due to up-regulation of ZO-1 Moreover, G-1 treatment did not alter the antitumor effect of 5-fluorouracil (Chen et al., 2021).

In a recent study, it was noted that levels of glutathione (GSH), a potent non-protein thiol antioxidant, and GSH/GSSG (glutathione disulfide) ratio (a potential indicator of cellular oxidative stress) in the small intestine mucosal tissue decreased compared to controls after 5-FU treatment. Therefore, modulating GSH synthesis can be explored as an intriguing strategy to reduce oxidative stress. Cystine and theanine, which is metabolized to glutamic acid in the liver, are essential precursors of GSH. In a recent study, Yoneda, Nishikawa, and Kurihara (2021) found that cystine and theanine (CT) administration decreased 5-FU-induced oxidative stress in the basal region of small intestinal crypts by increasing GSH levels and improving the GSH/GSSG ratio. CT-treated mice also exhibited reduced diarrhea and weight loss. Importantly CT did not affect the antitumor activity of 5-FU (Yoneda et al., 2021).

Ultimately, as has been discussed by many publications within the field of CIM, we feel that while all of these are individually exciting, progress has been slow and disappointing for a single solution to manage CIM effectively without ceasing or interrupting the chemotherapy regimen. A combinational therapeutic strategy which targets multiple or all of the individual components along the pathological pathway (ROS—Cytokines—Apoptosis) will likely yield the most promising clinical benefit (Cinausero et al., 2017; Dahlgren et al., 2021; Kwon, 2016; Sonis, 2004).

6. The gut and chemotherapy-induced peripheral neuropathy

Chemotherapy-induced peripheral neuropathy (CIPN) is a frequent and often dose-limiting complication of cancer therapy utilizing taxanes (docetaxel or paclitaxel), platinum compounds (cisplatin, carboplatin, and oxaliplatin) as well as other drugs such as vincristine, and bortezomib. Clinical manifestation of symptoms can vary greatly between patients, but often includes all or some of the following: thermal and mechanical hyperalgesia, tactile and thermal allodynia, spontaneous pain, paresthesia, dysesthesia as well as a variety of cognitive or motor impairments (Sałat, 2020). Currently no treatments exist that can reverse CIPN once it has developed, and pharmacological management of these neuropathies has been notoriously challenging with approximately 50% of patients being resistant to current pain management (Przewlocki & Przewlocka, 2005; Sałat, 2020). Over the last several years, rodent models of CIPN have been developed that demonstrate many of the neuropathological changes observed in patients undergoing cancer therapy as well as typical electrophysiological abnormalities associated with CIPN (Authier et al., 2009). Pre-clinical studies have shown long-lasting, bilateral neuropathy associated with a decrease in the density of peripheral intraepithelial nerve fibers (IENF), a hallmark of CIPN, as well as mechanical and cold allodynia utilizing a paclitaxel rodent model (Toma et al., 2017). While several explanations have been proposed for the development of CIPN, the mechanism(s) of paclitaxel-induced neuropathy and chronic pain are not well understood. However, there is growing evidence from animal models that paclitaxel and other cancer chemotherapeutic drugs can trigger an inflammatory response characterized by the production and release of pro-inflammatory cytokines, such as IL-1β and TNF-α (Shen et al., 2017; Vyas et al., 2014). Recent animal studies suggest that these cytokines that are released directly and indirectly by activated spinal glial cells within the dorsal root ganglia (DRG) are critical for the development of CIPN (Shen et al., 2017). Primary sensory afferents of the DRG innervate the intestinal mucosa and studies have recently demonstrated that intestinal inflammation and microbial dysbiosis can alter the excitability of DRG neurons (Kang et al., 2017; Mischel et al., 2018; Moore et al., 2002; Sessenwein et al., 2017). Therefore, inflammation emanating from the gut due to chemotherapy treatment might play an important role in the etiology of chemotherapy-induced peripheral hypersensitivity. Interestingly, several chemotherapeutic agents that are known to cause CIPN also induce gastro-intestinal mucositis and perturb the intestinal epithelial barrier. Disruption of the epithelial barrier is known to promote dysbiosis of the gut bacteria and bacterial translocation across the intestinal epithelium can induce inflammatory responses through activation of pattern recognition receptors. For example, pre-clinical studies have shown that changes in the gut microbiome can result in loss of epithelial integrity and promote an inflammatory response within the colonic wall through activation of Toll-like receptors (TLR) on enteric glia cells (Bhave et al., 2017).

A growing body of preclinical evidence has implicated the gut microbiome in chemotherapy-induced neuropathic pain (Ramakrishna et al., 2019; Shen et al., 2017). Ramakrishna et al. (Ramakrishna et al., 2019) recently reported that fecal microbial transplant from healthy donor mice attenuated chemotherapeutic-induced hypersensitivity in mice treated with paclitaxel. Additionally, oxaliplatin-induced mechanical hyperalgesia was reduced in germ-free and in antibiotic treated mice and restored by replenishing the dysbiotic microbiota (Shen et al., 2017). However, it is unclear how the alterations to microbiota, or epithelial barrier result in chemotherapy-induced hyperalgesia.

An important constituent of the commensal microbiome are bacterial species that ferment fibers through complex enzymatic pathways and produce short chain fatty acids, acetate, propionate, and butyrate (70–140 mM) (Venegas et al., 2019). Bonomo et al. (2020) recently reported that fecal microbial transplant from lean to obese, insulin-resistant mice increased serum butyrate levels. This was associated with reversal of peripheral neuropathy, mechanical allodynia, DRG hyperactivity, and transformation of macrophages in the DRG to an anti-inflammatory phenotype. Furthermore, butyrate administration decreased leukocyte population in the DRG and enhanced desensitization of TRPV1 channels (Bonomo et al., 2020). The gut epithelium plays a central role in preventing bacterial translocation and systemic inflammation. Butyrate is known to improve the integrity of the epithelium and to enhance some of the antimicrobial peptides preventing colonization by pathogenic bacteria (Schulthess et al., 2019; Zhao et al., 2018). Thus, loss of butyrate-producing commensal bacteria by way of direct effect from the chemotherapeutics or by indirectly altering the microenvironment may be one possible pathway through which the gut epithelial barrier is compromised and excitability of primary afferent neurons is altered to produce peripheral hypersensitivity.

7. Conclusion

Chemotherapy-induced gastrointestinal toxicities, including diarrhea and mucositis, are major hurdles in cancer treatment. Despite major advances in chemotherapy, treatment of gastrointestinal toxicity remains inadequate often resulting in hospitalization, cessation of treatment and decrease in quality of life. In this review, we highlighted our current understanding of the pathophysiology (Fig. 1) and potential pharmacological approaches to treat CID and CIM. An exciting prospect is the potential to approach future treatments based on the role the gut microbiome and

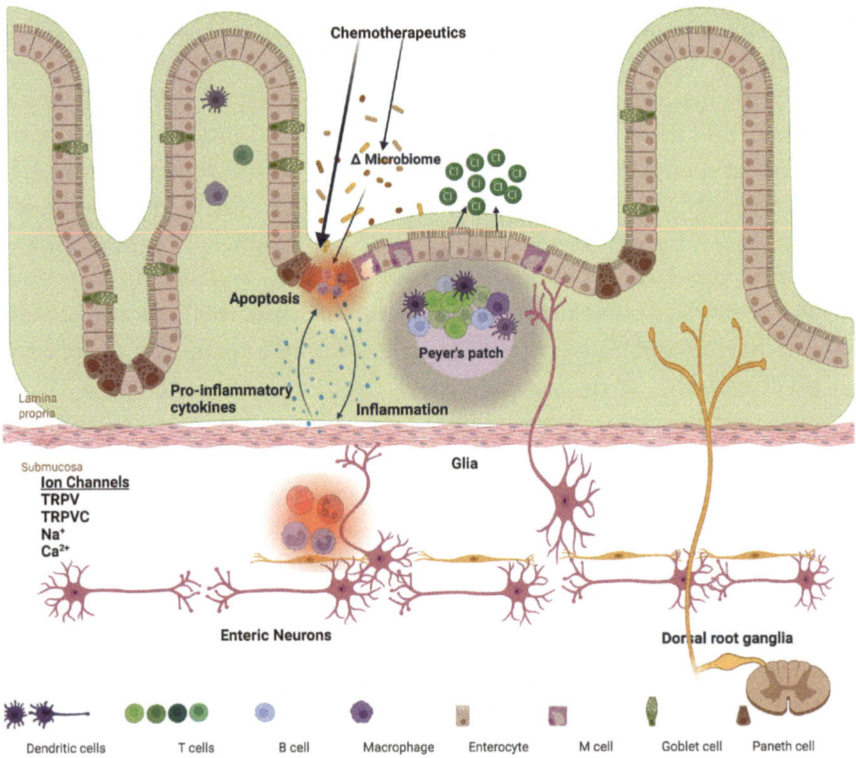

Fig. 1 A schematic presentation of mechanisms that underlie the chemotherapy-induced gastrointestinal toxicity. Chemotherapeutics can alter gut microbiome directly or through effects on the epithelial cells. Disruption of the gut barrier results in bacterial translocation and an inflammatory response within the lamina propria. This can affect ion channels in enteric and extrinsic sensory neurons projecting from the dorsal root ganglia. Inflammation also induces increase in epithelial secretions resulting in diarrhea. Injury to DNA of intestinal epithelial cells induce apoptosis.

metabolites may play on epithelial barrier function and enteric neurons. Future studies to dissect individual cell type specific responses in the gastrointestinal tract to chemotherapeutic agents will provide significant insights for treatment of gastrointestinal toxicities.

References

Abdel-Rahman, O., ElHalawani, H., & Essam-Eldin, S. (2016). S-1-based regimens and the risk of oral and gastrointestinal mucosal injury: A meta-analysis with comparison to other fluoropyrimidines. *Expert Opinion on Drug Safety*, *15*(1), 5–20. https://doi.org/10.1517/14740338.2016.1105959.

Akbarali, H. I. (2014). Oxidative stress and ion channels. In *Systems biology of free radicals and antioxidants* (pp. 355–373). Berlin Heidelberg: Springer. https://doi.org/10.1007/978-3-642-30018-9_12.

Allaire, J. M., Crowley, S. M., Law, H. T., Chang, S. Y., Ko, H. J., & Vallance, B. A. (2018). The intestinal epithelium: Central coordinator of mucosal immunity. *Trends in Immunology*, *39*(9), 677–696. https://doi.org/10.1016/j.it.2018.04.002.

Allais, L., De Smet, R., Verschuere, S., Talavera, K., Cuvelier, C. A., & Maes, T. (2017). Transient receptor potential channels in intestinal inflammation: What is the impact of cigarette smoking? *Pathobiology*, *84*(1), 1–15. https://doi.org/10.1159/000446568.

Andreyev, J., Ross, P., Donnellan, C., Lennan, E., Leonard, P., Waters, C., et al. (2014). Guidance on the management of diarrhoea during cancer chemotherapy. *The Lancet Oncology*, *15*(10), e447–e460. https://doi.org/10.1016/S1470-2045(14)70006-3.

Annunziato, L., Pannaccione, A., Cataldi, M., Secondo, A., Castaldo, P., Di Renzo, G., et al. (2002). Modulation of ion channels by reactive oxygen and nitrogen species: A pathophysiological role in brain aging? *Neurobiology of Aging*, *23*(5), 819–834. https://doi.org/10.1016/s0197-4580(02)00069-6.

Antonadou, D., Pepelassi, M., Synodinou, M., Puglisi, M., & Throuvalas, N. (2002). Prophylactic use of amifostine to prevent radiochemotherapy-induced mucositis and xerostomia in head-and-neck cancer. *International Journal of Radiation Oncology, Biology, Physics*, *52*(3), 739–747. https://doi.org/10.1016/S0360-3016(01)02683-9.

Aromolaran, K. A., & Goldstein, P. A. (2017). Ion channels and neuronal hyperexcitability in chemotherapy-induced peripheral neuropathy. *Molecular Pain*, *13*. https://doi.org/10.1177/1744806917714693. 174480691771469.

Authier, N., Balayssac, D., Marchand, F., Ling, B., Zangarelli, A., Descoeur, J., et al. (2009). Animal models of chemotherapy-evoked painful peripheral neuropathies. *Neurotherapeutics*, *6*(4), 620–629. https://doi.org/10.1016/j.nurt.2009.07.003.

Bajic, J. E., Johnston, I. N., Howarth, G. S., & Hutchinson, M. R. (2018). From the bottom-up: Chemotherapy and gut-brain axis dysregulation. *Frontiers in Behavioral Neuroscience*, *12*. https://doi.org/10.3389/fnbeh.2018.00104.

Basile, D., Di Nardo, P., Corvaja, C., Garattini, S. K., Pelizzari, G., Lisanti, C., et al. (2019). Mucosal injury during anti-cancer treatment: From pathobiology to bedside. *Cancers*, *11*(6), 857. https://doi.org/10.3390/cancers11060857.

Bensadoun, R. J., Schubert, M. M., Lalla, R. V., & Keefe, D. (2006). Amifostine in the management of radiation-induced and chemo-induced mucositis. *Supportive Care in Cancer*, *14*(6), 566–572. https://doi.org/10.1007/s00520-006-0047-4.

Bevins, C. L., & Salzman, N. H. (2011). Paneth cells, antimicrobial peptides and maintenance of intestinal homeostasis. *Nature Reviews Microbiology*, *9*(5), 356–368. https://doi.org/10.1038/nrmicro2546.

Bhatt, A. P., Pellock, S. J., Biernat, K. A., Walton, W. G., Wallace, B. D., Creekmore, B. C., et al. (2020). Targeted inhibition of gut bacterial β-glucuronidase activity enhances anticancer drug efficacy. *Proceedings of the National Academy of Sciences of the United States of America*, *117*(13), 7374–7381. https://doi.org/10.1073/pnas.1918095117.

Bhave, S., Gade, A., Kang, M., Hauser, K. F., Dewey, W. L., & Akbarali, H. I. (2017). Connexin-purinergic signaling in enteric glia mediates the prolonged effect of morphine on constipation. *FASEB Journal*, *31*(6), 2649–2660. https://doi.org/10.1096/fj.201601068R.

Bohórquez, D. V., Samsa, L. A., Roholt, A., Medicetty, S., Chandra, R., & Liddle, R. A. (2014). An enteroendocrine cell—Enteric glia connection revealed by 3D electron microscopy. *PLoS One*, *9*(2). https://doi.org/10.1371/journal.pone.0089881.

Bonomo, R. R., Cook, T. M., Gavini, C. K., White, C. R., Jones, J. R., Bovo, E., et al. (2020). Fecal transplantation and butyrate improve neuropathic pain, modify immune cell profile, and gene expression in the PNS of obese mice. *Proceedings of the National Academy of Sciences of the United States of America*, *117*(42). https://doi.org/10.1073/pnas.2006065117.

Boushey, R. P., Yusta, B., & Drucker, D. J. (2001). Glucagon-like peptide (GLP)-2 reduces chemotherapy-associated mortality and enhances cell survival in cells expressing a transfected GLP-2 receptor. *Cancer Research*, *61*(2), 687–693.

Boussios, S., Pentheroudakis, G., Katsanos, K., & Pavlidis, N. (2012). Systemic treatment-induced gastrointestinal toxicity: Incidence, clinical presentation and management. *Annals of Gastroenterology*, *25*(2), 106–118.

Bowen, J., Al-Dasooqi, N., Bossi, P., Wardill, H., Van Sebille, Y., Al-Azri, A., et al. (2019). The pathogenesis of mucositis: Updated perspectives and emerging targets. *Supportive Care in Cancer*, *27*(10), 4023–4033. https://doi.org/10.1007/s00520-019-04893-z.

Bowen, J. M., Gibson, R. J., Cummins, A. G., & Keefe, D. M. K. (2006). Intestinal mucositis: the role of the Bcl-2 family, p53 and caspases in chemotherapy-induced damage. *Supportive Care in Cancer*, *14*(7), 713–731. https://doi.org/10.1007/s00520-005-0004-7.

Bowen, J. M., Gibson, R. J., Cummins, A. G., Tsykin, A., & Keefe, D. M. K. (2007). Irinotecan changes gene expression in the small intestine of the rat with breast cancer. *Cancer Chemotherapy and Pharmacology*, *59*(3), 337–348. https://doi.org/10.1007/s00280-006-0275-9.

Brehmer, A. (2006). Structure of enteric neurons. In F. F. Beck, F. Clascá, D. E. Haines, H.-W. Korf, E. Marani, R. Putz, … K. Zilles (Eds.), *Vol. 186. Advances in anatomy, embryology, and cell biology*. Berlin Heidelberg: Springer.

Brown, I. A. M., McClain, J. L., Watson, R. E., Patel, B. A., & Gulbransen, B. D. (2016). Enteric glia mediate neuron death in colitis through purinergic pathways that require connexin-43 and nitric oxide. *Cellular and Molecular Gasteroenterology and Hepatology*, *2*(1). https://doi.org/10.1016/j.jcmgh.2015.08.007.

Brunton, L. L., Hilal-Dandan, R., & Knollmann, B. C. (2017). In J. F. Shanahan, & H. Lebowitz (Eds.), *Goodman & Gilman's: The pharmacological basis of therapeutics* (13 ed.). McGraw Hill.

Buckinx, R., Van Nassauw, L., Avula, L. R., Alpaerts, K., Adriaensen, D., & Timmermans, J. P. (2013). Transient receptor potential vanilloid type 1 channel (TRPV1) immunolocalization in the murine enteric nervous system is affected by the targeted C-terminal epitope of the applied antibody. *Journal of Histochemistry & Cytochemistry*, *61*(6), 421–432. https://doi.org/10.1369/0022155413484764.

Chai, S., Liu, K., Feng, W., Liu, T., Wang, Q., Zhou, R., et al. (2019). Activation of G protein–coupled estrogen receptor protects intestine from ischemia/reperfusion injury in mice by protecting the crypt cell proliferation. *Clinical Science*, *133*(3), 449–464. https://doi.org/10.1042/CS20180919.

Chang, C. W., Lee, H. C., Li, L. H., Chiang Chiau, J. S., Wang, T. E., Chuang, W. H., et al. (2020). Fecal microbiota transplantation prevents intestinal injury, upregulation of toll-like receptors, and 5-fluorouracil/oxaliplatin-induced toxicity in colorectal cancer. *International Journal of Molecular Sciences*, *21*(2), 386. https://doi.org/10.3390/ijms21020386.

Chang, C. W., Liu, C. Y., Lee, H. C., Huang, Y. H., Li, L. H., Chiau, J. S. C., et al. (2018). Lactobacillus casei variety rhamnosus probiotic preventively attenuates 5-fluorouracil/oxaliplatin-induced intestinal injury in a syngeneic colorectal cancer model. *Frontiers in Microbiology*, *9*(May). https://doi.org/10.3389/fmicb.2018.00983.

Chen, C., Tian, L., Zhang, M., Sun, Q., Zhang, X., Li, X., et al. (2013). Protective effect of amifostine on high-dose methotrexate-induced small intestinal mucositis in mice. *Digestive Diseases and Sciences*, *58*(11), 3134–3143. https://doi.org/10.1007/s10620-013-2826-3.

Chen, G., Zeng, H., Li, X., Liu, J., Li, Z., Xu, R., et al. (2021). Activation of G protein coupled estrogen receptor prevents chemotherapy-induced intestinal mucositis by inhibiting the DNA damage in crypt cell in an extracellular signal-regulated kinase 1- and 2-dependent manner. *Cell Death & Disease*, *12*(11), 1034. https://doi.org/10.1038/s41419-021-04325-z.

Chen, Y. K., Hou, H. A., Chow, J. M., Chen, Y. C., Hsueh, P. R., & Tien, H. F. (2011). The impact of oral herpes simplex virus infection and candidiasis on chemotherapy-induced oral mucositis among patients with hematological malignancies. *European Journal of Clinical Microbiology & Infectious Diseases*, *30*(6), 753–759. https://doi.org/10.1007/s10096-010-1148-z.

Chiu, I. M., Heesters, B. A., Ghasemlou, N., Von Hehn, C. A., Zhao, F., Tran, J., et al. (2013). Bacteria activate sensory neurons that modulate pain and inflammation. *Nature*, *501*(7465). https://doi.org/10.1038/nature12479.

Chuang, H. H., & Lin, S. (2009). Oxidative challenges sensitize the capsaicin receptor by covalent cysteine modification. *Proceedings of the National Academy of Sciences of the United States of America*, *106*(47), 20097–20102. https://doi.org/10.1073/pnas.0902675106.

Cinausero, M., Aprile, G., Ermacora, P., Basile, D., Vitale, M. G., Fanotto, V., et al. (2017). New frontiers in the pathobiology and treatment of cancer regimen-related mucosal injury. *Frontiers in Pharmacology*, *8*(Jun). https://doi.org/10.3389/fphar.2017.00354.

Coddou, C., Codocedo, J. F., Li, S., Lillo, J. G., Acuna-Castillo, C., Bull, P., et al. (2009). Reactive oxygen species potentiate the P2X2 receptor activity through intracellular Cys430. *Journal of Neuroscience*, *29*(39), 12284–12291. https://doi.org/10.1523/JNEUROSCI.2096-09.2009.

Colvin, L. A. (2019). Chemotherapy-induced peripheral neuropathy: Where are we now? *Pain*, *160*(1), S1–S10. https://doi.org/10.1097/j.pain.0000000000001540.

Costa, D. V. S., Bon-Frauches, A. C., Silva, A. M. H. P., Lima-Júnior, R. C. P., Martins, C. S., Leitão, R. F. C., et al. (2019). 5-Fluorouracil induces enteric neuron death and glial activation during intestinal mucositis via a S100B-RAGE-NFκB-dependent pathway. *Scientific Reports*, *9*(1). https://doi.org/10.1038/s41598-018-36878-z.

Crosnier, C., Stamataki, D., & Lewis, J. (2006). Organizing cell renewal in the intestine: Stem cells, signals and combinatorial control. *Nature Reviews. Genetics*, 7(5), 349–359. https://doi.org/10.1038/nrg1840.

da Silva Ferreira, A. R., Wardill, H. R., Tissing, W. J. E., & Harmsen, H. J. M. (2020). Pitfalls and novel experimental approaches to optimize microbial interventions for chemotherapy-induced gastrointestinal mucositis. *Current Opinion in Supportive & Palliative Care*, *14*(2), 127–134. https://doi.org/10.1097/SPC.0000000000000497.

Dahlgren, D., Sjöblom, M., Hellström, P. M., & Lennernäs, H. (2021). Chemotherapeutics-induced intestinal mucositis: Pathophysiology and potential treatment strategies. *Frontiers in Pharmacology*, *12*. https://doi.org/10.3389/fphar.2021.681417.

Daig, R. (2000). Human intestinal epithelial cells secrete interleukin-1 receptor antagonist and interleukin-8 but not interleukin-1 or interleukin-6. *Gut*, *46*(3), 350–358. https://doi.org/10.1136/gut.46.3.350.

de Man, F. M., Goey, A. K. L., van Schaik, R. H. N., Mathijssen, R. H. J., & Bins, S. (2018). Individualization of irinotecan treatment: A review of pharmacokinetics, pharmacodynamics, and pharmacogenetics. *Clinical Pharmacokinetics*, *57*(10), 1229–1254. https://doi.org/10.1007/S40262-018-0644-7.

de Miranda, J. A. L., Martins, C.d. S., Fideles, L.d. S., Barbosa, M. L. L., Barreto, J. E. F., Pimenta, H. B., et al. (2020). Troxerutin prevents 5-fluorouracil induced morphological changes in the intestinal mucosa: Role of cyclooxygenase-2 pathway. *Pharmaceuticals*, *13*(1), 10. https://doi.org/10.3390/ph13010010.

Deng, C., Deng, B., Jia, L., & Tan, H. (2017). Efficacy of long-acting release octreotide for preventing chemotherapy-induced diarrhoea: Protocol for a systematic review. *BMJ Open*, 7(6), e014916. https://doi.org/10.1136/bmjopen-2016-014916.

Denlinger, C. S., & Barsevick, A. M. (2009). The challenges of colorectal cancer survivorship. *Journal of the National Comprehensive Cancer Network*, 7(8), 883–894. https://doi.org/10.6004/jnccn.2009.0058.

Drokhlyansky, E., Smillie, C. S., Van Wittenberghe, N., Ericsson, M., Griffin, G. K., Eraslan, G., et al. (2020). The human and mouse enteric nervous system at single-cell resolution. *Cell*, *182*(6), 1606–1622.e23. https://doi.org/10.1016/j.cell.2020.08.003.

Elting, L. S., Cooksley, C., Chambers, M., Cantor, S. B., Manzullo, E., & Rubenstein, E. B. (2003). The burdens of cancer therapy. *Cancer*, *98*(7), 1531–1539. https://doi.org/10.1002/cncr.11671.

Engel, M. A., Khalil, M., Mueller-Tribbensee, S. M., Becker, C., Neuhuber, W. L., Neurath, M. F., et al. (2012). The proximodistal aggravation of colitis depends on substance P released from TRPV1-expressing sensory neurons. *Journal of Gastroenterology*, *47*(3), 256–265. https://doi.org/10.1007/s00535-011-0495-6.

Fraiser, L. H., Kanekal, S., & Kehrer, J. P. (1991). Cyclophosphamide toxicity: Characterizing and avoiding the problem. *Drugs*, *42*(5), 781–795. https://doi.org/10.2165/00003495-199142050-00005.

Furness, J. B. (2007). The enteric nervous System. In J. B. Furness (Ed.), *The enteric nervous system* Blackwell Publishing. https://doi.org/10.1002/9780470988756.

Gallo, R. L., & Hooper, L. V. (2012). Epithelial antimicrobial defence of the skin and intestine. *Nature Reviews. Immunology*, *12*(7), 503–516. https://doi.org/10.1038/nri3228.

Gershon, M. D., Kirchgessner, A. L., & Wade, P. R. (1994). Functional anatomy of the enteric nervous system. In L. Johnson, D. Alpers, E. Jacobson, & J. Walsh (Eds.), *Vol. 1. Physiology of the gastrointestinal tract* (3rd ed., pp. 381–422). New York: Raven Press.

Gibson, R. J., Bowen, J. M., Inglis, M. R. B., Cummins, A. G., & Keefe, D. M. K. (2003). Irinotecan causes severe small intestinal damage, as well as colonic damage, in the rat with implanted breast cancer. *Journal of Gastroenterology and Hepatology*, *18*(9), 1095–1100. https://doi.org/10.1046/j.1440-1746.2003.03136.x.

Gibson, R. J., Coller, J. K., Wardill, H. R., Hutchinson, M. R., Smid, S., & Bowen, J. M. (2016). Chemotherapy-induced gut toxicity and pain: Involvement of TLRs. *Supportive Care in Cancer*, *24*(5). https://doi.org/10.1007/s00520-015-3020-2.

Gibson, R. J., & Keefe, D. M. K. (2006). Cancer chemotherapy-induced diarrhoea and constipation: mechanisms of damage and prevention strategies. *Support Care Cancer*, *14*, 890–900. https://doi.org/10.1007/s00520-006-0040-y.

Grundmann, D., Loris, E., Maas-Omlor, S., Huang, W., Scheller, A., Kirchhoff, F., et al. (2019). Enteric glia: S100, GFAP, and beyond. *Anatomical Record*, *302*(8). https://doi.org/10.1002/ar.24128.

Guabiraba, R., Besnard, A. G., Menezes, G. B., Secher, T., Jabir, M. S., Amaral, S. S., et al. (2014). IL-33 targeting attenuates intestinal mucositis and enhances effective tumor chemotherapy in mice. *Mucosal Immunology*, 7(5), 1079–1093. https://doi.org/10.1038/mi.2013.124.

Han, X., Wu, Z., Di, J., Pan, Y., Zhang, H., Du, Y., et al. (2011). CXCL9 attenuated chemotherapy-induced intestinal mucositis by inhibiting proliferation and reducing apoptosis. *Biomedicine & Pharmacotherapy*, *65*(8), 547–554. https://doi.org/10.1016/j.biopha.2011.03.008.

Haramis, A. P. G., Begthel, H., van den Born, M., van Es, J., Jonkheer, S., Offerhaus, G. J. A., et al. (2004). De novo crypt formation and juvenile polyposis on BMP inhibition in mouse intestine. *Science*, *303*(5664), 1684–1686. https://doi.org/10.1126/science.1093587.

Hartmann, J. T., & Lipp, H. P. (2003). Toxicity of platinum compounds. *Expert Opinion on Pharmacotherapy*, *4*(6), 889–901. https://doi.org/10.1517/14656566.4.6.889.

Hayward, R., Hydock, D., Gibson, N., Greufe, S., Bredahl, E., & Parry, T. (2013). Tissue retention of doxorubicin and its effects on cardiac, smooth, and skeletal muscle function. *Journal of Physiology and Biochemistry*, *69*(2). https://doi.org/10.1007/s13105-012-0200-0.

He, X. C., Zhang, J., Tong, W. G., Tawfik, O., Ross, J., Scoville, D. H., et al. (2004). BMP signaling inhibits intestinal stem cell self-renewal through suppression of Wnt–β-catenin signaling. *Nature Genetics*, *36*(10), 1117–1121. https://doi.org/10.1038/ng1430.

Holzer, P. (2011). Transient receptor potential (TRP) channels as drug targets for diseases of the digestive system. *Pharmacology & Therapeutics*, *131*(1), 142–170. https://doi.org/10.1016/j.pharmthera.2011.03.006.

Ibiza, S., García-Cassani, B., Ribeiro, H., Carvalho, T., Almeida, L., Marques, R., et al. (2016). Glial-cell-derived neuroregulators control type 3 innate lymphoid cells and gut defence. *Nature*, *535*(7612), 440–443. https://doi.org/10.1038/nature18644.

Iguchi, N., Dönmez, M.İ., Carrasco, A., Wilcox, D. T., Pineda, R. H., Malykhina, A. P., et al. (2019). Doxorubicin induces detrusor smooth muscle impairments through myosin dysregulation, leading to a risk of lower urinary tract dysfunction. *American Journal of Physiology - Renal Physiology*, *317*(1). https://doi.org/10.1152/ajprenal.00090.2019.

Ikuno, N., Soda, H., Watanabe, M., & Oka, M. (1995). Irinotecan (CPT-11) and characteristic mucosal changes in the mouse ileum and cecum. *Journal of the National Cancer Institute*, *87*(24), 1876–1883. https://doi.org/10.1093/jnci/87.24.1876.

Jaćević, V., Dragojević-Simić, V., Tatomirović, Ž., Dobrić, S., Bokonjić, D., Kovačević, A., et al. (2018). The efficacy of amifostine against multiple-dose doxorubicin-induced toxicity in rats. *International Journal of Molecular Sciences*, *19*(8), 2370. https://doi.org/10.3390/ijms19082370.

Javle, M. M., Cao, S., Durrani, F. A., Pendyala, L., Lawrence, D. D., Smith, P. F., et al. (2007). Celecoxib and mucosal protection: Translation from an animal model to a phase I clinical trial of celecoxib, irinotecan, and 5-fluorouracil. *Clinical Cancer Research*, *13*(3), 965–971. https://doi.org/10.1158/1078-0432.CCR-06-0551.

Kamp, K., Gumz, B., Feelders, R. A., Kwekkeboom, D. J., Kaltsas, G., Costa, F. P., et al. (2013). Safety and efficacy of everolimus in gastrointestinal and pancreatic neuroendocrine tumors after 177Lu-octreotate. *Endocrine-Related Cancer*, *20*(6), 825–831. https://doi.org/10.1530/ERC-13-0254.

Kanarek, N., Grivennikov, S. I., Leshets, M., Lasry, A., Alkalay, I., Horwitz, E., et al. (2014). Critical role for IL-1β in DNA damage-induced mucositis. *Proceedings of the National Academy of Sciences of the United States of America*, *111*(6), E702–E711. https://doi.org/10.1073/pnas.1322691111.

Kang, M., Mischel, R. A., Bhave, S., Komla, E., Cho, A., Huang, C., et al. (2017). The effect of gut microbiome on tolerance to morphine mediated antinociception in mice. *Scientific Reports*, 7(1), 42658. https://doi.org/10.1038/srep42658.

Kang, M., Morsy, N., Jin, X., Lupu, F., & Akbarali, H. I. (2004). Protein and gene expression of Ca2+ channel isoforms in murine colon: Effect of inflammation. *Pflugers Archiv European Journal of Physiology*, *449*(3). https://doi.org/10.1007/s00424-004-1339-5.

Kang, M., Ross, G. R., & Akbarali, H. I. (2007). COOH-terminal association of human smooth muscle calcium channel Ca v1.2b with Src kinase protein binding domains: Effect of nitrotyrosylation. *American Journal of Physiology - Cell Physiology, 293*(6). https://doi.org/10.1152/ajpcell.00308.2007.

Kang, M., Ross, G. R., & Akbarali, H. I. (2010). The effect of tyrosine nitration of L-type Ca2+ channels on excitation-transcription coupling in colonic inflammation. *British Journal of Pharmacology, 159*(6). https://doi.org/10.1111/j.1476-5381.2009.00599.x.

Koizumi, N., Hatano, E., Nitta, T., Tada, M., Harada, N., Taura, K., et al. (2005). Blocking of PI3K/Akt pathway enhances apoptosis induced by SN-38, an active form of CPT-11, in human hepatoma cells. *International Journal of Oncology, 26*(5), 1301–1306. https://doi.org/10.3892/ijo.26.5.1301.

Kulkarni, S., Ganz, J., Bayrer, J., Becker, L., Bogunovic, M., & Rao, M. (2018). Advances in enteric neurobiology: The "Brain" in the gut in health and disease. *The Journal of Neuroscience, 38*(44), 9346–9354. https://doi.org/10.1523/JNEUROSCI.1663-18.2018.

Kulkarni, S., Micci, M. A., Leser, J., Shin, C., Tang, S. C., Fu, Y. Y., et al. (2017). Adult enteric nervous system in health is maintained by a dynamic balance between neuronal apoptosis and neurogenesis. *Proceedings of the National Academy of Sciences of the United States of America, 114*(18), E3709–E3718. https://doi.org/10.1073/pnas.1619406114.

Kwon, Y. (2016). Mechanism-based management for mucositis: Option for treating side effects without compromising the efficacy of cancer therapy. *OncoTargets and Therapy, 9*, 2007. https://doi.org/10.2147/OTT.S96899.

Lai, N. Y., Musser, M. A., Pinho-Ribeiro, F. A., Baral, P., Jacobson, A., Ma, P., et al. (2020). Gut-innervating nociceptor neurons regulate peyer's patch microfold cells and SFB levels to mediate Salmonella host defense. *Cell, 180*(1). https://doi.org/10.1016/j.cell.2019.11.014.

Lalla, R. V., Bowen, J., Barasch, A., Elting, L., Epstein, J., Keefe, D. M., et al. (2014). MASCC/ISOO clinical practice guidelines for the management of mucositis secondary to cancer therapy. *Cancer, 120*(10), 1453–1461. https://doi.org/10.1002/cncr.28592.

Lee, C. S. (2014). Gastro-intestinal toxicity of chemotherapeutics in colorectal cancer: The role of inflammation. *World Journal of Gastroenterology, 20*(14), 3751. https://doi.org/10.3748/wjg.v20.i14.3751.

Li, H. L., Lu, L., Wang, X. S., Qin, L. Y., Wang, P., Qiu, S. P., et al. (2017). Alteration of gut microbiota and inflammatory cytokine/chemokine profiles in 5-fluorouracil induced intestinal mucositis. *Frontiers in Cellular and Infection Microbiology*, 7(Oct). https://doi.org/10.3389/fcimb.2017.00455.

Li, Z., Chapleau, M. W., Bates, J. N., Bielefeldt, K., Lee, H.-C., & Abboud, F. M. (1998). Nitric oxide as an autocrine regulator of sodium currents in baroreceptor neurons. *Neuron, 20*(5), 1039–1049. https://doi.org/10.1016/S0896-6273(00)80484-5.

Liu, Y. A., Chung, Y. C., Pan, S. T., Shen, M. Y., Hou, Y. C., Peng, S. J., et al. (2013). 3-D imaging, illustration, and quantitation of enteric glial network in transparent human colon mucosa. *Neurogastroenterology and Motility, 25*(5). https://doi.org/10.1111/nmo.12115.

Logan, R. M., Gibson, R. J., Sonis, S. T., & Keefe, D. M. K. (2007). Nuclear factor-κB (NF-κB) and cyclooxygenase-2 (COX-2) expression in the oral mucosa following cancer chemotherapy. *Oral Oncology, 43*(4), 395–401. https://doi.org/10.1016/j.oraloncology.2006.04.011.

Luo, J., Qian, A., Oetjen, L. K., Yu, W., Yang, P., Feng, J., et al. (2018). TRPV4 channel signaling in macrophages promotes gastrointestinal motility via direct effects on smooth muscle cells. *Immunity, 49*(1). https://doi.org/10.1016/j.immuni.2018.04.021.

Lupi, L. A., Cucielo, M. S., Silveira, H. S., Gaiotte, L. B., Cesário, R. C., Seiva, F. R. F., et al. (2020). The role of Toll-like receptor 4 signaling pathway in ovarian, cervical, and endometrial cancers. *Life Sciences, 247*. https://doi.org/10.1016/j.lfs.2020.117435, 117435.

Madison, B. B., Braunstein, K., Kuizon, E., Portman, K., Qiao, X. T., & Gumucio, D. L. (2005). Epithelial hedgehog signals pattern the intestinal crypt-villus axis. *Development*, *132*(2), 279–289. https://doi.org/10.1242/dev.01576.

Maroun, J. A., Anthony, L. B., Blais, N., Burkes, R., Dowden, S. D., Dranitsaris, G., et al. (2007). Prevention and management of chemotherapy-induced diarrhea in patients with colorectal cancer: A consensus statement by the Canadian Working Group on Chemotherapy-Induced Diarrhea. *Current Oncology*, *14*(1), 13–20. https://doi.org/10.3747/co.2007.96.

Mason, H. S., Bourke, S., & Kemp, P. J. (2004). Selective modulation of ligand-gated P2X purinoceptor channels by acute hypoxia is mediated by reactive oxygen species. *Molecular Pharmacology*, *66*(6), 1525–1535. https://doi.org/10.1124/mol.104.000851.

Mayo, B. J., Stringer, A. M., Bowen, J. M., Bateman, E. H., & Keefe, D. M. (2017). Irinotecan-induced mucositis: The interactions and potential role of GLP-2 analogues. *Cancer Chemotherapy and Pharmacology*, *79*(2), 233–249. https://doi.org/10.1007/s00280-016-3165-9.

McQuade, R. M., Al Thaalibi, M., & Nurgali, K. (2020). Impact of chemotherapy-induced enteric nervous system toxicity on gastrointestinal mucositis. *Current Opinion in Supportive & Palliative Care*, *14*(3), 293–300. https://doi.org/10.1097/SPC.0000000000000515.

McQuade, R. M., Carbone, S. E., Stojanovska, V., Rahman, A., Gwynne, R. M., Robinson, A. M., et al. (2016). Role of oxidative stress in oxaliplatin-induced enteric neuropathy and colonic dysmotility in mice. *British Journal of Pharmacology*, *173*(24), 3502–3521. https://doi.org/10.1111/bph.13646.

McQuade, R. M., Stojanovska, V., Abalo, R., Bornstein, J. C., & Nurgali, K. (2016). Chemotherapy-induced constipation and diarrhea: Pathophysiology, current and emerging treatments. *Frontiers in Pharmacology*, 7(Nov), 414. Frontiers Media S.A. https://doi.org/10.3389/fphar.2016.00414.

Millington, K. A., Pascasio, J. M., Halligan, G. E., & De Chadarévian, J. P. (2012). Chemotherapy-related leiomyopathy: A suggested morphological explanation for the intestinal dysmotility affecting patients treated with anthracyclines. *Modern Pathology*, *25*(2). https://doi.org/10.1038/modpathol.2011.115.

Min, K., Kwon, O. S., Smuder, A. J., Wiggs, M. P., Sollanek, K. J., Christou, D. D., et al. (2015). Increased mitochondrial emission of reactive oxygen species and calpain activation are required for doxorubicin-induced cardiac and skeletal muscle myopathy. *Journal of Physiology*, *593*(8). https://doi.org/10.1113/jphysiol.2014.286518.

Mischel, R. A., Dewey, W. L., & Akbarali, H. I. (2018). Tolerance to morphine-induced inhibition of TTX-R sodium channels in dorsal root ganglia neurons is modulated by gut-derived mediators. *IScience*, *2*, 193–209. https://doi.org/10.1016/j.isci.2018.03.003.

Moore, B. A., Stewart, T. M. R., Hill, C., & Vanner, S. J. (2002). TNBS ileitis evokes hyperexcitability and changes in ionic membrane properties of nociceptive DRG neurons. *American Journal of Physiology-Gastrointestinal and Liver Physiology*, *282*(6), G1045–G1051. https://doi.org/10.1152/ajpgi.00406.2001.

Muchhala, K. H., Jacob, J. C., Alam, I., Hasan, S., Khan, A., Kang, M., et al. (2020). Rapid tolerance to morphine in the myenteric neurons of the small intestine is independent of β-arrestin-2 and mediated by PKC. *BioRxiv*, *2020*(07), 17.209437. https://doi.org/10.1101/2020.07.17.209437.

National Cancer Institute. (2017). *Common terminology criteria for adverse events (CTCAE) common terminology criteria for adverse events (CTCAE) v5.0*. https://ctep.cancer.gov.

Neunlist, M., Aubert, P., Bonnaud, S., Van Landeghem, L., Coron, E., Wedel, T., et al. (2007). Enteric glia inhibit intestinal epithelial cell proliferation partly through a TGF-β1-dependent pathway. *American Journal of Physiology - Gastrointestinal and Liver Physiology*, *292*(1). https://doi.org/10.1152/ajpgi.00276.2005.

Nogueira, L. T., Costa, D. V. S., Gomes, A. S., Martins, C. S., Silva, A. M. H. P., Coelho-Aguiar, J. M., et al. (2017). The involvement of mast cells in the irinotecan-induced enteric neurons loss and reactive gliosis. *Journal of Neuroinflammation, 14*(1). https://doi.org/10.1186/s12974-017-0854-1.

Pawolski, V., & Schmidt, M. H. H. (2020). Neuron–glia interaction in the developing and adult enteric nervous system. *Cells, 10*(1), 47. https://doi.org/10.3390/cells10010047.

Peterson, L. W., & Artis, D. (2014). Intestinal epithelial cells: Regulators of barrier function and immune homeostasis. *Nature Reviews. Immunology, 14*(3), 141–153. https://doi.org/10.1038/nri3608.

Prisciandaro, L. D., Geier, M. S., Butler, R. N., Cummins, A. G., & Howarth, G. S. (2011). Probiotic factors partially improve parameters of 5-fluorouracil-induced intestinal mucositis in rats. *Cancer Biology & Therapy, 11*(7), 671–677. https://doi.org/10.4161/cbt.11.7.14896.

Przewlocki, R., & Przewlocka, B. (2005). Opioids in neuropathic pain. *Current Pharmaceutical Design, 11*(23), 3013–3025. https://doi.org/10.2174/1381612054865055.

Qi, L., Luo, Q., Zhang, Y., Jia, F., Zhao, Y., & Wang, F. (2019). Advances in toxicological research of the anticancer drug cisplatin. *Chemical Research in Toxicology, 32*(8), 1469–1486. https://doi.org/10.1021/acs.chemrestox.9b00204.

Ramakrishna, C., Corleto, J., Ruegger, P. M., Logan, G. D., Peacock, B. B., Mendonca, S., et al. (2019). Dominant role of the gut microbiota in chemotherapy induced neuropathic pain. *Scientific Reports, 9*(1), 20324. https://doi.org/10.1038/s41598-019-56832-x.

Rao, M., Rastelli, D., Dong, L., Chiu, S., Setlik, W., Gershon, M. D., et al. (2017). Enteric glia regulate gastrointestinal motility but are not required for maintenance of the epithelium in mice. *Gastroenterology, 153*(4), 1068–1081.e7. https://doi.org/10.1053/j.gastro.2017.07.002.

Ribeiro, R. A., Wanderley, C. W. S., Wong, D. V. T., Mota, J. M. S. C., Leite, C. A. V. G., Souza, M. H. L. P., et al. (2016). Irinotecan- and 5-fluorouracil-induced intestinal mucositis: Insights into pathogenesis and therapeutic perspectives. *Cancer Chemotherapy and Pharmacology, 78*(5), 881–893. https://doi.org/10.1007/s00280-016-3139-y.

Robinson, A. M., Stojanovska, V., Rahman, A. A., McQuade, R. M., Senior, P. V., & Nurgali, K. (2016). Effects of oxaliplatin treatment on the enteric glial cells and neurons in the mouse ileum. *Journal of Histochemistry and Cytochemistry, 64*(9). https://doi.org/10.1369/0022155416656842.

Ross, G. R., Kang, M., & Akbarali, H. I. (2010). Colonic inflammation alters Src kinase-dependent gating properties of single Ca2+ channels via tyrosine nitration. *American Journal of Physiology - Gastrointestinal and Liver Physiology, 298*(6). https://doi.org/10.1152/ajpgi.00056.2010.

Ross, G. R., Kang, M., Shirwany, N., Malykhina, A. P., Drozd, M., & Akbarali, H. I. (2007). Nitrotyrosylation of Ca2+ channels prevents c-Src kinase regulation of colonic smooth muscle contractility in experimental colitis. *Journal of Pharmacology and Experimental Therapeutics, 322*(3). https://doi.org/10.1124/jpet.107.123075.

Said, H. M. (2018). *Physiology of the gastrointestinal tract* (6th ed., pp. 1–2). Elsevier.

Sałat, K. (2020). Chemotherapy-induced peripheral neuropathy: Part 1—current state of knowledge and perspectives for pharmacotherapy. *Pharmacological Reports, 72*(3), 486–507. https://doi.org/10.1007/s43440-020-00109-y.

Sanders, K. M., Koh, S. D., Ro, S., & Ward, S. M. (2012). Regulation of gastrointestinal motility-insights from smooth muscle biology. *Nature Reviews Gastroenterology and Hepatology, 9*(11). https://doi.org/10.1038/nrgastro.2012.168.

Sang, Q., Williamson, S., & Young, H. M. (1997). Projections of chemically identified myenteric neurons of the small and large intestine of the mouse. *Journal of Anatomy, 190*(2), 209–222. https://doi.org/10.1046/j.1469-7580.1997.19020209.x.

Schiller, L. R., Pardi, D. S., & Sellin, J. H. (2017). Chronic diarrhea: Diagnosis and management. *Clinical Gastroenterology and Hepatology*, *15*(2), 182–193.e3. https://doi.org/10.1016/j.cgh.2016.07.028.

Schilling, T., & Eder, C. (2009). Importance of the non-selective cation channel TRPV1 for microglial reactive oxygen species generation. *Journal of Neuroimmunology*, *216*(1–2), 118–121. https://doi.org/10.1016/j.jneuroim.2009.07.008.

Schulthess, J., Pandey, S., Capitani, M., Rue-Albrecht, K. C., Arnold, I., Franchini, F., et al. (2019). The short chain fatty acid butyrate imprints an antimicrobial program in macrophages. *Immunity*, *50*(2). https://doi.org/10.1016/j.immuni.2018.12.018.

Schwab, M., Zanger, U. M., Marx, C., Schaeffeler, E., Klein, K., Dippon, J., et al. (2008). Role of genetic and nongenetic factors for fluorouracil treatment-related severe toxicity: A prospective clinical trial by the German 5-FU toxicity study group. *Journal of Clinical Oncology*, *26*(13), 2131–2138. https://doi.org/10.1200/JCO.2006.10.4182.

Secombe, K. R., Van Sebille, Y. Z. A., Mayo, B. J., Coller, J. K., Gibson, R. J., & Bowen, J. M. (2020). Diarrhea induced by small molecule tyrosine kinase inhibitors compared with chemotherapy: Potential role of the microbiome. *Integrative Cancer Therapies*, *19*. https://doi.org/10.1177/1534735420928493. 153473542092849.

Seguella, L., & Gulbransen, B. D. (2021). Enteric glial biology, intercellular signalling and roles in gastrointestinal disease. *Nature Reviews. Gastroenterology & Hepatology*, *18*(8), 571–587. https://doi.org/10.1038/s41575-021-00423-7.

Sessenwein, J. L., Baker, C. C., Pradhananga, S., Maitland, M. E., Petrof, E. O., Allen-Vercoe, E., et al. (2017). Protease-mediated suppression of DRG neuron excitability by commensal bacteria. *Journal of Neuroscience*, *37*(48). https://doi.org/10.1523/JNEUROSCI.1672-17.2017.

Sharkey, K. A. (2015). Emerging roles for enteric glia in gastrointestinal disorders. *Journal of Clinical Investigation*, *125*(3). https://doi.org/10.1172/JCI76303.

Shen, S., Lim, G., You, Z., Ding, W., Huang, P., Ran, C., et al. (2017). Gut microbiota is critical for the induction of chemotherapy-induced pain. *Nature Neuroscience*, *20*(9), 1213–1216. https://doi.org/10.1038/nn.4606.

Soares, P. M. G., Mota, J. M. S. C., Gomes, A. S., Oliveira, R. B., Assreuy, A. M. S., Brito, G. A. C., et al. (2008). Gastrointestinal dysmotility in 5-fluorouracil-induced intestinal mucositis outlasts inflammatory process resolution. *Cancer Chemotherapy and Pharmacology*, *63*(1), 91–98. https://doi.org/10.1007/s00280-008-0715-9.

Sonis, S. T. (2004). The pathobiology of mucositis. *Nature Reviews Cancer*, *4*(4), 277–284. https://doi.org/10.1038/nrc1318.

Sougiannis, A. T., VanderVeen, B. N., Davis, J. M., Fan, D., & Murphy, E. A. (2021). Understanding chemotherapy-induced intestinal mucositis and strategies to improve gut resilience. *American Journal of Physiology - Gastrointestinal and Liver Physiology*, *320*(5). https://doi.org/10.1152/AJPGI.00380.2020.

Stein, A., Voigt, W., & Jordan, K. (2010). Chemotherapy-induced diarrhea: pathophysiology, frequency and guideline-based management. *Therapeutic Advances in Medical Oncology*, *2*(1), 51–63. https://doi.org/10.1177/1758834009355164.

Stojanovska, V., Sakkal, S., & Nurgali, K. (2015). Platinum-based chemotherapy: Gastrointestinal immunomodulation and enteric nervous system toxicity. *American Journal of Physiology. Gastrointestinal and Liver Physiology*, *308*, 223–232. https://doi.org/10.1152/ajpgi.00212.2014.-The.

Stojanovska, V., Stewart, M., McQuade, R., Timpani, C., Sorensen, J., Orbell, J., et al. (2016). Platinum accumulation and changes in mitochondrial function of the longitudinal muscle and myenteric plexus following oxaliplatin administration. *Neurogastroenterology & Motility*, *28*, 55.

Stringer, A. M., Gibson, R. J., Bowen, J. M., Logan, R. M., Ashton, K., Yeoh, A. S. J., et al. (2009). Irinotecan-induced mucositis manifesting as diarrhoea corresponds with an amended intestinal flora and mucin profile. *International Journal of Experimental Pathology*, *90*(5), 489–499. https://doi.org/10.1111/j.1365-2613.2009.00671.x.

Stringer, A. M., Gibson, R. J., Bowen, J. M., Logan, R. M., Yeoh, A. S. J., & Keefe, D. M. K. (2007). Chemotherapy-induced mucositis: The role of gastrointestinal microflora and mucins in the luminal environment. *The Journal of Supportive Oncology*, *5*(6), 259–267.

Stringer, A. M., Gibson, R. J., Logan, R. M., Bowen, J. M., Yeoh, A. S. J., & Keefe, D. M. K. (2008). Faecal microflora and β-glucuronidase expression are altered in an irinotecan-induced diarrhea model in rats. *Cancer Biology & Therapy*, 7(12), 1919–1925. https://doi.org/10.4161/cbt.7.12.6940.

Stringer, A. M., Gibson, R. J., Logan, R. M., Bowen, J. M., Yeoh, A. S. J., Laurence, J., et al. (2009). Irinotecan-induced mucositis is associated with changes in intestinal mucins. *Cancer Chemotherapy and Pharmacology*, *64*(1), 123–132. https://doi.org/10.1007/s00280-008-0855-y.

Stringer, A. M., & Logan, R. M. (2015). The role of oral flora in the development of chemotherapy-induced oral mucositis. *Journal of Oral Pathology and Medicine*, *44*(2), 81–87. Blackwell Publishing Ltd. https://doi.org/10.1111/jop.12152.

Sun, R., Zhu, L., Li, L., Song, W., Gong, X., Qi, X., et al. (2020). Irinotecan-mediated diarrhea is mainly correlated with intestinal exposure to SN-38: Critical role of gut Ugt. *Toxicology and Applied Pharmacology*, *398*, 115032. https://doi.org/10.1016/j.taap.2020.115032.

Takeba, Y., Kumai, T., Matsumoto, N., Nakaya, S., Tsuzuki, Y., Yanagida, Y., et al. (2007). Irinotecan activates p53 with its active metabolite, resulting in human hepatocellular carcinoma apoptosis. *Journal of Pharmacological Sciences*, *104*(3), 232–242. https://doi.org/10.1254/jphs.FP0070442.

Tao, G., Huang, J., Moorthy, B., Wang, C., Hu, M., Gao, S., et al. (2020). Potential role of drug metabolizing enzymes in chemotherapy-induced gastrointestinal toxicity and hepatotoxicity. *Expert Opinion on Drug Metabolism & Toxicology*, *16*(11), 1109–1124. https://doi.org/10.1080/17425255.2020.1815705.

Toma, W., Kyte, S. L., Bagdas, D., Alkhlaif, Y., Alsharari, S. D., Lichtman, A. H., et al. (2017). Effects of paclitaxel on the development of neuropathy and affective behaviors in the mouse. *Neuropharmacology*, *117*, 305–315. https://doi.org/10.1016/j.neuropharm.2017.02.020.

Touchefeu, Y., Montassier, E., Nieman, K., Gastinne, T., Potel, G., Bruley Des Varannes, S., et al. (2014). Systematic review: The role of the gut microbiota in chemotherapy- or radiation-induced gastrointestinal mucositis—Current evidence and potential clinical applications. *Alimentary Pharmacology & Therapeutics*, *40*(5). https://doi.org/10.1111/apt.12878.

Ustaris, F., Saura, C., Palma, J. D., Bryce, R., Moran, S., Neuman, L., et al. (2015). Effective management and prevention of neratinib-induced diarrhea. *American Journal of Hematology/Oncology*, *11*(11).

van der Flier, L. G., & Clevers, H. (2009). Stem cells, self-renewal, and differentiation in the intestinal epithelium. *Annual Review of Physiology*, *71*(1), 241–260. https://doi.org/10.1146/annurev.physiol.010908.163145.

van der Flier, L. G., van Gijn, M. E., Hatzis, P., Kujala, P., Haegebarth, A., Stange, D. E., et al. (2009). Transcription factor achaete scute-like 2 controls intestinal stem cell fate. *Cell*, *136*(5), 903–912. https://doi.org/10.1016/j.cell.2009.01.031.

Venegas, D. P., De La Fuente, M. K., Landskron, G., González, M. J., Quera, R., Dijkstra, G., et al. (2019). Short chain fatty acids (SCFAs)mediated gut epithelial and immune regulation and its relevance for inflammatory bowel diseases. *Frontiers in Immunology*, *10*(Mar). https://doi.org/10.3389/fimmu.2019.00277.

Vicentini, F. A., Keenan, C. M., Wallace, L. E., Woods, C., Cavin, J.-B., Flockton, A. R., et al. (2021). Intestinal microbiota shapes gut physiology and regulates enteric neurons and glia. *Microbiome, 9*(1), 210. https://doi.org/10.1186/s40168-021-01165-z.

Villa, A., & Sonis, S. T. (2015). Mucositis. *Current Opinion in Oncology, 27*(3), 159–164. https://doi.org/10.1097/CCO.0000000000000180.

Villarreal, A., Aviles Reyes, R. X., Angelo, M. F., Reinesf, A. G., & Ramos, A. J. (2011). S100B alters neuronal survival and dendrite extension via RAGE-mediated NF-κB signaling. *Journal of Neurochemistry, 117*(2). https://doi.org/10.1111/j.1471-4159.2011.07207.x.

Vyas, D., Laput, G., & Vyas, A. K. (2014). Chemotherapy-enhanced inflammation may lead to the failure of therapy and metastasis. *OncoTargets and Therapy, 7*. https://doi.org/10.2147/OTT.S60114.

Wadler, S., Benson, A. B., Engelking, C., Catalano, R., Field, M., Kornblau, S. M., et al. (1998). Recommended guidelines for the treatment of chemotherapy-induced diarrhea. *Journal of Clinical Oncology, 16*(9), 3169–3178. https://doi.org/10.1200/JCO.1998.16.9.3169.

Wang, Q., Li, Z., Liu, K., Liu, J., Chai, S., Chen, G., et al. (2021). Activation of the G protein–coupled estrogen receptor prevented the development of acute colitis by protecting the crypt cell. *Journal of Pharmacology and Experimental Therapeutics, 376*(2), 281–293. https://doi.org/10.1124/jpet.120.000216.

Wei, L., Wen, X. S., & Xian, C. J. (2021). Chemotherapy-induced intestinal microbiota dysbiosis impairs mucosal homeostasis by modulating toll-like receptor signaling pathways. *International Journal of Molecular Sciences, 22*(17), 9474. https://doi.org/10.3390/ijms22179474.

Wu, C. H., Ko, J. L., Liao, J. M., Huang, S. S., Lin, M. Y., Lee, L. H., et al. (2019). D-methionine alleviates cisplatin-induced mucositis by restoring the gut microbiota structure and improving intestinal inflammation. *Therapeutic Advances in Medical Oncology, 11*. https://doi.org/10.1177/1758835918821021. 175883591882102.

Wu, J., Gan, Y., Li, M., Chen, L., Liang, J., Zhuo, J., et al. (2020). Patchouli alcohol attenuates 5-fluorouracil-induced intestinal mucositis via TLR2/MyD88/NF-kB pathway and regulation of microbiota. *Biomedicine & Pharmacotherapy, 124*, 109883. https://doi.org/10.1016/j.biopha.2020.109883.

Wu, Y., Wu, J., Lin, Z., Wang, Q., Li, Y., Wang, A., et al. (2021). Administration of a probiotic mixture ameliorates cisplatin-induced mucositis and pica by regulating 5-HT in rats. *Journal of Immunology Research, 2021*, 1–16. https://doi.org/10.1155/2021/9321196.

Wu, Z., Han, X., Qin, S., Zheng, Q., Wang, Z., Xiang, D., et al. (2010). Interleukin 1 receptor antagonist reduces lethality and intestinal toxicity of 5-fluorouracil in a mouse mucositis model. *Biomedicine & Pharmacotherapy, 64*(9), 589–593. https://doi.org/10.1016/j.biopha.2010.06.006.

Wu, Z., Han, X., Qin, S., Zheng, Q., Wang, Z., Xiang, D., et al. (2011). Interleukin 1 receptor antagonist reduces lethality and intestinal toxicity of 5-Fluorouracil in a mouse mucositis model. *Biomedicine & Pharmacotherapy, 65*(5), 339–344. https://doi.org/10.1016/j.biopha.2011.04.013.

Yamamoto, H., Ishihara, K., Takeda, Y., Koizumi, W., & Ichikawa, T. (2013). Changes in the mucus barrier during cisplatin-induced intestinal mucositis in rats. *BioMed Research International, 2013*, 1–8. https://doi.org/10.1155/2013/276186.

Yeung, C. Y., Chan, W. T., Jiang, C. B., Cheng, M. L., Liu, C. Y., Chang, S. W., et al. (2015). Amelioration of chemotherapy-induced intestinal mucositis by orally administered probiotics in a mouse model. *PLoS One, 10*(9), e0138746. https://doi.org/10.1371/journal.pone.0138746.

Yoneda, J., Nishikawa, S., & Kurihara, S. (2021). Oral administration of cystine and theanine attenuates 5-fluorouracil-induced intestinal mucositis and diarrhea by suppressing both

glutathione level decrease and ROS production in the small intestine of mucositis mouse model. *BMC Cancer, 21*(1), 1343. https://doi.org/10.1186/s12885-021-09057-z.
Zhao, Y., Chen, F., Wu, W., Sun, M., Bilotta, A. J., Yao, S., et al. (2018). GPR43 mediates microbiota metabolite SCFA regulation of antimicrobial peptide expression in intestinal epithelial cells via activation of mTOR and STAT3. *Mucosal Immunology, 11*(3), 752–762. https://doi.org/10.1038/mi.2017.118.
Zidan, J., Haim, N., Beny, A., Stein, M., Gez, E., & Kuten, A. (2001). Octreotide in the treatment of severe chemotherapy-induced diarrhea. *Annals of Oncology, 12*(2), 227–229. https://doi.org/10.1023/A:1008372228462.

CHAPTER FIVE

Cardiac complications of cancer therapies

Adolfo G. Mauro, Katherine Hunter, and Fadi N. Salloum*
Division of Cardiology, Department of Internal Medicine, Virginia Commonwealth University, Pauley Heart Center, Richmond, VA, United States
*Corresponding author: e-mail address: fadi.salloum@vcuhealth.org

Contents

Abstract

The quest of defeating cancer and improving prognosis in survivors has generated remarkable strides forward in research and have advanced the development of new antineoplastic therapies. These achievements, combined with rapid screening and early detection, have considerably extended the life expectancy of patients surviving multiple types of malignancies. Consequently, chemotherapy-related toxicity in several organ systems, especially the cardiovascular system, has surfaced as one of the leading causes of morbidity and mortality among cancer survivors. Recent evidence classifies chemotherapy-induced cardiotoxicity as the second-leading cause of morbidity and mortality, closely comparing with secondary cancer malignancies. While a certain

Advances in Cancer Research, Volume 155
ISSN 0065-230X
https://doi.org/10.1016/bs.acr.2022.03.006

degree of cardiotoxicity has been reported to accompany most chemotherapies, including anthracyclines, anti-metabolites, and alkylating agents, even the latest targeted cancer therapies such as immune checkpoint inhibitors and tyrosine kinase inhibitors have been associated with acute and chronic cardiac sequelae. In this chapter, we focus on describing the principal mechanism(s) for each class of chemotherapeutic agents that lead to cardiotoxicity and the innovative translational research approaches that are currently being explored to prevent or treat cancer therapy-induced cardiotoxicity and related cardiac complications.

Abbreviations

ABC adenosine triphosphate binding cassettes
AIF apoptosis-inducing factor
APC antigen-presenting cells
ATP adenosine triphosphate
Bax BCL2 associated X, apoptosis regulator
BCR breakpoint cluster region
CBR carbonyl reductases
CHF congestive heart failure
CML chronic myeloid leukemia
CTLA4 cytotoxic T lymphocytes-associated protein 4
CYBA cytochrome *b*-245, alpha polypeptide
DNA deoxyribonucleic acid
DOX doxorubicin
EDTA ethylenediaminetetraacetic acid
EF ejection fraction
EGFR epidermal growth factor receptor
ER endoplasmic reticulum
ETC electron transport chain
FAK focal adhesion kinase
FGFR fibroblast growth factor receptor
GDP guanosine diphosphate
GLS global longitudinal strain
GPX4 protein glutathione peroxidase 4
GTP guanosine triphosphate
Gy gray
HER2 human epidermal growth factor receptor 2
hERG ether-ago-go channel
HF heart failure
HIF1-α hypoxia inducible factor alpha
HO-1 heme oxygenase 1
HSF-1 heat shock factor-1
ICI immune checkpoint inhibitors
IFN-γ interferon gamma
IKr acting delayed rectifier potassium channel
IL-1β interleukin-1 beta
IL-6 interleukin-6

IRP	iron regulatory protein
IRP	iron regulatory proteins
IκB	Ikappab kinase
JAK	Janus kinase
LVEF	left ventricular ejection fraction
mAbs	monoclonal antibodies
MAP	mitogen-activated protein
MHC	major histocompatibility complex
MRI	magnetic resonance imaging
mtDNA	mitochondrial DNA
NADPH	nicotinamide adenine dinucleotide NAD + phosphate (P) + hydrogen (H)
NCF4	neutrophil cytosolic factor 4
NFAT	nuclear factor of activated T-lymphocytes
NF-κB	nuclear factor-Kb
NLRP3	NOD-, LRR- and pyrin domain-containing protein 3
nrf-2	nuclear factor erythroid 2-related factor 2
NRTK	non-receptor tyrosine kinase
NSLCLC	non-small cell-lung carcinoma
PD-1	programmed cell death 1
PDGFR	platelet-derived growth factor receptor
RAC2	Ras-related C3 botulinum toxin substrate 2
RIPC	remote ischemic preconditioning
RNA	ribonucleic acid
RNS	reactive nitrogen species
ROS	reactive oxygen species
RT	radiation therapy
RTK	receptor tyrosine kinase
SERCA	sarco/endoplasmic reticulum Ca^{2+} ATPase
TCR	T cell receptor
TGF-β	transforming growth factor beta
TKI	tyrosine kinase inhibitors
TLRs	toll-like receptors
TNF-α	tumor necrosis factor alpha
TopII	topoisomerase II
VEGFR	vascular endothelial growth factor receptor

1. Introduction

Defeating cancer and improving the prognosis of patients with cancer malignancies is the primary goal for the oncology field, as cancer remains the second leading cause of mortality in the United States

(Siegel, Miller, & Jemal, 2019). A continued effort in advancing research and developing new anticancer therapies, combined with early screening, has significantly extended longevity and improved quality of life for a majority of cancer patients. However, the progressively increasing number of people living with a history of cancer; i.e., cancer survivors, presents challenges associated with long-term treatment-related toxicities, which have become a substantive concern among long-term survivors (Siegel, Miller, & Jemal, 2019).

Cardiotoxicity resulting as a secondary consequence of chemotherapy is a form of cardiovascular system toxicity occurring after the administration of drugs targeting tumor growth and promoting patient survival. The way cardiotoxicity has been defined is changing over time (Pardo Sanz & Zamorano, 2020). One of the initial assessments of cancer-related cardiotoxicity is a marked decrease in left ventricular ejection fraction (LVEF). However, due to the limitations of this measurement (i.e., intra- and inter-observer variability), new cardiac biomarkers and new imaging techniques are being studied to detect early signs of cardiac injury.

Cardiotoxicity is divided into two distinct categories: (i) type I cardiac dysfunction, when irreversible structural myocardial damage occurs, and (ii) type II ventricular dysfunction, which does not manifest with myocardial rearrangements and is reversible after treatment discontinuation (Herrmann, 2020; Herrmann et al., 2014; Madeddu et al., 2016).

The timeline in which the side effects occur varies significantly; increased acute coronary syndromes and myopericarditis as well as different manifestations of ventricular arrhythmias can become evident as early as a few weeks from the initiation of the antineoplastic regimen. On the other hand, long-term survivors, although asymptomatic through the entire remission phase, can experience clear signs of systolic and diastolic dysfunction, which create concerns regarding a possible progression to a more severe congestive heart failure (CHF) phenotype (Herrmann, 2020).

Currently, toxicity to the cardiovascular system is becoming the leading cause of morbidity and mortality among patients surviving a cancer malignancy (Cadeddu et al., 2016). Several factors need to be considered with respect to the incidence of cardiotoxicity, including the antineoplastic agent(s) used, the dosage, and the chosen treatment regimen (Table 1). Furthermore, age, preexisting cardiovascular pathologies, and combined treatments, such as the inclusion of radiation therapy, raise the risk for each individual (Herrmann, 2020; Minotti, Salvatorelli, & Menna, 2010).

Table 1 Summary of cardiac complications of cancer therapies.

Chemotherapeutic agent	Summary	Molecular targets	Cardiotoxic effects
Anthracyclines	Anthracyclines are fungal-derived, cytostatic antibiotics that are widely used against solid tumors (e.g., lymphomas, sarcomas, breast cancers, and hematologic cancers). They exhibit a dose-dependent cardiotoxicity.	Anthracyclines inhibit DNA and RNA synthesis to stop replication by binding to topoisomerase II. This interaction affects the ability to relax supercoils in DNA using double-stranded breaks (Cadeddu et al., 2016)	Cardiomyopathy
Tyrosine Kinase Inhibitors (TKIs)	Tyrosine Kinases are enzymes that phosphorylate specific tyrosine residues on proteins to transduce signals through the cellular membrane (Pardo Sanz & Zamorano, 2020; R. L. Siegel et al., 2019; Weiss, Sarosy, Clagett-Carr, Russo, & Leyland-Jones, 1986) Cancer cells often have mutated tyrosine kinases that are continuously activated, altering pathways controlled by TKs like those controlling cellular death and proliferation. (Mondal, Jain, Aronow, & Frishman, 2019; Nitiss, 2009) Tyrosine kinase inhibitors (TKIs) are irreversible inhibitors of tyrosine kinases	TKIs allosterically inhibit the enzyme by covalently binding to the ATP-binding site of the tyrosine kinase. There are five classes of TKIs according to their inhibition strategy: Type I inhibits by binding to the ATP pocket of TKs, Type II binds a hydrophobic pocket adjacent to the ATP pocket of the inactive kinase, Type III and IV are noncompetitive, allosteric inhibitors, and Type V span two regions of the kinase.(Mercuro et al., 2007)	Cardiomyopathy, Arterial vascular disease, Arrhythmia, Venous thromboembolism, Systemic hypertension, Pulmonary hypertension

Continued

Table 1 Summary of cardiac complications of cancer therapies.—cont'd

Chemotherapeutic agent	Summary	Molecular targets	Cardiotoxic effects
Fluorouracil and antimetabolites	Antimetabolites are the oldest class of cancer-treating drugs. To exert their antitumor effect, they target rapidly dividing cells and promote apoptosis (Mondal et al., 2019)	These molecules mimic cell metabolites used in genetic replication and induce apoptosis by causing irreparable DNA damage	Cardiomyopathy, Arterial vascular disease, Pericardial disease
Immune therapies	Immune therapies are used to combat the immune evasion employed by cancer cells. These therapies can take many forms, including the use of cytokines, immune checkpoint inhibitor blockades, adoptive cell therapies, vaccines that trigger immune responses, and targeted oncolytic viruses (Lyon, Yousaf, Battisti, Moslehi, & Larkin, 2018)	For example, immune checkpoint inhibitors like CTLA4 are molecules that act as an antagonist to antigen-presenting cell ligands, preventing the activation of T cells. To oppose this immune suppression, monoclonal antibodies like Ipilimumab and Tremelimumab target CTLA4 to promote T cell activity	Cardiomyopathy, Arterial vascular disease, Arrhythmia, Venous thromboembolism, Pulmonary hypertension, Pericardial disease
Radiation therapy	Radiation is a prevalent cancer therapy that is used in over 60% of cancer patients, often in combination therapy (Stewart, Seemann, Hoving, & Russell, 2013)	Radioactive particles are directed to tumor cells that cause DNA damage and immediate cell death (Groarke et al., 2014)	Cardiomyopathy, Arterial vascular disease, Arrhythmia, Pulmonary hypertension, Pericardial disease, Valvular heart disease

Alkylating agents	Alkylating agents are nonspecific antitumor molecules that interrupt replication. Due to their generality, they cause significant side effects to other tissues, particularly to the GI mucosa, bone marrow, and reproductive system.	Alkylating agents permanently cross-link DNA strands by creating an aziridinium cation on guanine, preventing transcription and replication. The inability of the cell to separate DNA strands leads to apoptosis.	Cardiomyopathy, Arterial vascular disease, Arrhythmia
Proteasome inhibitors	Proteasomes are protein complexes that cleave malfunctioning proteins, and inhibitors of these complexes display antitumor activity. While the mechanism of this treatment is not fully understood, it may be due to the degradation of signaling proteins that would ordinarily promote cell survival and proliferation	Proteasome inhibitors bind to the catalytic subunits of proteasomes irreversibly.	Cardiomyopathy, Arterial vascular disease, Systemic hypertension

2. Cancer therapeutic agents

2.1 Anthracyclines

In investigating new antineoplastic drugs from soil-based microbes, an Italian pharmaceutical company in the early sixties identified, near the region of Castel del Monte (Apulia), a new bacterial strain in the genus Streptomyces, the *Streptomyces paucities*, able to produce a novel antibiotic called Daunorubicin (Cassinelli, 2016). The effort to overcome side effects and rapid evolution of drug-resistance has culminated in the discovery of other anthracycline family members such as Epirubicin and Idarubicin (Broxterman, Gotink, & Verheul, 2009; Shaul et al., 2013). Currently, anthracyclines continue to be utilized extensively due to their broad-spectrum cytotoxicity against solid tumors such as lymphomas, sarcomas, and breast cancer, and hematologic cancers including myeloid and lymphoid leukemias. Anthracyclines directly affect the ability of topoisomerase II to produce cleaved single and double-stranded DNA during the transcription and replication of genomic material (Nitiss, 2009). Anthracycline derivatives differ from the initial chemical structure by substituting patterns on the anthraquinone segment or the carbohydrate unit (Weiss et al., 1986). Unfortunately, this entire class of compounds shares the same limitations, including their dose-dependent cardiotoxicity, which compromises their therapeutic benefit.

The timeline over which anthracycline cardiotoxicity manifests can vary tremendously, and the effects of anthracyclines on the cardiovascular system are broad and often categorized based on the time of presentation of symptoms. Acute cardiotoxicity symptoms can manifest within hours or days from administration of the chemotherapy, often evidenced by disturbances in the heart electrical conduction system such as repolarization changes and atrial and ventricular arrhythmias. Many of the acute sequelae, including myopericarditis, have a benign course and tend to resolve spontaneously (McGowan et al., 2017; Mondal et al., 2019). In contrast, long-term chronic toxicity often presents as asymptomatic and subclinical for years. Particularly for oncology patients in the pediatric age group, cardiac damage may not become apparent until years or even decades after a patient has received a cardiotoxic chemotherapeutic treatment (Bansal, Amdani, Lipshultz, & Lipshultz, 2017). The ostensible reason for these delayed effects resides in the fact that the heart possesses significant functional reserve, which might limit the detection of cardiac damage as appreciable alterations of the systolic or diastolic function. This translates into patients who carry out an everyday life style without any symptoms for decades after the initial treatment that can abruptly progress to cardiomyopathy and overt CHF (Cadeddu et al., 2016; McGowan et al., 2017; Mondal et al., 2019).

The most critical determinant of anthracycline toxicity is the cumulative dose. The incidence of CHF following doxorubicin (DOX) administration is nearly 4–7% when patients receive 400–550 mg/m^2, 18% in those receiving up to 700 mg/m^2, and 36% in patients receiving more than 701 mg/m^2 (Cadeddu et al., 2016; Mercuro et al., 2007; Swain, Whaley, & Ewer, 2003). With a progressive increase in cumulative dose, irreversible myocyte injury becomes even more evident and results in progressive LV adverse remodeling and impaired contractility until heart failure symptoms appear (Fig. 1) (Schwartz et al., 1987).

Several additional factors can detrimentally affect the risk of cardiotoxicity, including sex (i.e., being female), pre-existing cardiac complications (particularly hypertension and obesity), age (patients 65 years and older experience a higher incidence of cardiac complications), and concomitant administration of

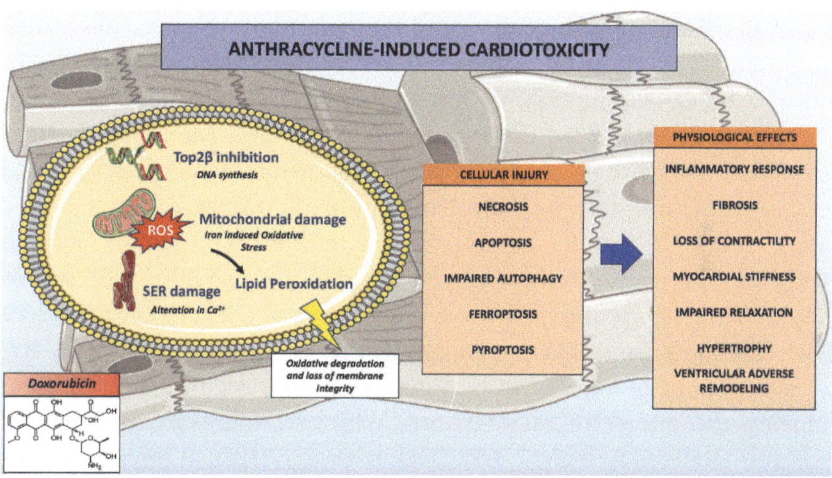

Fig. 1 Mechanisms of cardiotoxicity induced by anthracycline administration. Schematic showing the principal myocardial cytotoxic effects following the administration of anthracyclines. Anthracyclines, such as doxorubicin, induce inhibition of topoisomerase IIβ, contribute to mitochondrial dysfunction, and generate reactive oxygen species (ROS). The oxidative stress generated, impairs the smooth endoplasmic reticulum (SER) with consequent intracellular altered calcium handling and lipid peroxidation. The cellular injuries accumulated post-anthracycline exposure, activate several cell death pathways with detrimental effects on the total heart physiology.

mediastinal radiotherapy or further pharmacological chemotherapy agents. Genetic polymorphisms may also increase the susceptibility of the individual patient to anthracycline-induced cardiotoxicity (Higgins, O'Halloran, & Chang, 2015; Mondal et al., 2019). Multiple lines of evidence support the conclusion that modifications in genes coding for several adenosine triphosphate binding cassettes (ABC) correlate with an increased risk of cardiotoxicity (Leong, Chaiyakunapruk, & Lee, 2017). Additionally, a number of small nuclear polymorphisms in Carbonyl reductases (CBR), Cytochrome b-245, alpha polypeptide (CYBA), Neutrophil cytosolic factor 4 (NCF4), and Ras-Related C3 Botulinum Toxin Substrate 2 (RAC2) genes contribute to the variability among individuals in developing cardiac complications following anthracycline treatment (Leong et al., 2017). Unfortunately, individual variability and differences in drug metabolism create difficulties in determining and generalizing the maximum permitted effective dosage of the anthracyclines necessary to achieve the optimal antineoplastic effects without harming the cardiovascular system.

2.1.1 Mechanisms of anthracycline cardiotoxicity

2.1.1.1 ROS and oxidative damage

The most widely cited and recognized mechanism of anthracycline-induced cardiotoxicity involves the generation of reactive oxygen species (ROS)—and to a lesser extent, reactive nitrogen species (RNS)—and subsequent oxidative stress promoted through several pathways (McSweeney, Bozza, Alterovitz, & Zhang, 2019). ROS/RNS accumulation results in the death, dysfunction, and chronic inflammation of cardiomyocytes that exemplifies cardiotoxicity, by causing mitochondrial dysfunction, damaging DNA and proteins, weakening cell membranes via lipid peroxidation, and inducing apoptotic pathways (Fabiani et al., 2021; Octavia et al., 2012; Wenningmann, Knapp, Ande, Vaidya, & Ait-Oudhia, 2019). ROS can also alter gene expression by interacting with regulatory proteins and inducing the release of calcium ions, which activate calpains to initiate apoptosis (Kalivendi et al., 2005; Octavia et al., 2012).

One source of ROS generation results directly from the redox cycling of anthracyclines, catalyzed by several enzymes such as NADH dehydrogenase (Complex I), xanthine oxidases, NADPH oxidases, uncoupled nitric oxide synthases, and peroxisomes; many of which are found along the electron transport chain (ETC) in the inner mitochondrial membrane (Cappetta et al., 2017; Ju et al., 2006; Wenningmann et al., 2019). Anthracyclines contain a quinone moiety—key to its redox cycling—which can be reduced into an intermediate semiquinone free radical, which then forms the superoxide radical in the presence of oxygen and restores the original quinone; therefore, small levels of anthracycline can produce much larger amounts of ROS in cardiomyocytes (Octavia et al., 2012; Vejpongsa & Yeh, 2014). Redox cycling can also be catalyzed by manganese superoxide dismutase to produce hydrogen peroxide, which can be converted to reactive hydroxyl radicals in the presence of iron or copper (Octavia et al., 2012). The interaction of anthracyclines with these enzymes is aggravated by the localization of anthracyclines to the inner mitochondrial membrane due to a strong association between anthracyclines and cardiolipin, a phospholipid of the inner mitochondrial membrane (Octavia et al., 2012; Wenningmann et al., 2019). This association interferes with complexes I, III, and IV of the electron transport chain, which all require cardiolipin for optimal activity, and the redox cycling of doxorubicin when bound to cardiolipin strengthens their association (Wenningmann et al., 2019). Since mitochondrial DNA (mtDNA) is responsible for producing ETC components, ROS-mediated damage to mtDNA can inhibit ETC protein production, contributing to mitochondrial dysfunction (Wenningmann et al., 2019). ROS accumulation and mitochondrial dysfunction leads to the release of cytochrome *c* and other apoptotic factors, leading to cardiomyocyte death (Wenningmann et al., 2019). Cardiomyocytes are especially susceptible to this oxidative damage, because they have a high volume of mitochondria, occupying approximately 30% of the cell (Bernstein, 2018).

A non-enzymatic pathway exists that produces ROS directly from anthracyclines using intracellular iron as a reducing agent (McSweeney et al., 2019; Octavia et al., 2012). Normally, free iron levels in the cell are very low, but doxorubicinol (a metabolite of doxorubicin) forms complexes with the Fe-S group of a cytoplasmic iron regulatory protein (aconitase/iron regulatory proteins (IRP-1)), thereby preventing the translation of iron sequestration proteins and increasing free iron for more ROS production (Octavia et al., 2012; Wenningmann et al., 2019). Iron-mediated ROS generation has led to the use of dexrazoxane, an iron-chelator, as a treatment for cardiotoxicity (Russo, Bono, & Ghigo, 2021; Vejpongsa & Yeh, 2014).

Due to the consequences of ROS generation in anthracycline-induced cardiotoxicity, antioxidants were anticipated to prevent to some degree oxidative damage when used as a treatment, and they did show some promise both in vitro and in vivo (Vejpongsa & Yeh, 2014). However, these results have been inconsistent among chronic animal models (Octavia et al., 2012; Vejpongsa & Yeh, 2014; Zhang et al., 2012).

In addition to mitochondrial dysfunction, anthracycline-generated ROS can disrupt further cellular processes as well, including the induction of autophagy, senescence, and inflammation (Fabiani et al., 2021; Hou et al., 2021; Russo et al., 2021). Anthracyclines have been shown to increase the expression and activation of several inflammatory proteins and factors, including nuclear factor-kappa B (NF-κB) and tumor necrosis factor alpha (TNF-α) as well as toll-like receptors (TLRs) that are involved in the initiation of the innate immune response and cytokine secretion (Fabiani et al., 2021). Other anti-inflammatory factors, like cyclooxygenases and lipoxygenases, are repressed by anthracyclines (Fabiani et al., 2021). A related process is senescence, or cellular aging, which is characterized by halted cellular replication and production of proinflammatory cytokines, culminating in chronic inflammation (Hou et al., 2021).

2.1.1.2 Topoisomerase II β

It is established that anthracyclines induce their cytotoxic effects by binding to topoisomerase, an enzyme that regulates DNA over/under-winding using temporary double stranded breaks, but more recent studies have also implicated the enzyme in anthracycline-induced cardiotoxicity (Henriksen, 2018; Hou et al., 2021; Zhang et al., 2012). Doxorubicin binds to both topoisomerase II (TopII) and DNA to form a TopII-doxorubicin-DNA cleavage complex causing permanent double-stranded breaks in the DNA to trigger cell death (Zhang et al., 2012). Other studies emphasize an anthracycline-TopII interaction that alters gene expression. There are two types of topoisomerase enzymes: TopIIα, which is found in proliferating cells and is overexpressed in tumor cells, and TopIIβ, which is found in quiescent cells like cardiomyocytes (Vejpongsa & Yeh, 2014; Zhang et al., 2012). Binding to TopIIα was thought to be the reason why anthracyclines target cancer cells, but they also appear to target and inhibit TopIIβ (Henriksen, 2018). Zhang et al. (2012) created a mouse model with a cardiomyocyte-specific TopIIβ knockout, $Top2\beta^{\Delta/\Delta}$, which was then treated with doxorubicin. After 16h, the mice were sacrificed, and gene expression was determined; and this analysis was repeated after 72h. Treatment with DOX at the 16-h time point led to the activation of the p53 tumor suppressor pathway and the beta-adrenergic pathway in the $Top2\beta^{+/+}$ mice, but not in the $Top2\beta^{\Delta/\Delta}$ strain. Also, only in the control mice were there increases in pro-apoptotic transcripts, like Trp53inp1, Apaf1, Bax, and Fas, potentially due to the double-stranded breaks caused by the DOX-TopII-DNA complex. In contrast to the shorter-term exposure, after the 72-h treatment mitochondrial function and oxidative phosphorylation pathways were suppressed in the control, but not in the Top2ß$^{\Delta/\Delta}$ group (Zhang et al., 2012). Other changes that only affected the control group included visible mitochondrial damage observed by electron microscopy, a disruption in mitochondrial membrane potential, and a reduced rate of oxygen consumption. These findings indicate that mitochondrial damage after DOX exposure may be due to transcriptome changes rather than ROS generation, as ROS generation was reduced by 70% in the knockout mice compared to controls. A chronic treatment study applying DOX over 5 weeks indicated a decrease in ejection fraction (EF) from 53% to 43% in the control mice while the knockout mice showed no change in EF. Mitochondrial abnormalities could be due to the down-regulation of *Ppargc1a* and *Ppargc1b* which are necessary for mitochondrial biogenesis and function (Ju et al., 2006). It is possible that the p53 pathway (activated by the double stranded breaks) causes this effect, but it could also be caused by a p53-independent pathway. Another study found that at low concentrations, DOX stabilizes the TopII-DNA cleavage complex and blocks ligation, but at high concentrations it acts as an intercalating agent which prevents TopII from binding to DNA, leading to transcriptional arrest and DNA damage (Vejpongsa & Yeh, 2014). However, it was found that TopII beta was not necessary for adult cardiomyocyte homeostasis (Zhang et al., 2012).

2.1.1.3 Apoptosis

Anthracycline-induced cardiotoxicity has been linked to both intrinsic, mitochondria-mediated apoptosis, and extrinsic, Fas/Fas L-mediated apoptosis (Kalivendi et al., 2005; Wenningmann et al., 2019). In the intrinsic pathway, the upregulation of proteins like Bax induces caspase-9 to activate the effector caspase-3 (Wenningmann et al., 2019). One study found that DOX-induced oxidative stress initiates the p53 pathways by activating heat shock factor-1 (HSF-1), leading to more HSP25, which stabilizes p53 (McSweeney et al., 2019; Octavia et al., 2012). This would show that anthracycline exposure can activate the p53 pathway both through the double-stranded DNA breaks caused by the DOX-TopII-DNA complex and through a ROS-mediated pathway originating from the redox cycling of anthracyclines. Moreover, McSweeney et al. (2019) conducted an in vitro study using human induced pluripotent stem cell-derived cardiomyocytes to study transcriptomic changes after DOX exposure and found that p53 and other members of the extrinsic apoptotic pathway were the most significantly upregulated gene cluster (McSweeney et al., 2019). Another source of apoptosis activation is due to anthracycline-induced intracellular calcium dysregulation (Kalivendi et al., 2005; Octavia et al., 2012; Wenningmann et al., 2019). DOX increases the probability that calcium-release channels in the sarcoplasmic reticulum will be in the open state, increasing intracellular calcium levels (Octavia et al., 2012). The presence of ROS also increases intracellular calcium (Kalivendi et al., 2005). The excess of calcium induces calpains (calcium-dependent proteases) to cleave caspase-12 which initiates apoptosis. Calpain activation also degrades titin (an important structural protein for muscle elasticity) which may be the cause of myofibrillar deterioration associated with DOX cardiotoxicity (Octavia et al., 2012). Calcium release also causes the calcineurin-dependent activation of the nuclear factor of activated T-lymphocytes (NFAT), a transcription factor that advances transcription of Fas L to induce an extrinsic apoptotic cascade by binding to the Fas receptor on the cell surface (Kalivendi et al., 2005). Also, doxorubicinol (the hydroxyl metabolite of DOX) affects the sarco/endoplasmic reticulum Ca^{2+} ATPase (SERCA) and the Na^{+}/K^{+} exchanger on the sarcolemma, causing an increase in intracellular calcium (Wenningmann et al., 2019). While caspase inhibitors would seem like an effective prevention strategy to inhibit apoptosis in anthracycline-treated cardiomyocytes, they do not completely prevent apoptosis due to a caspase-independent pathway involving mitochondrial apoptosis-inducing factor (AIF) (Wenningmann et al., 2019). Apoptosis appears to be an acute effect of anthracycline treatment: in a study by McSweeney et al. (2019), over 75% of the genes dysregulated by DOX exposure in cardiomyocytes in vitro returned to normal transcription levels after 7 days of DOX washout (Arola et al., 2000; McSweeney et al., 2019; Wenningmann et al., 2019).

In addition to cardiomyocytes, other cell types - which make up ⅔ of cells in the heart - are damaged by anthracyclines, contributing to cardiotoxicity (Cappetta et al., 2017). Cardiac progenitor cells are among the few cell types of the heart that replicate for tissue homeostasis and repair (Cianflone et al., 2020). ROS-induced DNA damage, cell cycle arrest, cellular senescence, and apoptosis occur in cardiac progenitor cells after treatment with anthracyclines, which impairs the heart's ability for repair (Cappetta et al., 2017). DOX has been shown to affect vascular cells—increasing their permeability—and smooth muscle cells—decreasing vessel relaxation. ROS generated by anthracyclines can induce cardiac fibroblast cells to stimulate cardiac fibrosis, an outcome that contributes to heart failure. Other DOX-affected cells include mesenchymal stem cells and endothelial progenitor cells, which warrant further research to define their potential contributions to DOX-induced cardiotoxicity (Cappetta et al., 2017).

2.1.1.4 Ferroptosis

Ferroptosis has been described as one of the mechanisms of cell death following anthracycline administration. This highly regulated, iron-dependent cell death process is characterized by polyunsaturated fatty acid peroxidation, constituting the membrane phospholipid tails (Hirschhorn & Stockwell, 2019). DOX can disturb the iron homeostasis increasing intracellular iron by directly inhibiting the activity of iron regulatory proteins 1 and 2 (IRP1 and IRP2) (Kwok & Richardson, 2002; Minotti, Salvatorelli, Menna, Ronchi, & Cairo, 2001). The inactivated form of these proteins can function as transcriptional factors regulating the synthesis of transferrin receptor 1 and ferritin, essential for the uptake and storage of the iron in cells. DOX can also directly bind to iron by forming DOX-iron(III) complexes, which lead to further ROS formation and cellular damage (Hirschhorn & Stockwell, 2019; Xu, Persson, & Richardson, 2005).

Animals fed iron-rich chow display a significant increase in cardiotoxicity following DOX administration (Panjrath et al., 2007). DOX was also found to increase iron content in cardiac mitochondria of patients (Ichikawa et al., 2014). Another protein affected by anthracyclines is the nuclear factor erythroid 2-related factor 2 (nrf-2). The activation of Nrf-2 is involved in regulating heme oxygenase 1 (HO-1), a protein essential for the heme group metabolism (Fang et al., 2019). Following DOX treatment, HO-1 can increase the labile iron pool increasing the mitochondrial membrane lipid peroxidation. Additionally, DOX reduces the level of the antioxidant protein glutathione peroxidase 4 (GPX4), which induces lipid peroxidation through the DOX–Fe^{2+} complex at the mitochondrial level (Tadokoro et al., 2020).

2.1.1.5 Pyroptosis

Pyroptosis is a proinflammatory type of cellular death that exhibits both the features of apoptosis and necrosis (Toldo & Abbate, 2018). Pyroptosis is mediated by caspase-1 activation and is associated with the signaling of the NOD-, LRR- and pyrin domain-containing protein 3 (NLRP3), which induces the release of proinflammatory interleukins such as interleukin-1 beta (IL-1β) and IL-18 through gasdermin D pores (Toldo, Mauro, Cutter, & Abbate, 2018). Evidence confirming the link between DOX and NLRP3 are still scarce; however, DOX administration facilitates the upregulation of NLRP3 inflammasome components and subsequent pyroptotic cell death both in vivo and in vitro (Dargani & Singla, 2019; Meng et al., 2019). Treatment with an inflammasome inhibitor has proven beneficial in a model of DOX-induced cardiac dysfunction, lowering myocardial fibrosis and preventing the decline in cardiac function (Marchetti et al., 2015).

2.2 Tyrosine kinase inhibitors

Tyrosine kinases are essential enzymes mediating several intracellular pathways, including cell proliferation, differentiation, migration, metabolism, and programmed cell death. There are more than 500 protein kinase genes in the human genome, representing one of the largest families of enzymes (Manning, Whyte, Martinez, Hunter, & Sudarsanam, 2002). Genetic alterations of these kinases are often responsible for tumor progression and development. Tyrosine kinases catalyze the phosphorylation of specific tyrosine amino acid residues on targeted proteins to initiate and translate intracellular signals (Force, Krause, & Van Etten, 2007).

There are three central components in cell signaling: reception, transduction, and response (Force et al., 2007). Specific ligands activate their corresponding receptor tyrosine kinase (RTK), which typically induces receptor dimerization (Lemmon & Schlessinger, 2010). The stable dimerized receptor will then induce trans-phosphorylation of tyrosine in the cytoplasmatic portion of the receptor, propagating the signal through the plasma membrane (Lemmon & Schlessinger, 2010). The most common signaling propagation pathway involves the triggered mitogen-activated protein (MAP) kinase cascade, which activates the small G protein Ras (Santarpia, Lippman, & El-Naggar, 2012). Ras is attached to the inner layer of the plasma membrane and, in its inactive state, is bound to a molecule of guanosine diphosphate (GDP). When SH2-containing proteins join with activated RTKs, Ras activates and binds guanosine triphosphate (GTP) in place of GDP. The GTP-bound Ras phosphorylates the first serine-threonine kinase in the MAP kinase cascade, leading to the phosphorylation cascade of downstream MAP kinases. As a final step, the phosphorylation of the targeted transcription factor leads to a change in gene transcription (Degirmenci, Wang, & Hu, 2020; Force et al., 2007; Madhusudan & Ganesan, 2004; Santarpia et al., 2012).

Insulin-like growth factors do not involve the activation of MAP kinases, but instead activate the phosphoinositide 3-kinase pathway, which phosphorylates inositol phospholipids in the cell membrane (Hakuno & Takahashi, 2018). Signal transducers and activators of transcription (STAT) are proteins involved in cytokine hormonal signaling, binding directly to the phosphorylated tyrosine in the receptors (Ma, Qin, & Li, 2020). Once activated, STAT proteins directly access the nucleus, inducing changes in transcription (Ma et al., 2020).

Tyrosine kinases are classified primarily by the specificity of their receptors and ligands. Extracellular domain receptors kinases include EGFR, PDGFR, FGFR, and VEGFR, while SRC, ABL, the focal adhesion kinase (FAK), and Janus belong to the family of non-receptor tyrosine kinase (NRTK), which are intracellular proteins with regulatory activities (Paul & Mukhopadhyay, 2012). The role of tyrosine kinases in regulating essential cellular functions, such as cell proliferation, differentiation, and migration, makes these specific proteins highly critical in the mutagenesis toward a neoplastic phenotype. Unfortunately, Tyrosine kinase proteins are often genetically altered in cancer cells, thus conferring a selective advantage to newly formed cancer cells (Erickson, 2010; Lemmon & Schlessinger, 2010).

Multicellular organisms have genetically evolved to possess a tightly controlled system capable of controlling each tyrosine activation with a system of antagonistic tyrosine kinase and tyrosine phosphatases. However, cancers often acquire phenotypical mutations, disrupting the regulatory function of cell proliferation and cellular death pathways, leading to constitutive activation of select tyrosine kinases (Paul & Mukhopadhyay, 2012).

Several mechanisms of gain of function have been identified in the tyrosine kinase pathways (Madhusudan & Ganesan, 2004). For example, mutations in the extracellular or intracellular domain, e.g., EGFRv III, have been recognized in glioblastomas, ovarian tumors, and non-small cell lung carcinoma (Paul & Mukhopadhyay, 2012). These mutations confer constitutive activity to the receptor despite the ligand's presence, resulting in uncontrolled proliferation (Nishikawa et al., 1994; Zwick, Bange, & Ullrich, 2002). Abnormal chromosomal translocations are also involved in cancer etiology (Castel, Rauen, & McCormick, 2020). In chronic myeloid leukemia (CML), the translocation of the portion of chromosome-22 onto chromosome-9 creates a new chromosome called Philadelphia (Harvey & Tasian, 2020). This chromosome abnormality fuses the Breakpoint cluster region (BCR) sequences with the tyrosine-protein kinase ABL. The corresponding protein generated, BCR-ABL, has uncontrolled tyrosine kinase activity, which is responsible for 95% of CML cases (Harvey & Tasian, 2020). Other examples of chromosome translocations are: the TEL-ABL tyrosine kinase, the product of the fusion between chromosome-9 and chromosome-12, leading to a constitutively phosphorylated kinase; and TEL-PDGF, a fusion product from the chromosome-2 and 5 in anaplastic large cell lymphoma (Lamore, Kohnken, Peters, & Kolaja, 2020; Metibemu et al., 2019).

The fusion protein is not the sole modality of kinase-gained function in cancer cells. The autocrine-paracrine loop can occur in the event of an uncontrolled overexpression of an RTK or its cognate ligand (Paul & Mukhopadhyay, 2012). Overexpressed EGFR, PDGFR, and IGF receptors are the culprits for a variety of human cancers. EGFR and EGF overexpression are responsible for bladder cancer, breast cancer, glioblastoma multiforme, and non-small cell lung cancer (NSCLC) (Paul & Mukhopadhyay, 2012). Overexpression of PDGFR, PDGF-A, and PDGF-B is responsible for many astrocytic brain tumors and gliomas, while Insulin-like growth factor receptors (IGFR) are implicated in the pathogenesis of prostate cancer, NSLCLC, and breast cancer (Paul & Mukhopadhyay, 2012).

There are two different strategies to inhibit targeted tyrosine kinases: small molecule inhibitors and monoclonal antibodies. Tyrosine kinase inhibitors (TKI) are molecules developed to covalently bind and block the ATP site or induce an enzymatic conformational change allosterically, resulting in irreversible inhibition of the tyrosine kinase activity. Monoclonal antibodies, instead, are humanized, human chimeric, or bispecific antibodies targeting the extracellular domain of the RTKs (Tong & Seeliger, 2015).

The Food and Drug Administration has approved more than 50 large and small molecule inhibitors of TKIs, and the majority are utilized in cancer treatment (Lamore et al., 2020). Based on the kinase inhibition modality, TKIs are classified as five types of inhibitors: (i) Type-I inhibitors are molecules that bind actively to the TKI's adenosine triphosphate (ATP) pocket, inhibiting access to the catalytic units of the enzyme. (ii) Type-II inhibitors instead bind to an inactive kinase occupying a hydrophobic pocket adjacent to the ATP binding site. These compounds are often more specific than type-I inhibitors; (iii) Type-III and Type-IV are non-competitive inhibitors, respectively binding within and outside the cleft between the small and large lobes of the enzyme, therefore acting as allosteric inhibitors. (iv) Type-V inhibitors instead are bivalent molecules that stretch through two regions of the protein kinase (Bhullar et al., 2018; Pottier et al., 2020). Despite a targeted approach, small molecule TKIs and inhibiting antibodies frequently carry unwanted side effects, including cardiac toxicity, significantly impacting patients' quality of life (Force et al., 2007).

2.2.1 Mechanisms of tyrosine kinase inhibitor cardiotoxicity

TKI is a broad class of inhibitors that target different pathways; therefore, it is not surprising that the side effects reported during their clinical use can manifest as a plethora of diverse cardiovascular complications. The major cardiovascular adverse events registered include QT prolongation, reduced cardiac contractility, hypertension, congestive heart failure, and increased incidence of myocardial infarction (Force et al., 2007; Kerkelä et al., 2006; Lamore et al., 2020). The mechanism proposed for cardiac toxicity does not always correlate directly with the intended target, but it is often specific for the single drug used, which is related to the rather high incidence of off-target effects of TKI (Lamore et al., 2020).

TKIs blocking those kinases essential for the upkeep of cardiomyocyte homeostasis or for the cells constituting the vasculature bed, will have sides effect specific for the target. For example, TKI targeting VEGF, VEGFR, HER2, and Bcr-Abl are responsible for a higher incidence of hypertension, left ventricular dysfunction, and heart failure (Chen, Kerkelä, & Force, 2008; Gharwan & Groninger, 2016; Touyz & Herrmann, 2018). Meanwhile, TKIs inhibiting c-Jun N-terminal kinase, pyruvate dehydrogenase kinase, and protein kinase A have been instead linked to cardiomyocyte-specific mitochondrial dysfunction leading to cellular abnormalities and myocardial apoptotic cell death (Force et al., 2007; Thomson, 2002).

2.2.1.1 QT prolongation

QT prolongation refers to delayed ventricular repolarization predisposing the myocardium to the development of an early after-depolarization, which can cause lethal cardiac arrhythmias and torsade de pointes (Malik & Camm, 2001). QT prolongation is an example of off-targeted side effects of TKI administration, which inhibit the activity of the ether-ago-go channel (hERG), an essential pore subunit responsible for the acting delayed rectifier potassium channel (IKr) and involved in cardiac repolarization (Chen, Sampson, & Kass, 2016). TKIs that have been proven to cause QT prolongation include Dasatinib, an inhibitor of Bcr-Abl, c-Kit, platelet-derived growth factor receptor (PDGFR), and members of the SCR family (Force et al., 2007). Dasatinib is approved for treating chronic myeloid leukemia and Philadelphia chromosome-positive acute lymphoblastic leukemia (Force et al., 2007). Dasatinib was found to interfere with hERG with an indirect effect on prolonging the QT interval (Force et al., 2007; Malato et al., 2012). Nilotinib, which targets PDGFRα/β, c-kit, and Bcr-Abl, increased QT intervals by more than 500 ms as compared to the patients pre-chemotherapy. Interestingly, in subsequent clinical trials for nilotinib and dasatinib, the exclusion in the enrollment of patients with present impaired cardiac function resulted in no signs of QT prolongation in the cohort of patients studied (Breccia & Alimena, 2013; Kantarjian et al., 2011). This result may potentially prove useful in identifying patients with a higher risk of cardiac complications based on their progressed health history. Sunitinib, a TKI targeting VEGFR1–3, PDGFRα/β, and c-kit, has proven to be disruptive to the trafficking movement of the hERG channel from the endoplasmic reticulum (ER). Treatment with Sorafenib, an inhibitor of BRAF, VEGFR, the PDGFR family, c-kit, clinically showed evidence of electrocardiogram abnormalities in patients (Tolcher et al., 2011). Ponatinib, a pan-Abl inhibitor, and Vemurafenib, which targets BRAF, were also associated with arrhythmias in treated patients (Herrmann, 2020). Additionally, Trastuzumab, a monoclonal antibody binding the human epidermal growth factor receptor 2 (HER2), approved for the adjuvant treatment of patients with HER2-positive early breast cancer, was also found responsible for QT prolongation, possibly as a consequence of increased oxidative stress (Dirican et al., 2014).

2.2.1.2 Left ventricular dysfunction

Left ventricular dysfunction is an impairment in the ability of the heart to effectively pump blood systemically in the body. The echocardiogram is the elected, noninvasive technique to assess cardiac contractility by measuring the percentage of LVEF, an indirect measurement of the amount of blood pumped from heart into the system each cardiac cycle. Impaired contractility leads to a decline in LVEF, approximately in the range of 55%–75% in an average individual (Khouri et al., 2012). In the field of cardio-oncology, a reduction $\geq$10% compared to baseline would prompt investigation and monitoring. A decline in LVEF is an early sign of a possible progression to cardiac failure (Cardinale, Iacopo, & Cipolla, 2020). A declined cardiac contractility was evident in patients treated with imatinib, an inhibitor of the PDGFRα/β, Bcr-Abl and c-kit (Kerkelä et al., 2006). Imatinib is approved by the FDA for the treatment of CML (Lamore et al., 2020). Imatinib showed a reduction in LVEF of approximately 25% compared to before initiating treatment. This cardiac complication seems to be caused by excessive unfolded protein accumulation into the endoplasmic reticulum (Chen et al., 2008). Reversible decline in LVEF was observed following the treatment with HER2 inhibitors;

Trastuzumab and lapatinib (Nemeth, Varga, Wu, & Pacher, 2017). To increase the cytotoxic effect in women with breast cancer, Trastuzumab was administered with an anthracycline. Unfortunately, excessive rates of cardiac toxicity were observed due to the combination of both drugs, thus exacerbating cardiotoxic sides effects (Montemurro, 2014).

2.2.1.3 Hypertension

Higher arterial mean pressure > 140 mmHg (systolic) and >90 mmHg (diastolic) can induce deleterious damage to the cardiocirculatory system. Bevacizumab, a monoclonal antibody targeting VEGF-A, regorafenib, an inhibitor of the VEGFR2 and sorafenib, have been associated with increased blood pressure in more than 30% of treated patients who experienced systolic blood pressure ≥ 160 mmHg or diastolic ≥100 mmHg, which then required pharmacological intervention (Lamore et al., 2020). Severe hypertension was reported after using sunitinib, an inhibitor of VEGFR, PDGF-R and c-kit used in patients affected by renal cell carcinoma and imatinib-resistant gastrointestinal stromal tumors. Sunitinib caused hypertension in 47% of the patients treated (León-Mateos, Mosquera, & Antón Aparicio, 2015). Both Trastuzumab and momelotinib, a Janus kinase (JAK)1/2 inhibitor, were associated with an increased risk of developing hypertension in more than 20% of the patients (Chen et al., 2008; Verstovsek et al., 2017).

2.3 Fluorouracil and other anti-metabolites

Anti-metabolites are the oldest class of largely cytotoxic antitumor drugs. Their mechanism of action generally involves small molecules that mimic metabolites used by tumor cells to replicate their genetic material (Mondal et al., 2019; Tiwari, 2012). Once incorporated, these molecules inhibit DNA synthesis, activating intracellular signaling for apoptotic cell death (Tiwari, 2012). Anti-metabolites currently in clinical use include 6-mercaptopurine, fludarabine, 5-fluorouracil, gemcitabine, cytarabine, methotrexate and pemetrexed. However, the most commonly used are 5-fluorouracil and 5-fluorodeoxyuridine to treat head and neck and gastrointestinal tumors (Sara et al., 2018). 5-fluorouracil and 5-fluorodeoxyuridine are thymidylate synthase inhibitors that inhibit DNA synthesis and arrest the cell cycle in the S-phase during DNA synthesis (Tiwari, 2012).

2.3.1 Fluorouracil and anti-metabolites induced cardiac toxicity

The significant use of anti-metabolites to treat cancer makes fluoropyrimidines the second most common cause of chemotherapy-related cardiotoxicity (Saif, Shah, & Shah, 2009). The 5-fluorouracil, one of the most commonly used antimetabolites, was found to increase the incidence of coronary thrombosis, arterial inflammation, and vasospasms. The incidence of cardiotoxicity following the use of 5-fluorouracil varies considerably, ranging from 2 to 4%, especially in patients with pre-existing conditions (Sara et al., 2018). The risk of experiencing cardiac-related side effects also seems to increase under the co-administration of other chemotherapies, including cisplatin (Madeddu et al., 2016). The mechanism of toxicity of 5-fluorouracil appears to be related to the enzyme involved in the drug's metabolism: namely, thymidine phosphorylase. Thymidine phosphorylase is responsible for the upregulation of endothelin 1, a potent vasoconstrictor involved in the arterial stiffening and responsible for vasospasms and ischemia (Zanoli et al., 2019). 5-fluorouracil and capecitabine are also potent inducers of oxidative stress and lipidic peroxidation, with consequent activation of apoptotic and necroptotic pathways in the myocardium (Focaccetti et al., 2015; Macdonald, 1999). Anti-metabolites have been studied in animal models, revealing lysosomal membrane leakage in cardiomyocytes and impairment of the energy-producing Krebs cycle while depleting high-energy phosphate production (Kosmas et al., 2008; Sara et al., 2018).

2.4 Immune therapies

Inhibiting the mechanisms used by tumors to evade activated T lymphocytes has recently generated great interest. These therapies have proven to change the care of cancer patients dramatically, especially those with challenging malignancies previously proven difficult to treat (Stein-Merlob, Rothberg, Ribas, & Yang, 2021). Several types of immunotherapies exist, including the use of cytokines (i.e., Interleukin-2 and Interferons), immune checkpoint inhibitors (ICI) blockade, adoptive cell therapies using activated chimeric T cells, or natural killer cells, vaccines to trigger an immune response, and targeted oncolytic viruses (Zhang, Reynolds, Lyon, Palaskas, & Neilan, 2021).

The concept of immune therapy as a potential cure for cancer has its origins back in the 1980s with the first approved studies on interferons for the treatment of hairy cell leukemia performed in 1986. These studies were followed by consideration of interferons as an adjuvant therapy for melanoma in 1995 (Berraondo et al., 2019; Herndon et al., 2012). As part of the evolution of the disease, cancer cells adapt and are able to evade the immune system by suppressing anti-tumor T cell activity (O'Donnell, Teng, & Smyth, 2019). T cells are generated in the bone marrow from hematopoietic precursors and are characterized by a T cell receptor (TCR) formed by two glycoprotein chains: α- and β- (Courtney, Lo, & Weiss, 2018; Waldman, Fritz, & Lenardo, 2020). TCR is responsible for recognizing a particular antigen, and it is unique for each T cells clone. A multitude of α- and β-glycoproteins isoforms are formed by rearrangement of specific gene segments, which ultimately leads to the generation of an incredible repertoire of T cells clonotypes capable of recognizing a virtually unlimited number of antigens (Waldman et al., 2020). Naïve T cells then migrate to the thymus, where reactive T cells towards specific self-antigens are actively depleted to build a tolerant immune system (Waldman et al., 2020). TCR is responsible for activating T cells following interaction with the major histocompatibility complex (MHC) present on antigen-presenting cells (APC). A second co-stimulatory signal is essential for the transduction of the TCR and involves the coreceptor CD28, which interacts with the ligands B7–1 and B7–2 expressed on activated APC. The absence of CD28 colligation induces a non-responsive state or lymphocyte cell death (Rudd, Taylor, & Schneider, 2009).

Inhibitory "checkpoint" molecules exist to prevent hyperactivation of T cells, such as cytotoxic T lymphocytes-associated protein 4 (CTLA4) and the programmed cell death 1 (PD-1) (Zhang et al., 2021). The primary purpose of these molecules is to control the hyperactivity of naïve T cells, therefore, limiting T cell activation and proliferation (Fig. 2) (Waldman et al., 2020).

CTLA4 is induced following T cell activation, except for T regulatory cells, which constitutively express high levels of this molecule. CTLA has high sequence homology with CD28 and antagonizes the binding of CD28 with coreceptors B7-1 and B7-2 on APCs, therefore, preventing the restimulation between T cells and APC (Waldman et al., 2020). In an effort to increase anti-tumor activity of T cells against cancer, monoclonal antibodies (mAbs) against CTLA4 have been developed, such as Ipilimumab and Tremelimumab. Ipilimumab has been shown clinically to exert anti-tumoral activity on the progression of stage III/IV non-resectable melanoma, increasing short-term survival (Hodi et al., 2003, 2010). PD-1 is another ligand that shares more than 15% homology with CTLA4 and CD28, binding instead to the B7 homologs ligand PD-L1 and PD-L2

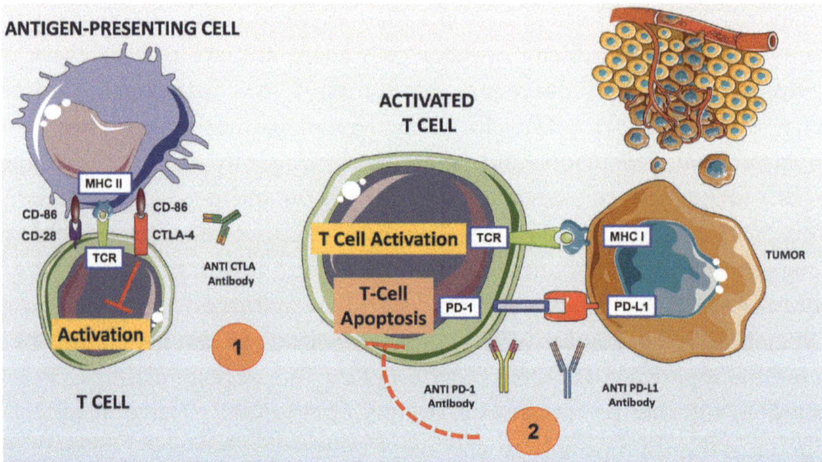

Fig. 2 T cell response to immune checkpoint inhibitors. Representation of the mechanisms used by tumor cells to evade immune systems by suppressing antitumor T cells. (1) Antigen-presenting cells activate T cells against tumor antigen presentation. Blockade of CTLA-4 co-stimulation by specific monoclonal antibodies increases the antitumor activity of T cells. (2) Tumor cells induce T cell apoptosis by exposing the programmed cell death (PD-1) and PD ligand 1 (PD-L1). The activation of the programmed cell death pathway limits T cell activation and proliferation.

(Zhang et al., 2021). While CTLA4 interaction is held predominantly in lymphoid organs, PD-1/PD-L1 contact occurs at the peripheral tissue level, i.e., later in the maturation of T cells. PD-1 interaction with its ligand induces a state of exhaustion in T cells characterized by inhibitory stimuli affecting both survival and proliferation (Wherry, 2011; Wherry & Kurachi, 2015). Unfortunately, tumor cells can cause exhaustion in T cells by using the PD-L1 to escape their anti-tumor activity. The engineering of monoclonal antibodies (mAbs) blocking the PD-1 axis has shown exciting results in enhancing tumor cytotoxicity from T cells (Iwai et al., 2002).

Pembrolizumab and nivolumab are two of the FDA-approved anti-human PD-1 mAbs to treat unresectable melanoma, with a remarkable increased survival for 72.9% of patients at 1 year compared to the standard of treatment (Robert et al., 2015). Pembrolizumab and nivolumab were also approved for the treatment of neck and head squamous cell carcinomas, gastric/esophageal cancers, and Hodgkin lymphoma (Waldman et al., 2020). From 2016 forward, humanized anti-PD-L1 mAbs were tested clinically, including atezoliumab, avelumab, and durvalumab, which were ultimately approved for urothelial carcinoma, non-small-cell lung cancer, triple-negative breast cancer and advanced renal carcinoma (Waldman et al., 2020).

2.4.1 IC-associated cardiotoxicities

Clinical adverse events from ICI therapy are rare, and the risk factors for immune-related adverse events are still poorly understood. Almost 70 to 90% of cancer patients treated with ICI experience side effects following the treatment, including encephalitis, hepatitis, and myositis (including myocarditis) (Mahmood et al., 2018; Zhang et al., 2021). The timeline and onset of symptoms varies, but typically, they occur acutely within 12 weeks from therapy initiation, with 15% percent of the patients experiencing severe manifestation and a significant impact on their quality of life (Zhang et al., 2021). The most threatening adverse event that arises from ICI therapy is myocarditis (Zhang et al., 2021). Myocarditis is characterized by an intense inflammatory reaction within the cardiac muscle. Pathologically, myocarditis is characterized by the presence of myocyte necrosis and mononuclear infiltrates. These infiltrates are constituted mainly of a lymphocytic phenotype with CD3, CD4 and CD8 positive cells and CD68 positive macrophages (Jain et al., 2018). The incidence of myocarditis in patients receiving ICI is not yet entirely determined and appears to be increasing; however, IC-associated myocarditis is an incredibly life-threatening complication with a mortality rate approaching 50% (Stein-Merlob et al., 2021). Compared with controls, patients who have been treated with a combination of ICI therapies are more likely to experience myocardial inflammation. The higher risk was associated with the combination of anti-CTLA4 and anti-PD-1 antibodies (Mahmood et al., 2018). From a study conducted on 20,594 patients, the incidence of myocarditis was reported to be 0.06% in patients receiving nivolumab as monotherapy, while patients that received a combination of therapies had an increased risk up to 0.27% (Johnson et al., 2016). In a concomitant clinical trial, other reports have found an increased incidence of myocarditis up to 1.14% in patients receiving both mono- and combined ICI therapies, with a median onset time of 34 days after receiving the treatment (Mahmood et al., 2018).

Unfortunately, ICI administration has been linked to other cardiovascular manifestations; from a meta-analysis study involving 22 clinical trials, it appears that monotherapy with PD-1 or PD-1 blockers increase the incidence of heart failure up to 2%, while 1% of the total patients experienced cardiac arrest and ventricular arrhythmias (Hu et al., 2017). Pericarditis and pericardial effusions were also reported, with an associated case fatality of 13%, and 0.38% of patients with pericardial effusion needed pericardiocentesis to relieve symptoms (Palaskas et al., 2019; Zhang et al., 2021).

Along with immune checkpoint inhibitors, CAR T-cell therapy has been associated with cardiac toxicity manifestation. Upon recognition with the tumor antigen, the engineered T-cells elicit a strong immune response, leading to strong cytokine syndrome induced by the release of tumor necrosis factor alpha (TNF- α), Interferon gamma (IFN-γ) and Interleukin-6 (IL-6) (Stein-Merlob et al., 2021). CAR T-cell therapy was found to increase the incidence of arrhythmias up to 77.6%, heart failure in 14.3% of the cases and, myocardial infarction in 0.5% of patients treated (Guha et al., 2020).

2.5 Radiation therapy

Radiation therapy (RT) is considered a mainstay of cancer treatment, with more than 60% of cancer patients receiving radiation during the course of their disease (Abshire & Lang, 2018). Radiotherapy can be used both alone and in combination with other chemotherapy agents, thereby increasing overall increased survival of the patients (Gardner, Kim, & Chetty, 2019). Linear accelerators, powered mostly by cobalt and cesium, are used to generate radiation in the form of alpha, beta or gamma particles; a collimator then shapes, and modulates the beam's intensity to precisely direct it towards the area affected by the neoplasia (Abshire & Lang, 2018). The selection of the radioisotope is determined based upon the application and therefore the dosage rate delivered; high dose rate >12 Gray (Gy)/h, low dose rate (0.4–2 Gy/h), and very low dose rate (<0.4 Gy/h) (Abshire & Lang, 2018; Gardner et al., 2019; Otter, Stewart, & Devlin, 2019). Cancers treated by radiation generally display immediate cell death caused by the irreparable DNA breakage sustained from the treatment. RT is often used to decrease the size of tumors before surgery, as shown in the treatment of breast cancer, where RT as an adjuvant prevents tumor reoccurrence by almost 75% (Taunk, Haffty, Kostis, & Goyal, 2015). Radiation can also be delivered within the tumor tissue by brachytherapy (Otter et al., 2019; Taunk et al., 2015). Brachytherapy uses very low dose radiation delivered within the tumor or the tumor bed after surgical resection in the treatment of cervical, rectal and prostate cancers (Otter et al., 2019).

2.5.1 Cardiotoxicity due to radiation therapy

Unfortunately, RT to the breast and chest has been associated with increased risk of cardiac toxicity, which can limit the benefit of this treatment. Radiation-induced cardiomyopathy is characterized by impairment in left ventricular resting systolic function, contractile reserve, compliance, and filling pattern, and is associated with interstitial fibrosis, pericardial and valvular thickening, degeneration of the conduction tissue, and premature or accelerated atherosclerosis (Groarke et al., 2014; Min & Wierzbicki, 2017).

The cellular injury immediately following radiation therapy causes endothelial activation and consequent recruitment of inflammatory cells (Stewart et al., 2013). Proinflammatory mediators are acutely released, including TNF-α, and Interleukin-1, -6 and -8 followed by pro-fibrotic mediators including PDGF, TGF-β and basic fibroblast growth factor (bFGF) (Mezzaroma et al., 2015; Van Der Meeren, Squiban, Gourmelon, Lafont, & Gaugler, 1999). Additionally, damage to the microcirculation fosters the induction of coagulation with a resultant increase of thromboembolic events (Kennedy, Maity, & Sanzari, 2016). Radiation therapy induces a chronic activation of the nuclear factor-κB (NF-κB), which mediates the upregulation of hundreds of proinflammatory genes including hypoxia inducible factor α (HIF1-α) and Transforming growth factor beta (TGF- β) (Herrmann, 2020).

The pathological changes affecting the irradiated arterial district morphologically resemble the pathogenesis of atherosclerosis, including vasculitis, subendothelial fibrosis and accumulation of lipid and foam cells under the intima layer (Stewart et al., 2013). The risk of the concomitant presence of atherosclerotic plaque before the radiation treatment increases the risk of plaque rupture and thrombosis, leading to devastating consequences including myocardial infarction and ischemic stroke (Arthurs, Hanna, Zaza, Peng, & Hall, 2016; Jacobse et al., 2019; Taunk et al., 2015). Microvasculature damage is the initial injury at the myocardial level, with more than 40% of patients subjected to mediastinal radiation experiencing perfusion defects (Taunk et al., 2015). The resulting inflammatory process drives an increase in myocardial cell death, which fosters the accumulation of reparative fibrous tissue with consequent ventricular stiffening and impaired elasticity and cardiac compliance (Taunk et al., 2015). Valvulopathy has been reported following RT, with more than 80% of patients who receive at least 35 Gy experiencing direct damage of the valve, including thickening, increased calcification, and leaflet fibrosis (Brosius, Waller, & Roberts, 1981; Taunk et al., 2015).

Increased evidence of pericardial disease was experienced by almost 90% of the patients exposed to significant mediastinal radiation (Groarke et al., 2014). The pericardium is a double membrane that protects and lubricates the heart, facilitating its movements in the chest cavity. Radiation to the pericardium increases the chances of thickening of the membranes and is associated with the more life-threatening cardiac tamponade and fibrinous deposition. Following RT, the pericardial fat is likely to be substituted by fibrous tissue, decreasing the elasticity of the serous membranes, which could restrain the cardiac movements and result in constrictive pericarditis (Groarke et al., 2014).

Fibrosis deposition and directed damage to the conductive system (sino-atrial and atrioventricular nodes) by RT can increase the incidence of arrhythmias in patients. Patients treated with a cumulative dose of more than 40 Gy are more likely to experience partial or complete right bundle branch block and ventricular ectopic beats (Groarke et al., 2014).

2.6 Alkylating agents

Alkylating agents exert antitumor activity either by cross-.linking the two strands of DNA, preventing a complete separation (difunctional alkylators), or by binding to the DNA, which results in DNA damage as the cells attempt to remove this lesion (monofunctional alkylators) (Singh, Kumar, Prasad, & Bhardwaj, 2018). DNA binding to the N7 nitrogen on the guanine base is often favored, resulting in a highly reactive and aziridinium cation. The alkylating agent family includes the nitrogen mustards (busulfan, chlorambucil, and melphalan); the oxazsaphosphorines (cyclophosphamide and ifosfamide); hydrazine (temozolomide) and vinyl-quinazolinone (Madeddu et al., 2016). Alkylating agents have been widely employed in the treatment of blood tumors (i.e., leukemia, lymphomas, and myelomas) and solid tumors, including breast, lung, ovarian, prostate, sarcomas, and Hodgkin's disease (Higgins et al., 2015). Unfortunately, the mechanism of action of alkylating agents is not specific; as is the case for most of the classical/conventional chemotherapeutic agents, the alkylators can injure normal cells with deleterious sides effect, especially for gastrointestinal mucosa, bone marrow, and the reproductive system (Mondal et al., 2019).

2.6.1 Cardiotoxicity of alkylating agents

One of the most commonly used alkylating agents is cyclophosphamide, which is used to treat several tumors including non-Hodgkin's lymphoma, Burkitt's lymphoma, multiple myeloma, endometrial cancer, lung cancer, breast cancer, sarcoma, and neuroblastoma (Iqubal et al., 2019). The use of cyclophosphamide is not only relegated to tumor suppression but it can also be applied, as an immunosuppressor, in many immune-mediated diseases, including lupus erythematosus, multiple sclerosis, and bone marrow transplantation (Elmariah et al., 2018; Iqubal et al., 2019). The pharmacological effect exerted by cyclophosphamide depends on the dosage administered; when administered at a low dose (<5 mg/kg), the drug functions as an immunosuppressor with no known cardiotoxic effects (Goldberg, Antin, Guinan, & Rappeport, 1986; Gottdiener, Appelbaum, Ferrans, Deisseroth, & Ziegler, 1981). Unfortunately, especially in the treatment of solid and hematological tumors and bone marrow transplantation, higher doses of cyclophosphamide are used (>200 mg/kg) with documented side effects on the cardiovascular system (Goldberg et al., 1986; Steinherz et al., 1981). The incidence of cardiotoxicity from cyclophosphamide treatment varies, with 7%–28% of patients experiencing an acute onset within 21 days from the initiation of the therapy (Atalay, Gulmez, & Ozsancak Ugurlu, 2014; Goldberg et al., 1986; Higgins et al., 2015).

Clinically, cyclophosphamide is often used to augment the cytotoxic effects of other chemotherapies, including agents that possess intrinsic cardiotoxicities such as doxorubicin, TKIs, platinum analogs, antimetabolites, and taxanes (Iqubal et al., 2019). Cyclophosphamide is metabolized by the liver to 4-hydroxy cyclophosphamide and further reduced to aldophosphamide. Aldophophamide is then further metabolized to carboxyphospamide, nitrogen mustard, and acrolein. Acrolein seems to be responsible for the cardiotoxic effects of cyclophosphamide, increasing the oxidative and nitrate stress in the heart. Acrolein was also found to increase cardiomyocyte apoptotic rate while increasing activation of NF-κB and thereby fostering a proinflammatory environment (Moghe et al., 2015; Song et al., 2016). Cyclophosphamide and its metabolites were responsible for activating p53 and p38 mitogen-activated protein kinase pathways, causing cardiac apoptosis and hypertrophy. Additionally, recent evidence confirms an association of the use of cyclophosphamide with adverse ventricular remodeling by impairing mitochondrial function and inducing endoplasmic reticulum stress (Iqubal et al., 2019).

2.7 Platinum-based agents

Approved in 1978, platinum-based agents (cisplatin, carboplatin, and oxaliplatin) are a component of a broad spectrum of antineoplastic treatments (Ghosh, 2019). Like alkylating agents, platinum derivatives bind to DNA forming a covalent bond within two nucleotides generating cross-links on both the same and opposite DNA strands (Rottenberg, Disler, & Perego, 2021). One of the most common clinical applications of platinum-based agents is the treatment of small and non-small cell lung cancer. In 1978, the FDA approved the use of Cisplatin for treating ovarian and testicular cancer, including seminoma and non-seminoma cancers. Other applications of cisplatin include the treatment of pediatric brain tumors, breast cancer, gastric cancer, oropharyngeal cancer, leukemias, and refractory non-Hodgkin's lymphomas (Ghosh, 2019; Parmar et al., 2021).

2.7.1 Cardiotoxicity of platinum-based agents

Cisplatin use has been limited by its cardiotoxic complications, including congestive heart failure, myocardial infarction, myocarditis, hypertension, electrical abnormalities, and an increased risk of thromboembolism (Dugbartey, Peppone, & de Graaf, 2016; Madeddu et al., 2016; Yeh & Bickford, 2009). Cisplatin has also been linked to increased production of Von Willebrand factor and consequent endothelial activation (Dugbartey et al., 2016). Cisplatin has been shown to cause cardiotoxicity by depleting glutathione, the antioxidant cellular defense molecule, increasing cellular oxidative stress with negative consequences on mitochondrial respiration (El-Awady, Moustafa, Abo-Elmatty, & Radwan, 2011). In preclinical models, cisplatin markedly induced mitochondrial and DNA damage, leading to ultrastructural abnormalities and increased apoptosis and ventricular dysfunction (El-Awady et al., 2011; Ma et al., 2010; Madeddu et al., 2016).

2.8 Proteasome inhibitors

Proteasome inhibitors are drugs that target multi-catalytic protein complexes delegated to "dispose" of malfunctioning proteins by cleaving them into single peptides, which contribute to 80% of the protein turnover. Different proteasomes exist depending on their catalytic subunits (Fricker, 2020). Proteasomes share a 700 kDa common core (the 20S subunit) containing 14α and 14β subunits disposed into four rings. The active catalytic unit is included in each inner circle and has a characteristic structure formed by three β subunits: β1, β2, and β5 (Fricker, 2020). Bortezomib, Carfilzomib, and Ixazomib are the three clinically approved irreversible proteasome inhibitors targeting the catalytic subunits of the proteasomes. Marizomib, oprozomib, and delanzomib are newer compounds currently being tested in clinical trials (Fricker, 2020). The therapeutic effect of proteasome inhibitors is not entirely elucidated, however, targeting protein turnover can directly affect the downstream pathways regulated by these proteins, by changing their intracellular levels. As an example, NF-κB signaling has been implicated in tumor growth. Upon phosphorylation, the IkappaB kinase (IκB) kinase proteins, negative regulators of the nuclear factor NF-κB, are degraded by proteasomes (Grisham et al., 1999). NF-κB is free to translocate into the nucleus, promoting the transcription of many genes involved with cell survival (Didonato, Mercurio, & Karin, 2012). Additionally, the inhibition of the degradation of cyclins and cyclin-dependent kinases, proteins involved in controlling cellular cycle, can result in uncontrolled tumor cell division as hypothesized in mantle cell lymphoma, which is caused by the overexpression of cyclin D1 (Musgrove, Caldon, Barraclough, Stone, & Sutherland, 2011).

2.8.1 Cardiotoxicity from proteasome inhibitors

Bortezomib is the first proteasome inhibitor to be approved by the FDA to treat newly diagnosed multiple myeloma (Cole & Frishman, 2018; Fricker, 2020). It has become the standard of care therapy for the antitumor pharmacological regimen of high-dose chemotherapy autologous stem cell transplantation in patients with melanoma (Cole & Frishman, 2018; Fricker, 2020). A small number of published case reports have demonstrated cardiotoxicity following bortezomib treatment, including arrhythmias, ischemic heart disease, and heart failure (Xiao, Yin, Wei, & Shang, 2014). However, a more complex and extensive stratified-analysis involving more than 5000 patients have found no significant correlation between the use of bortezomib and increased incidence of all grades of cardiovascular adverse events (Xiao et al., 2014).

Carfilzomib, approved in 2012, is considered a second-generation proteasome inhibitor for the treatment of bortezomib-resistant malignancies (Wu, Oren, Gertz, & Yang, 2020). Increased cardiotoxicities like hypertension, ischemic cardiomyopathy, heart failure, and sudden cardiac death have been associated with the use of carfilzomib (Wu et al., 2020). Integrated safety data pooled from four different phase II studies reveal an increase in the incidence of atrial fibrillation by 13%, with 7.2% of patients experiencing heart failure and 3% developing ischemic heart disease (Siegel et al., 2013). When given in combination with another chemotherapeutic drug (lenalidomide and dexamethasone), carfilzomib had an increased incidence of almost 4% for both hypertension and heart failure; while the increase in ischemic heart disease was reported to be higher than 3% (Wu et al., 2020). More extensive observational studies enrolled 95 patients with relapsed and refractory multiple myeloma that were followed for 18 months. Of those 95 patients, 65 received carfilzomib, while the remaining 30 were treated with bortezomib. Cardiovascular events were recorded in 51% of patients treated with carfilzomib, while only 17% of patients treated with bortezomib experienced cardiovascular side effects (Cornell et al., 2019). Higher natriuretic peptides measured after the first mid-cycle, correlated with a higher risk of cardiovascular events in the carfilzomib cohorts; however, the patients were not required to discontinue the therapy (Cornell et al., 2019). This last trial highlighted an increased incidence of major cardiovascular events compared with the previous clinical trials, possibly because, contrary to the previous study, the cardiovascular events were part of the end-points of the study, ensuring that all events were promptly reported and analyzed (Cornell et al., 2019; Wu et al., 2020). In an effort to monitor the use of carfilzomib in the general population, a retrospective study recently published on 635 patients revealed a 22% incidence of hypertension and up to 14% increased risk of heart failure (Fakhri, Fiala, Shah, Vij, & Wildes, 2020).

3. Current and promising therapies to prevent chemotherapy-induced cardiotoxicity

As described in the previous sections, the use of antineoplastic drugs is often limited by their cardiotoxicity, although there are a spectrum of other toxicities to normal tissues such as renal damage by platinum-based compounds, peripheral neuropathy by taxane, platinum-based drugs and proteosome inhibitors, arthralgia and chemobrain associated with the utilization of aromatase inhibitors in breast cancer and severe delayed onset diarrhea from irinotecan in colorectal cancer, with many additional examples. The widespread use of anthracyclines as chemotherapeutic agents and the devastating effect on the cardiovascular system have encouraged extensive research to find a possible strategy to mitigate and delay the onset of cardiac dysfunction.

Initial efforts to decrease toxicity from anthracyclines involved establishing the optimal therapeutic dose. It is well established that the cumulative amount of anthracyclines received by the patient increases the risk of cardiotoxicity, especially in patients with unmodifiable risk factors such as pre-existing cardiac complications, age (pediatric or elderly), and concomitant administration of other chemotherapies (including radiation therapy). For the reasons mentioned above, attempts to modify drug chemistry (i.e., pegylated liposomal doxorubicin) or delivery (slow infusion vs. bolus dose) were some of the earliest explored clinical options (Olivieri et al., 2017; Payne & Nohria, 2017).

3.1 Dexrazoxane

One of the mechanisms of anthracycline-induced cytotoxicity is the generation of ROS through a Fenton reaction between molecular iron and hydrogen peroxide (Shaul et al., 2013). Dexrazoxane is a chelating agent derived from ethylenediaminetetraacetic acid (EDTA), but with specificity for iron. Dexrazoxane has been shown to reduce ROS generation when administered concomitantly with anthracyclines, without reducing the antineoplastic effects (Bansal et al., 2019). Dexrazoxane is the only FDA and European Medicine Agency-approved drug to prevent anthracycline-induced cardiotoxicity in patients with advanced breast cancer (Bansal et al., 2019).

In a multicenter clinical study involving 534 randomized patients, dexrazoxane significantly decreased the LVEF decline compared to placebo (Swain et al., 1997). A subsequent study was able to determine the efficacy of dexrazoxane even after delayed administration in patients that received 300 mg/m^2 of anthracycline (Swain et al., 1997; Swain, Whaley, Gerber, Weisberg, et al., 1997). Unfortunately, dexrazoxane use was also linked to several side effects including neutropenia, leukopenia, granulocytopenia, and thrombocytopenia. Additionally, nausea, increased renal excretion of iron and zinc, hair loss, mucositis, inconsistent increases in liver enzymes and reactions at the injection site were reported following dexrazoxane use (Mack, 2007).

3.2 Neurohormonal blockade

The therapies antagonizing neurohormonal activation are considered a mainstay in the treatment of heart failure (Metra & Teerlink, 2017). The application of neurohormonal inhibitors in the treatment of chemotherapy-induced cardiotoxicity was explored based on preclinical evidence supporting the role of Angiotensin II and endothelin in the pathogenesis of anthracycline-induced cardiomyopathy.(Payne & Nohria, 2017) Additionally, the antioxidant property of beta-blockers has accelerated further research efforts to explore their potential clinical activities in cancer patients (Metra & Teerlink, 2017).

The prophylactic administration of the beta-blocker, carvedilol, was explored in a randomized study of 25 participants treated with anthracyclines. Carvedilol was administered daily for 6 months. Although the number of patients analyzed was small, carvedilol appeared to reduce both systolic and diastolic dysfunction markers (Kalay et al., 2006). Two additional studies involving 40 and 80 women with breast cancer and anthracycline therapy were conducted using carvedilol as a preventative strategy. Although there were improvements, both studies only marginally supported a benefit of carvedilol prophylaxis in reducing the degree of dysfunction measured by strain rate analysis (Elitok et al., 2014; Tashakori Beheshti et al., 2016). The use of carvedilol was then explored in a more extensive study, the CECCY trial, which enrolled 192 HER2+ breast cancer patients. The use of carvedilol in these settings failed to reduce the incidence of LVEF reduction. However, carvedilol administration reduced the cardiomyocyte injury measured as a decreased plasmatic release of Cardiac troponin I (cTnI), a protein responsible for myocardial contractility and widely used as a biomarker for heart damage. The reduction of cardiotoxicity by carvedilol administration preserved the decline of diastolic function, improving heart relaxation (Avila et al., 2018).

The combined therapy of enalapril, an angiotensin-converting enzyme inhibitor, and carvedilol was also clinically tested in the OVERCOME trial, a randomized controlled study with 90 participants (Bosch et al., 2013). Enalapril and carvedilol were administered in parallel with anthracyclines in recently-diagnosed acute leukemia patients and autologous hematopoietic stem cell transplantation for hematologic malignancies. This study revealed that the combined therapy prevented the decline in LVEF while significantly reducing heart failure and all-cause mortality compared to the placebo-treated group (6.7 vs. 22%; $P=0.036$) (Bosch et al., 2013). Although promising, more extensive studies with a larger population are needed to confirm the clinical relevance of carvedilol as a therapeutic option to lower anthracycline-mediated cardiotoxicity.

Neurohormonal blockade was also explored as a strategy for the mitigation of trastuzumab-induced cardiac dysfunction. In the MANTICORE trial, a double-blinded, placebo-controlled trial, patients with HER2+ breast cancer were randomized 1:1 to perindopril, bisoprolol, and placebo for the duration of trastuzumab therapy, beginning 1 week before chemotherapy. (Pituskin et al., 2017) The results revealed that bisoprolol, but not perindopril, marginally attenuated the decline of LVEF compared to placebo and perindopril. Both drugs failed to preserve adverse ventricular remodeling measured by LV end-diastolic volume (Pituskin et al., 2017).

The combination of candesartan, an angiotensin receptor blocker, and metoprolol was explored in HER2+ breast cancer patients to contrast cardiotoxicity induced by trastuzumab and anthracycline therapy. This clinical study, named the PRADA trial, was a two × two factorial, randomized, placebo-controlled, double-blind trial involving 130 women with early breast cancer and no other comorbidities (Gulati et al., 2016). The primary outcome measured was LVEF change by cardiac magnetic resonance imaging (MRI). Candesartan was the only drug that showed significant prevention in LVEF decline compared to placebo (Gulati et al., 2016).

Finally, spironolactone, an aldosterone inhibitor, was clinically evaluated in a small study involving 83 women with breast cancer and treated with anthracyclines. Daily treatment with spironolactone prevented LVEF decline compared to the control group when simultaneously administered during chemotherapy (Akpek et al., 2015). If confirmed in larger clinical trials, spironolactone can then be considered as a promising protective agent against cardiotoxicity when simultaneously administered with anthracyclines.

3.3 Statins

Statins are a family of drugs used to treat patients with high levels of cholesterol. Recent evidence has, however, explored other pleiotropic benefits associated with the use of statins aside from lowering cholesterol. Statins were found to reduce inflammation, apoptosis, and the risk of stroke and heart failure (HF) (Lee, Sattar, McMurray, & Packard, 2019). The pleiotropic effects of statins to reduce anthracycline-induced cardiomyopathy were investigated in two small prospective studies. In the first study, 40 patients undergoing anthracycline treatment were randomized either to statin therapy or placebo before the initiation of chemotherapy (Acar et al., 2011). Patients were then treated with statins for an additional 6 months. Statin therapy was able to significantly prevent LVEF decline following anthracycline treatment when compared with the placebo group. Similar results were obtained in the second study enrolling 51 participants (Acar et al., 2011; Chotenimitkhun et al., 2015). Cardiac MRI images were recorded before and 6 months after anthracycline administration. Whereas patients that were not treated with statins experienced a decline in LVEF over 6 months, the use of statins prevented the early onset of cardiac dysfunction following anthracycline therapy (Acar et al., 2011; Chotenimitkhun et al., 2015). Further validations are needed since the number of participants in this study was limited. To this point, the ongoing PREVENT study will evaluate if treatment with atorvastatin could prevent LVEF decline in patients receiving anthracyclines. The study was designed as double-blind, placebo-controlled, randomized clinical trial of 0 or 40 mg of atorvastatin/day in 250 women with breast cancer (NCT01988571). In this study LVEF, myocardial strain, fibrosis, aortic pulse wave velocity and wall thickness will be measured as well as serum biomarkers for the presence of oxidative/nitrosative stress, systemic inflammation and circulating neurohormones.

3.4 Exercise therapy

As mentioned above, the ASCO recommends lifestyle changes and increased exercise activity for all patients undergoing cancer treatment. Exercise capacity is predictive of better prognosis in patients with heart failure, and its use as an intervention has been increasingly explored in the setting of cardiovascular diseases. The link between exercise intervention and anthracycline cardiomyopathy was explored in a prospective trial involving 2973 women with metastatic breast cancer (Payne & Nohria, 2017). Women who exercised for ≥ 9 metabolic equivalent task (MET)-hours/week significantly reduced the risk of developing cardiovascular events, including heart failure by 23% (Jones et al., 2016). Unfortunately, the exercise therapy as an adjuvant to trastuzumab therapy did not show benefits compared to the control group in reducing adverse ventricular remodeling (Haykowsky, Mackey, Thompson, Jones, & Paterson, 2009). These discordant results highlight the need for larger, randomized, double-blinded control trials to assess the potential benefits of exercise intervention to prevent or treat therapy-induced cardiotoxicity and heart failure in cancer patients.

3.5 Remote ischemic preconditioning intervention

Ischemic pre-conditioning refers to controlled short periods of ischemia designed to prepare an organ to a more extended ischemic event (Heusch & Gersh, 2017; Yellon & Downey, 2003). Early animal models showed that transient occlusions of a coronary artery could be beneficial in the event of a myocardial infarction. On the other hand, remote ischemic preconditioning (RIPC) refers to a distal vasculature bed that is subjected to transient ischemia that results in the release of soluble protective mediators that deliver such protective effects in a distant organ. RIPC has been explored in a preclinical model treated with a high dose of DOX. The results showed a remarkable improvement in animal survival, as well as reduction in myocardial apoptosis, fibrosis deposition, and consequent preservation of LV mass (Gertz et al., 2019). RIPC effects on cancer patients are now being explored in the ERIC-ONC clinical trial (NCT02471885). This pilot study will demonstrate whether RIPC, delivered through repeated inflations and deflations of a limb blood pressure cuff, can reduce subclinical myocardial injury from anthracycline administration. Patients will be monitored up to 1 year after anthracycline administration. LV function will be monitored with global longitudinal strain (GLS) while cardiac troponin, NT-proBNP, and micro-RNA levels will be measured at different time points. Major adverse clinical cardiovascular events and incidence of arrhythmias will also be recorded.

4. Conclusions

This chapter provides an extensive overview of the central mechanisms involved in the cardiotoxicity mediated by antineoplastic treatments. Research and development of new cancer treatments coupled with the multidisciplinary approach in managing cancer patients have markedly increased the chances of a patient surviving cancer. Unfortunately, the consequent decline in mortality is counteracted by an increase in cardiotoxicity secondary to cancer therapy, which is now the leading cause of morbidity and mortality among cancer survivors. While CHF is a significant cause of late morbidity and mortality after DOX chemotherapy, the combination of cancer and cardiotoxicity due to cancer therapies comprises a 'dangerous intersection' with increased mortality compared to each disease alone. Several strategies to mitigate anthracycline and other cancer therapy-related cardiotoxicity have been tested so far with limited success. Therefore, enhancing our understanding of the pathophysiology of chemo- and radiotherapy-induced cardiotoxicity represents an urgent unmet clinical need to refine the current therapeutic strategies and improve clinical outcomes in cancer survivors.

Acknowledgments

This work was supported by National Heart, Lung, and Blood Institute (R35 HL155651) to Dr. Salloum.

Figs. 1 and 2 were drawn using the image bank of Servier Medical Art (http://smart.servier.com/).

References

Abshire, D., & Lang, M. K. (2018). The evolution of radiation therapy in treating cancer. *Seminars in Oncology Nursing*. https://doi.org/10.1016/j.soncn.2018.03.006.

Acar, Z., Kale, A., Turgut, M., Demircan, S., Durna, K., Demir, S., et al. (2011). Efficiency of atorvastatin in the protection of anthracycline-induced cardiomyopathy. *Journal of the American College of Cardiology*, *58*(9), 988–989. https://doi.org/10.1016/j.jacc.2011.05.025.

Akpek, M., Ozdogru, I., Sahin, O., Inanc, M., Dogan, A., Yazici, C., et al. (2015). Protective effects of spironolactone against anthracycline-induced cardiomyopathy. *European Journal of Heart Failure*, *17*(1), 81–89. https://doi.org/10.1002/ejhf.196.

Arola, O. J., Saraste, A., Pulkki, K., Kallajoki, M., Parvinen, M., & Voipio-Pulkki, L. M. (2000). Acute doxorubicin cardiotoxicity involves cardiomyocyte apoptosis. *Cancer Research*, *60*(7), 1789–1792.

Arthurs, E., Hanna, T. P., Zaza, K., Peng, Y., & Hall, S. F. (2016). Stroke after radiation therapy for head and neck cancer: What is the risk? *International Journal of Radiation Oncology Biology Physics*, *96*(3), 589–596. https://doi.org/10.1016/j.ijrobp.2016.07.007.

Atalay, F., Gulmez, O., & Ozsancak Ugurlu, A. (2014). Cardiotoxicity following cyclophosphamide therapy: A case report. *Journal of Medical Case Reports*, *8*(1). https://doi.org/10.1186/1752-1947-8-252.

Avila, M. S., Ayub-Ferreira, S. M., de Barros Wanderley, M. R., das Dores Cruz, F., Gonçalves Brandão, S. M., Rigaud, V. O. C., et al. (2018). Carvedilol for prevention of chemotherapy-related cardiotoxicity: The CECCY trial. *Journal of the American College of Cardiology*, *71*(20), 2281–2290. https://doi.org/10.1016/j.jacc.2018.02.049.

Bansal, N., Adams, M. J., Ganatra, S., Colan, S. D., Aggarwal, S., Steiner, R., et al. (2019). Strategies to prevent anthracycline-induced cardiotoxicity in cancer survivors. *Cardio-Oncology*, *5*(1), 1–22. https://doi.org/10.1186/s40959-019-0054-5.

Bansal, N., Amdani, S., Lipshultz, E. R., & Lipshultz, S. E. (2017). Chemotherapy-induced cardiotoxicity in children. *Expert Opinion on Drug Metabolism and Toxicology*, *13*(8), 817–832. https://doi.org/10.1080/17425255.2017.1351547.

Bernstein, D. (2018). Anthracycline cardiotoxicity. *Circulation Research*, *122*(2), 188–190. https://doi.org/10.1161/CIRCRESAHA.117.312395.

Berraondo, P., Sanmamed, M. F., Ochoa, M. C., Etxeberria, I., Aznar, M. A., Pérez-Gracia, J. L., et al. (2019). Cytokines in clinical cancer immunotherapy. *British Journal of Cancer*, *120*(1), 6–15. https://doi.org/10.1038/s41416-018-0328-y.

Bhullar, K. S., Lagarón, N. O., McGowan, E. M., Parmar, I., Jha, A., Hubbard, B. P., et al. (2018). Kinase-targeted cancer therapies: Progress, challenges and future directions. *Molecular Cancer*. https://doi.org/10.1186/s12943-018-0804-2.

Bosch, X., Rovira, M., Sitges, M., Domènech, A., Ortiz-Pérez, J. T., de Caralt, T. M., et al. (2013). Enalapril and carvedilol for preventing chemotherapy-induced left ventricular systolic dysfunction in patients with malignant hemopathies. *Journal of the American College of Cardiology*, *61*(23), 2355–2362. https://doi.org/10.1016/j.jacc.2013.02.072.

Breccia, M., & Alimena, G. (2013). Occurrence and current management of side effects in chronic myeloid leukemia patients treated frontline with tyrosine kinase inhibitors. *Leukemia Research*, *37*(6), 713–720. https://doi.org/10.1016/j.leukres.2013.01.021.

Brosius, F. C., Waller, B. F., & Roberts, W. C. (1981). Radiation heart disease. Analysis of 16 young (aged 15 to 33 years) necropsy patients who received over 3,500 rads to the heart. *The American Journal of Medicine*, *70*(3), 519–530. https://doi.org/10.1016/0002-9343(81)90574-X.

Broxterman, H. J., Gotink, K. J., & Verheul, H. M. W. (2009). Understanding the causes of multidrug resistance in cancer: A comparison of doxorubicin and sunitinib. *Drug Resistance Updates*, *12*(4–5), 114–126. https://doi.org/10.1016/j.drup.2009.07.001.

Cadeddu, C., Mercurio, V., Spallarossa, P., Nodari, S., Triggiani, M., Monte, I., et al. (2016). Preventing antiblastic drug-related cardiomyopathy: Old and new therapeutic strategies. *Journal of Cardiovascular Medicine*, *17*, S64–S75. https://doi.org/10.2459/JCM.0000000000000382.

Cappetta, D., De Angelis, A., Sapio, L., Prezioso, L., Illiano, M., Quaini, F., et al. (2017). Oxidative stress and cellular response to doxorubicin: A common factor in the complex milieu of anthracycline cardiotoxicity. *Oxidative Medicine and Cellular Longevity*, *2017*. https://doi.org/10.1155/2017/1521020.

Cardinale, D., Iacopo, F., & Cipolla, C. M. (2020). Cardiotoxicity of anthracyclines. *Frontiers in Cardiovascular Medicine*. https://doi.org/10.3389/fcvm.2020.00026.

Cassinelli, G. (2016). The roots of modern oncology: From discovery of new antitumor anthracyclines to their clinical use. *Tumori, 102*(3), 226–235. https://doi.org/10.5301/tj.5000507.

Castel, P., Rauen, K. A., & McCormick, F. (2020). The duality of human oncoproteins: Drivers of cancer and congenital disorders. *Nature Reviews Cancer, 20*(7), 383–397. https://doi.org/10.1038/s41568-020-0256-z.

Chen, M. H., Kerkelä, R., & Force, T. (2008). Mechanisms of cardiac dysfunction associated with tyrosine kinase inhibitor cancer therapeutics. *Circulation, 118*(1), 84–95. https://doi.org/10.1161/CIRCULATIONAHA.108.776831.

Chen, L., Sampson, K. J., & Kass, R. S. (2016). Cardiac delayed rectifier potassium channels in health and disease. *Cardiac Electrophysiology Clinics, 8*(2), 307–322. https://doi.org/10.1016/j.ccep.2016.01.004.

Chotenimitkhun, R., D'Agostino, R., Lawrence, J. A., Hamilton, C. A., Jordan, J. H., Vasu, S., et al. (2015). Chronic statin administration may attenuate early anthracycline-associated declines in left ventricular ejection function. *Canadian Journal of Cardiology, 31*(3), 302–307. https://doi.org/10.1016/j.cjca.2014.11.020.

Cianflone, E., Torella, M., Biamonte, F., De Angelis, A., Urbanek, K., Costanzo, F. S., et al. (2020). Targeting cardiac stem cell senescence to treat cardiac aging and disease. *Cell, 9*(6), 1558. https://doi.org/10.3390/cells9061558.

Cole, D. C., & Frishman, W. H. (2018). Cardiovascular complications of proteasome inhibitors used in multiple myeloma. *Cardiology in Review, 26*(3), 122–129. https://doi.org/10.1097/CRD.0000000000000183.

Cornell, R. F., Ky, B., Weiss, B. M., Dahm, C. N., Gupta, D. K., Du, L., et al. (2019). Prospective study of cardiac events during proteasome inhibitor therapy for relapsed multiple myeloma. *Journal of Clinical Oncology, 37*(22), 1946–1955. https://doi.org/10.1200/JCO.19.00231.

Courtney, A. H., Lo, W. L., & Weiss, A. (2018). TCR signaling: Mechanisms of initiation and propagation. *Trends in Biochemical Sciences*. https://doi.org/10.1016/j.tibs.2017.11.008.

Dargani, Z. T., & Singla, D. K. (2019). Embryonic stem cell-derived exosomes inhibit doxorubicin-induced TLR4-NLRP3-mediated cell death-pyroptosis. *American Journal of Physiology - Heart and Circulatory Physiology, 317*(2), H460–H471. https://doi.org/10.1152/ajpheart.00056.2019.

Degirmenci, U., Wang, M., & Hu, J. (2020). Targeting aberrant RAS/RAF/MEK/ERK signaling for cancer therapy. *Cell, 9*(1). https://doi.org/10.3390/cells9010198.

Didonato, J. A., Mercurio, F., & Karin, M. (2012). NF-κB and the link between inflammation and cancer. *Immunological Reviews, 246*(1), 379–400. https://doi.org/10.1111/j.1600-065X.2012.01099.x.

Dirican, A., Levent, F., Alacacioglu, A., Kucukzeybek, Y., Varol, U., Kocabas, U., et al. (2014). Acute cardiotoxic effects of adjuvant trastuzumab treatment and its relation to oxidative stress. *Angiology, 65*(10), 944–949. https://doi.org/10.1177/0003319714523112.

Dugbartey, G. J., Peppone, L. J., & de Graaf, I. A. M. (2016). An integrative view of cisplatin-induced renal and cardiac toxicities: Molecular mechanisms, current treatment challenges and potential protective measures. *Toxicology, 371*, 58–66. https://doi.org/10.1016/j.tox.2016.10.001.

El-Awady, E. S. E., Moustafa, Y. M., Abo-Elmatty, D. M., & Radwan, A. (2011). Cisplatin-induced cardiotoxicity: Mechanisms and cardioprotective strategies. *European Journal of Pharmacology, 650*(1), 335–341. https://doi.org/10.1016/j.ejphar.2010.09.085.

Elitok, A., Oz, F., Ahmet, Y., Kilic, L., Ciftci, R., Sen, F., et al. (2014). Effect of carvedilol on silent anthracycline-induced cardiotoxicity assessed by strain imaging: A prospective randomized controlled study with six-month follow-up. *Cardiology Journal, 21*(5), 509–515. https://doi.org/10.5603/CJ.a2013.0150.

Elmariah, H., Kasamon, Y. L., Zahurak, M., Macfarlane, K. W., Tucker, N., Rosner, G. L., et al. (2018). Haploidentical bone marrow transplantation with post-transplant cyclophosphamide using non–first-degree related donors. *Biology of Blood and Marrow Transplantation, 24*(5), 1099–1102. https://doi.org/10.1016/j.bbmt.2018.02.005.

Erickson, R. P. (2010). Somatic gene mutation and human disease other than cancer: An update. *Mutation research—Reviews in mutation research*. https://doi.org/10.1016/j.mrrev.2010.04.002.

Fabiani, I., Aimo, A., Grigoratos, C., Castiglione, V., Gentile, F., Saccaro, L. F., et al. (2021). Oxidative stress and inflammation: Determinants of anthracycline cardiotoxicity and possible therapeutic targets. *Heart Failure Reviews, 26*(4), 881–890. https://doi.org/10.1007/s10741-020-10063-9.

Fakhri, B., Fiala, M. A., Shah, N., Vij, R., & Wildes, T. M. (2020). Measuring cardiopulmonary complications of carfilzomib treatment and associated risk factors using the SEER-Medicare database. *Cancer, 126*(4), 808–813. https://doi.org/10.1002/cncr.32601.

Fang, X., Wang, H., Han, D., Xie, E., Yang, X., Wei, J., et al. (2019). Ferroptosis as a target for protection against cardiomyopathy. *Proceedings of the National Academy of Sciences, 116*(7), 2672–2680. https://doi.org/10.1073/pnas.1821022116.

Focaccetti, C., Bruno, A., Magnani, E., Bartolini, D., Principi, E., Dallaglio, K., et al. (2015). Effects of 5-fluorouracil on morphology, cell cycle, proliferation, apoptosis, autophagy and ROS production in endothelial cells and cardiomyocytes. *PLoS One, 10*(2). https://doi.org/10.1371/journal.pone.0115686.

Force, T., Krause, D. S., & Van Etten, R. A. (2007). Molecular mechanisms of cardiotoxicity of tyrosine kinase inhibition. *Nature Reviews Cancer*, 7(5), 332–344. https://doi.org/10.1038/nrc2106.

Fricker, L. D. (2020). Proteasome inhibitor drugs. *Annual Review of Pharmacology and Toxicology, 60*, 457–476. https://doi.org/10.1146/annurev-pharmtox-010919-023603.

Gardner, S. J., Kim, J., & Chetty, I. J. (2019). Modern radiation therapy planning and delivery. *Hematology/Oncology Clinics of North America, 33*(6), 947–962. https://doi.org/10.1016/j.hoc.2019.08.005.

Gertz, Z. M., Cain, C., Kraskauskas, D., Devarakonda, T., Mauro, A. G., Thompson, J., et al. (2019). Remote ischemic pre-conditioning attenuates adverse cardiac remodeling and mortality following doxorubicin administration in mice. *JACC: CardioOncology, 1*(2), 221–234. https://doi.org/10.1016/j.jaccao.2019.11.004.

Gharwan, H., & Groninger, H. (2016). Kinase inhibitors and monoclonal antibodies in oncology: Clinical implications. *Nature Reviews Clinical Oncology, 13*(4), 209–227. https://doi.org/10.1038/nrclinonc.2015.213.

Ghosh, S. (2019). Cisplatin: The first metal based anticancer drug. *Bioorganic Chemistry, 88*, 102925. https://doi.org/10.1016/j.bioorg.2019.102925.

Goldberg, M. A., Antin, J. H., Guinan, E. C., & Rappeport, J. M. (1986). Cyclophosphamide cardiotoxicity: An analysis of dosing as a risk factor. *Blood, 68*(5), 1114–1118. https://doi.org/10.1182/blood.v68.5.1114.1114.

Gottdiener, J. S., Appelbaum, F. R., Ferrans, V. J., Deisseroth, A., & Ziegler, J. (1981). Cardiotoxicity associated with high-dose cyclophosphamide therapy. *Archives of Internal Medicine, 141*(6), 758–763. https://doi.org/10.1001/archinte.1981.00340060066015.

Grisham, M. B., Palombella, V. J., Elliott, P. J., Conner, E. M., Brand, S., Wong, H. L., et al. (1999). Inhibition of NF-κB activation in vitro and in vivo: Role of 26S proteasome. *Methods in Enzymology, 300*, 345–363. https://doi.org/10.1016/S0076-6879(99)00140-8.

Groarke, J. D., Nguyen, P. L., Nohria, A., Ferrari, R., Cheng, S., & Moslehi, J. (2014). Cardiovascular complications of radiation therapy for thoracic malignancies: The role for non-invasive imaging for detection of cardiovascular disease. *European Heart Journal, 35*(10), 612–623. https://doi.org/10.1093/eurheartj/eht114.

Guha, A., Addison, D., Jain, P., Gutierrez, J. M., Ghosh, A., Roddie, C., et al. (2020). Cardiovascular events associated with chimeric antigen receptor T cell therapy: Cross-sectional FDA adverse events reporting system analysis: Cardiovascular events with CAR-T therapy. *Biology of Blood and Marrow Transplantation, 26*(12), 2211–2216. https://doi.org/10.1016/j.bbmt.2020.08.036.

Gulati, G., Heck, S. L., Ree, A. H., Hoffmann, P., Schulz-Menger, J., Fagerland, M. W., et al. (2016). Prevention of cardiac dysfunction during adjuvant breast cancer therapy (PRADA): A 2 × 2 factorial, randomized, placebo-controlled, double-blind clinical trial of candesartan and metoprolol. *European Heart Journal, 37*(21), 1671–1680. https://doi.org/10.1093/eurheartj/ehw022.

Hakuno, F., & Takahashi, S. I. (2018). 40 years of IGF1: IGF1 receptor signaling pathways. *Journal of Molecular Endocrinology, 61*(1), T69–T86. https://doi.org/10.1530/JME-17-0311.

Harvey, R. C., & Tasian, S. K. (2020). Clinical diagnostics and treatment strategies for Philadelphia chromosome-like acute lymphoblastic leukemia. *Blood Advances, 4*(1), 218–228. https://doi.org/10.1182/bloodadvances.2019000163.

Haykowsky, M. J., Mackey, J. R., Thompson, R. B., Jones, L. W., & Paterson, D. I. (2009). Adjuvant trastuzumab induces ventricular remodeling despite aerobic exercise training. *Clinical Cancer Research, 15*(15), 4963–4967. https://doi.org/10.1158/1078-0432.CCR-09-0628.

Henriksen, P. A. (2018). Anthracycline cardiotoxicity: An update on mechanisms, monitoring and prevention. *Heart, 104*(12), 971–977. https://doi.org/10.1136/heartjnl-2017-312103.

Herndon, T. M., Demko, S. G., Jiang, X., He, K., Gootenberg, J. E., Cohen, M. H., et al. (2012). U.S. food and drug administration approval: Peginterferon-alfa-2b for the adjuvant treatment of patients with melanoma. *The Oncologist, 17*(10), 1323–1328. https://doi.org/10.1634/theoncologist.2012-0123.

Herrmann, J. (2020). Adverse cardiac effects of cancer therapies: Cardiotoxicity and arrhythmia. *Nature Reviews Cardiology, 17*(8), 474–502. https://doi.org/10.1038/s41569-020-0348-1.

Herrmann, J., Lerman, A., Sandhu, N. P., Villarraga, H. R., Mulvagh, S. L., & Kohli, M. (2014). Evaluation and management of patients with heart disease and cancer: Cardio-oncology. *Mayo Clinic Proceedings, 89*(9), 1287–1306. https://doi.org/10.1016/j.mayocp.2014.05.013.

Heusch, G., & Gersh, B. J. (2017). The pathophysiology of acute myocardial infarction and strategies of protection beyond reperfusion: A continual challenge. *European Heart Journal, 38*(11), 774–784. https://doi.org/10.1093/eurheartj/ehw224.

Higgins, A. Y., O'Halloran, T. D., & Chang, J. D. (2015). Chemotherapy-induced cardiomyopathy. *Heart Failure Reviews, 20*(6), 721–730. https://doi.org/10.1007/s10741-015-9502-y.

Hirschhorn, T., & Stockwell, B. R. (2019). The development of the concept of ferroptosis. *Free Radical Biology and Medicine, 133*, 130–143. https://doi.org/10.1016/j.freeradbiomed.2018.09.043.

Hodi, F. S., Mihm, M. C., Soiffer, R. J., Haluska, F. G., Butler, M., Seiden, M. V., et al. (2003). Biologic activity of cytotoxic T lymphocyte-associated antigen 4 antibody blockade in previously vaccinated metastatic melanoma and ovarian carcinoma patients. *Proceedings of the National Academy of Sciences of the United States of America, 100*(8), 4712–4717. https://doi.org/10.1073/pnas.0830997100.

Hodi, F. S., O'Day, S. J., McDermott, D. F., Weber, R. W., Sosman, J. A., Haanen, J. B., et al. (2010). Improved survival with ipilimumab in patients with metastatic melanoma. *New England Journal of Medicine, 363*(8), 711–723. https://doi.org/10.1056/nejmoa1003466.

Hou, Y., Wei, Y., Lautrup, S., Yang, B., Wang, Y., Cordonnier, S., et al. (2021). NAD^+ supplementation reduces neuroinflammation and cell senescence in a transgenic mouse model of Alzheimer's disease via cGAS-STING. *Proceedings of the National Academy of Sciences of the United States of America, 118*(37). https://doi.org/10.1073/pnas.2011226118.

Hu, Y.-B., Zhang, Q., Li, H.-J., Michot, J. M., Liu, H.-B., Zhan, P., et al. (2017). Evaluation of rare but severe immune related adverse effects in PD-1 and PD-L1 inhibitors in non-small cell lung cancer: A meta-analysis. *Translational Lung Cancer Research, 6*(S1), S8–S20. https://doi.org/10.21037/tlcr.2017.12.10.

Ichikawa, Y., Ghanefar, M., Bayeva, M., Wu, R., Khechaduri, A., Naga Prasad, S. V., et al. (2014). Cardiotoxicity of doxorubicin is mediated through mitochondrial iron accumulation. *Journal of Clinical Investigation, 124*(2), 617–630. https://doi.org/10.1172/JCI72931.

Iqubal, A., Iqubal, M. K., Sharma, S., Ansari, M. A., Najmi, A. K., Ali, S. M., et al. (2019). Molecular mechanism involved in cyclophosphamide-induced cardiotoxicity: Old drug with a new vision. *Life Sciences*. https://doi.org/10.1016/j.lfs.2018.12.018.

Iwai, Y., Ishida, M., Tanaka, Y., Okazaki, T., Honjo, T., & Minato, N. (2002). Involvement of PD-L1 on tumor cells in the escape from host immune system and tumor immunotherapy by PD-L1 blockade. *Proceedings of the National Academy of Sciences of the United States of America, 99*(19), 12293–12297. https://doi.org/10.1073/pnas.192461099.

Jacobse, J. N., Duane, F. K., Boekel, N. B., Schaapveld, M., Hauptmann, M., Hooning, M. J., et al. (2019). Radiation dose-response for risk of myocardial infarction in breast cancer survivors. *International Journal of Radiation Oncology Biology Physics, 103*(3), 595–604. https://doi.org/10.1016/j.ijrobp.2018.10.025.

Jain, V., Mohebtash, M., Rodrigo, M. E., Ruiz, G., Atkins, M. B., & Barac, A. (2018). Autoimmune myocarditis caused by immune checkpoint inhibitors treated with antithymocyte globulin. *Journal of Immunotherapy, 41*(7), 332–335. https://doi.org/10.1097/CJI.0000000000000239.

Johnson, D. B., Balko, J. M., Compton, M. L., Chalkias, S., Gorham, J., Xu, Y., et al. (2016). Fulminant myocarditis with combination immune checkpoint blockade. *New England Journal of Medicine, 375*(18), 1749–1755. https://doi.org/10.1056/nejmoa1609214.

Jones, L. W., Habel, L. A., Weltzien, E., Castillo, A., Gupta, D., Kroenke, C. H., et al. (2016). Exercise and risk of cardiovascular events in women with nonmetastatic breast cancer. *Journal of Clinical Oncology, 34*(23), 2743–2749. https://doi.org/10.1200/JCO.2015.65.6603.

Ju, B. G., Lunyak, V. V., Perissi, V., Garcia-Bassets, I., Rose, D. W., Glass, C. K., et al. (2006). A topoisomerase IIβ-mediated dsDNA break required for regulated transcription. *Science, 312*(5781), 1798–1802. https://doi.org/10.1126/science.1127196.

Kalay, N., Basar, E., Ozdogru, I., Er, O., Cetinkaya, Y., Dogan, A., et al. (2006). Protective effects of carvedilol against anthracycline-induced cardiomyopathy. *Journal of the American College of Cardiology, 48*(11), 2258–2262. https://doi.org/10.1016/j.jacc.2006.07.052.

Kalivendi, S. V., Konorev, E. A., Cunningham, S., Vanamala, S. K., Kaji, E. H., Joseph, J., et al. (2005). Doxorubicin activates nuclear factor of activated T-lymphocytes and Fas ligand transcription: Role of mitochondrial reactive oxygen species and calcium. *Biochemical Journal, 389*(2), 527–539. https://doi.org/10.1042/BJ20050285.

Kantarjian, H. M., Hochhaus, A., Saglio, G., De Souza, C., Flinn, I. W., Stenke, L., et al. (2011). Nilotinib versus imatinib for the treatment of patients with newly diagnosed chronic phase, Philadelphia chromosome-positive, chronic myeloid leukaemia: 24-month minimum follow-up of the phase 3 randomised ENESTnd trial. *The Lancet Oncology*, *12*(9), 841–851. https://doi.org/10.1016/S1470-2045(11)70201-7.

Kennedy, A. R., Maity, A., & Sanzari, J. K. (2016). A review of radiation-induced coagulopathy and new findings to support potential prevention strategies and treatments. *Radiation Research*, *186*(2), 121–140. https://doi.org/10.1667/RR14406.1.

Kerkelä, R., Grazette, L., Yacobi, R., Iliescu, C., Patten, R., Beahm, C., et al. (2006). Cardiotoxicity of the cancer therapeutic agent imatinib mesylate. *Nature Medicine*, *12*(8), 908–916. https://doi.org/10.1038/nm1446.

Khouri, M. G., Douglas, P. S., Mackey, J. R., Martin, M., Scott, J. M., Scherrer-Crosbie, M., et al. (2012). Cancer therapy-induced cardiac toxicity in early breast cancer addressing the unresolved issues. *Circulation*, *126*(23), 2749–2763. https://doi.org/10.1161/CIRCULATIONAHA.112.100560.

Kosmas, C., Kallistratos, M. S., Kopterides, P., Syrios, J., Skopelitis, H., Mylonakis, N., et al. (2008). Cardiotoxicity of fluoropyrimidines in different schedules of administration: A prospective study. *Journal of Cancer Research and Clinical Oncology*, *134*(1), 75–82. https://doi.org/10.1007/s00432-007-0250-9.

Kwok, J. C., & Richardson, D. R. (2002). Unexpected anthracycline-mediated alterations in iron-regulatory protein-RNA-binding activity: The iron and copper complexes of anthracyclines decrease RNA-binding activity. *Molecular Pharmacology*, *62*(4), 888–900. https://doi.org/10.1124/mol.62.4.888.

Lamore, S. D., Kohnken, R. A., Peters, M. F., & Kolaja, K. L. (2020). Cardiovascular toxicity induced by kinase inhibitors: Mechanisms and preclinical approaches. *Chemical Research in Toxicology*. https://doi.org/10.1021/acs.chemrestox.9b00387.

Lee, M. M. Y., Sattar, N., McMurray, J. J. V., & Packard, C. J. (2019). Statins in the prevention and treatment of heart failure: A review of the evidence. *Current Atherosclerosis Reports*, *21*(10). https://doi.org/10.1007/s11883-019-0800-z.

Lemmon, M. A., & Schlessinger, J. (2010). Cell signaling by receptor tyrosine kinases. *Cell*, *141*(7), 1117–1134. https://doi.org/10.1016/j.cell.2010.06.011.

Leong, S. L., Chaiyakunapruk, N., & Lee, S. W. H. (2017). Candidate gene association studies of anthracycline-induced cardiotoxicity: A systematic review and meta-analysis. *Scientific Reports*, 7(1). https://doi.org/10.1038/s41598-017-00075-1.

León-Mateos, L., Mosquera, J., & Antón Aparicio, L. (2015). Treatment of sunitinib-induced hypertension in solid tumor by nitric oxide donors. *Redox Biology*, *6*, 421–425. https://doi.org/10.1016/j.redox.2015.09.007.

Lyon, A. R., Yousaf, N., Battisti, N. M. L., Moslehi, J., & Larkin, J. (2018). Immune checkpoint inhibitors and cardiovascular toxicity. *The Lancet. Oncology*, *19*(9), e447–e458. https://doi.org/10.1016/S1470-2045(18)30457-1.

Ma, H., Jones, K. R., Guo, R., Xu, P., Shen, Y., & Ren, J. (2010). Cisplatin compromises myocardial contractile function and mitochondrial ultrastructure: Role of endoplasmic reticulum stress. *Clinical and Experimental Pharmacology and Physiology*, *37*(4), 460–465. https://doi.org/10.1111/j.1440-1681.2009.05323.x.

Ma, J. H., Qin, L., & Li, X. (2020). Role of STAT3 signaling pathway in breast cancer. *Cell Communication and Signaling*, *18*(1), 1–13. https://doi.org/10.1186/s12964-020-0527-z.

Macdonald, J. S. (1999). Toxicity of 5-fluorouracil. *Oncology*, *13*(7 Suppl), 33–34. https://doi.org/10.2174/187152506775268785.

Mack, J. T. (2007). Dexrazoxane. In *XPharm: The comprehensive pharmacology reference* (pp. 1–7). https://doi.org/10.1016/B978-008055232-3.63698-0.

Madeddu, C., Deidda, M., Piras, A., Cadeddu, C., Demurtas, L., Puzzoni, M., et al. (2016). Pathophysiology of cardiotoxicity induced by nonanthracycline chemotherapy. *Journal of Cardiovascular Medicine*, *17*(type II), S12–S18. https://doi.org/10.2459/JCM.0000000000000376.

Madhusudan, S., & Ganesan, T. S. (2004). Tyrosine kinase inhibitors in cancer therapy. *Clinical Biochemistry*. https://doi.org/10.1016/j.clinbiochem.2004.05.006.
Mahmood, S. S., Fradley, M. G., Cohen, J. V., Nohria, A., Reynolds, K. L., Heinzerling, L. M., et al. (2018). Myocarditis in patients treated with immune checkpoint inhibitors. *Journal of the American College of Cardiology*, *71*(16), 1755–1764. https://doi.org/10.1016/j.jacc.2018.02.037.
Malato, A., Saccullo, G., Fazio, G., Vergara, B., Raso, S., Paolo Guarneri, G., et al. (2012). Drug-related cardiotoxicity for the treatment of haematological malignancies in elderly. *Current Pharmaceutical Design*, *16*(26), 2872–2879. https://doi.org/10.2174/13816121 0793176446.
Malik, M., & Camm, A. J. (2001). Evaluation of drug-induced QT interval prolongation: Implications for drug approval and labelling. *Drug Safety*, *24*(5), 323–351. https://doi.org/10.2165/00002018-200124050-00001.
Manning, G., Whyte, D. B., Martinez, R., Hunter, T., & Sudarsanam, S. (2002). The protein kinase complement of the human genome. *Science*. https://doi.org/10.1126/science.1075762.
Marchetti, C., Toldo, S., Chojnacki, J., Mezzaroma, E., Liu, K., Salloum, F. N., et al. (2015). Pharmacologic inhibition of the NLRP3 inflammasome preserves cardiac function after ischemic and nonischemic injury in the mouse. *Journal of Cardiovascular Pharmacology*, *66*(1), 1–8. https://doi.org/10.1097/FJC.0000000000000247.
McGowan, J. V., Chung, R., Maulik, A., Piotrowska, I., Walker, J. M., & Yellon, D. M. (2017). Anthracycline chemotherapy and cardiotoxicity. *Cardiovascular Drugs and Therapy*, *31*(1), 63–75. https://doi.org/10.1007/s10557-016-6711-0.
McSweeney, K. M., Bozza, W. P., Alterovitz, W. L., & Zhang, B. (2019). Transcriptomic profiling reveals p53 as a key regulator of doxorubicin-induced cardiotoxicity. *Cell Death Discovery*, *5*(1), 1–11. https://doi.org/10.1038/s41420-019-0182-6.
Meng, L., Lin, H., Zhang, J., Lin, N., Sun, Z., Gao, F., et al. (2019). Doxorubicin induces cardiomyocyte pyroptosis via the TINCR-mediated posttranscriptional stabilization of NLR family pyrin domain containing 3. *Journal of Molecular and Cellular Cardiology*, *136*, 15–26. https://doi.org/10.1016/j.yjmcc.2019.08.009.
Mercuro, G., Cadeddu, C., Piras, A., Dessì, M., Madeddu, C., Deidda, M., et al. (2007). Early epirubicin-induced myocardial dysfunction revealed by serial tissue doppler echocardiography: Correlation with inflammatory and oxidative stress markers. *The Oncologist*, *12*(9), 1124–1133. https://doi.org/10.1634/theoncologist.12-9-1124.
Metibemu, D. S., Akinloye, O. A., Akamo, A. J., Ojo, D. A., Okeowo, O. T., & Omotuyi, I. O. (2019). Exploring receptor tyrosine kinases-inhibitors in cancer treatments. *Egyptian Journal of Medical Human Genetics*. https://doi.org/10.1186/s43042-019-0035-0.
Metra, M., & Teerlink, J. R. (2017). Heart failure. *The Lancet*, *390*(10106), 1981–1995. https://doi.org/10.1016/S0140-6736(17)31071-1.
Mezzaroma, E., Mikkelsen, R. B., Toldo, S., Mauro, A. G., Sharma, K., Marchetti, C., et al. (2015). Role of Interleukin-1 in radiation-induced cardiomyopathy. *Molecular Medicine*, *21*(1), 210–218. https://doi.org/10.2119/molmed.2014.00243.
Min, S. S., & Wierzbicki, A. S. (2017). Radiotherapy, chemotherapy and atherosclerosis. *Current Opinion in Cardiology*. https://doi.org/10.1097/HCO.0000000000000404.
Minotti, G., Salvatorelli, E., & Menna, P. (2010). Pharmacological foundations of cardio-oncology. *Journal of Pharmacology and Experimental Therapeutics*, *334*(1), 2–8. https://doi.org/10.1124/jpet.110.165860.
Minotti, G., Salvatorelli, E., Menna, P., Ronchi, R., & Cairo, G. (2001). Doxorubicin irreversibly inactivates iron regulatory proteins 1 and 2 in cardiomyocytes: Evidence for distinct metabolic pathways and implications for iron-mediated cardiotoxicity of antitumor therapy. *Cancer Research*, *61*(23), 8422–8428.

Moghe, A., Ghare, S., Lamoreau, B., Mohammad, M., Barve, S., McClain, C., et al. (2015). Molecular mechanisms of acrolein toxicity: Relevance to human disease. *Toxicological Sciences, 143*(2), 242–255. https://doi.org/10.1093/toxsci/kfu233.

Mondal, P., Jain, D., Aronow, W. S., & Frishman, W. H. (2019). Cardiotoxicity of cancer therapies. *Cardiology in Review*. https://doi.org/10.1097/CRD.0000000000000239.

Montemurro, F. (2014). Should trastuzumab be administered concomitantly with anthracycline in women with early, HER2-positive breast cancer? *Translational Cancer Research, 3*(6), 541–546. https://doi.org/10.3978/j.issn.2218-676X.2014.06.09.

Musgrove, E. A., Caldon, C. E., Barraclough, J., Stone, A., & Sutherland, R. L. (2011). Cyclin D as a therapeutic target in cancer. *Nature Reviews Cancer, 11*(8), 558–572. https://doi.org/10.1038/nrc3090.

Nemeth, B. T., Varga, Z. V., Wu, W. J., & Pacher, P. (2017). Trastuzumab cardiotoxicity: From clinical trials to experimental studies. *British Journal of Pharmacology, 174*(21), 3727–3748. https://doi.org/10.1111/bph.13643.

Nishikawa, R., Ji, X. D., Harmon, R. C., Lazar, C. S., Gill, G. N., Cavenee, W. K., et al. (1994). A mutant epidermal growth factor receptor common in human glioma confers enhanced tumorigenicity. *Proceedings of the National Academy of Sciences of the United States of America, 91*(16), 7727–7731. https://doi.org/10.1073/pnas.91.16.7727.

Nitiss, J. L. (2009). DNA topoisomerase II and its growing repertoire of biological functions. *Nature Reviews Cancer*. https://doi.org/10.1038/nrc2608.

Octavia, Y., Tocchetti, C. G., Gabrielson, K. L., Janssens, S., Crijns, H. J., & Moens, A. L. (2012). Doxorubicin-induced cardiomyopathy: From molecular mechanisms to therapeutic strategies. *Journal of Molecular and Cellular Cardiology, 52*(6), 1213–1225. https://doi.org/10.1016/j.yjmcc.2012.03.006.

O'Donnell, J. S., Teng, M. W. L., & Smyth, M. J. (2019). Cancer immunoediting and resistance to T cell-based immunotherapy. *Nature Reviews Clinical Oncology*. https://doi.org/10.1038/s41571-018-0142-8.

Olivieri, J., Perna, G. P., Bocci, C., Montevecchi, C., Olivieri, A., Leoni, P., et al. (2017). Modern management of anthracycline-induced cardiotoxicity in lymphoma patients: Low occurrence of cardiotoxicity with comprehensive assessment and tailored substitution by nonpegylated liposomal doxorubicin. *The Oncologist, 22*(4), 422–431. https://doi.org/10.1634/theoncologist.2016-0289.

Otter, S. J., Stewart, A. J., & Devlin, P. M. (2019). Modern brachytherapy. *Hematology/Oncology Clinics of North America, 33*(6), 1011–1025. https://doi.org/10.1016/j.hoc.2019.08.011.

Palaskas, N., Morgan, J., Daigle, T., Banchs, J., Durand, J. B., Hong, D., et al. (2019). Targeted cancer therapies with pericardial effusions requiring pericardiocentesis focusing on immune checkpoint inhibitors. *American Journal of Cardiology, 123*(8), 1351–1357. https://doi.org/10.1016/j.amjcard.2019.01.013.

Panjrath, G. S., Patel, V., Valdiviezo, C. I., Narula, N., Narula, J., & Jain, D. (2007). Potentiation of doxorubicin cardiotoxicity by iron loading in a rodent model. *Journal of the American College of Cardiology, 49*(25), 2457–2464. https://doi.org/10.1016/j.jacc.2007.02.060.

Pardo Sanz, A., & Zamorano, J. L. (2020). Cardiotoxicity': Time to define new targets? *European Heart Journal*. https://doi.org/10.1093/eurheartj/ehaa013.

Parmar, A., Macluskey, M., Mc Goldrick, N., Conway, D. I., Glenny, A. M., Clarkson, J. E., et al. (2021). Interventions for the treatment of oral cavity and oropharyngeal cancer: Chemotherapy. *Cochrane Database of Systematic Reviews, 2021*(12). https://doi.org/10.1002/14651858.CD006386.pub4.

Paul, M. K., & Mukhopadhyay, A. K. (2012). Tyrosine kinase—Role and significance in cancer. *International Journal of Medical Sciences, 1*(2), 101–115. https://doi.org/10.7150/ijms.1.101.

Payne, D. L., & Nohria, A. (2017). Prevention of chemotherapy induced cardiomyopathy. *Current Heart Failure Reports, 14*(5), 398–403. https://doi.org/10.1007/s11897-017-0353-9.

Pituskin, E., Mackey, J. R., Koshman, S., Jassal, D., Pitz, M., Haykowsky, M. J., et al. (2017). Multidisciplinary approach to novel therapies in cardio-oncology research (MANTICORE 101-breast): A randomized trial for the prevention of trastuzumab-associated cardiotoxicity. *Journal of Clinical Oncology, 35*(8), 870–877. https://doi.org/10.1200/JCO.2016.68.7830.

Pottier, C., Fresnais, M., Gilon, M., Jérusalem, G., Longuespée, R., & Sounni, N. E. (2020). Tyrosine kinase inhibitors in cancer: Breakthrough and challenges of targeted therapy. *Cancers, 12*(3), 731. https://doi.org/10.3390/cancers12030731.

Robert, C., Long, G. V., Brady, B., Dutriaux, C., Maio, M., Mortier, L., et al. (2015). Nivolumab in previously untreated melanoma without BRAF mutation. *New England Journal of Medicine, 372*(4), 320–330. https://doi.org/10.1056/nejmoa1412082.

Rottenberg, S., Disler, C., & Perego, P. (2021). The rediscovery of platinum-based cancer therapy. *Nature Reviews Cancer, 21*(1), 37–50. https://doi.org/10.1038/s41568-020-00308-y.

Rudd, C. E., Taylor, A., & Schneider, H. (2009). CD28 and CTLA-4 coreceptor expression and signal transduction. *Immunological Reviews*. https://doi.org/10.1111/j.1600-065X.2009.00770.x.

Russo, M., Bono, E., & Ghigo, A. (2021). The interplay between autophagy and senescence in anthracycline cardiotoxicity. *Current Heart Failure Reports, 18*(4), 180–190. https://doi.org/10.1007/s11897-021-00519-w.

Saif, M. W., Shah, M. M., & Shah, A. R. (2009). Fluoropyrimidine-associated cardiotoxicity: Revisited. *Expert Opinion on Drug Safety, 8*(2), 191–202. https://doi.org/10.1517/14740330902733961.

Santarpia, L., Lippman, S. M., & El-Naggar, A. K. (2012). Targeting the MAPKRASRAF signaling pathway in cancer therapy. *Expert Opinion on Therapeutic Targets, 16*(1), 103–119. https://doi.org/10.1517/14728222.2011.645805.

Sara, J. D., Kaur, J., Khodadadi, R., Rehman, M., Lobo, R., Chakrabarti, S., et al. (2018). 5-fluorouracil and cardiotoxicity: A review. *Therapeutic Advances in Medical Oncology, 10*. https://doi.org/10.1177/1758835918780140.

Schwartz, R. G., McKenzie, W. B., Alexander, J., Sager, P., D'Souza, A., Manatunga, A., et al. (1987). Congestive heart failure and left ventricular dysfunction complicating doxorubicin therapy. Seven-year experience using serial radionuclide angiocardiography. *The American Journal of Medicine, 82*(6), 1109–1118. https://doi.org/10.1016/0002-9343(87)90212-9.

Shaul, P., Frenkel, M., Goldstein, E. B., Mittelman, L., Grunwald, A., Ebenstein, Y., et al. (2013). The structure of anthracycline derivatives determines their subcellular localization and cytotoxic activity. *ACS Medicinal Chemistry Letters, 4*(3), 323–328. https://doi.org/10.1021/ml3002852.

Siegel, D., Martin, T., Nooka, A., Harvey, R. D., Vij, R., Niesvizky, R., et al. (2013). Integrated safety profile of single-agent carfilzomib: Experience from 526 patients enrolled in 4 phase II clinical studies. *Haematologica, 98*(11), 1753–1761. https://doi.org/10.3324/haematol.2013.089334.

Siegel, R. L., Miller, K. D., & Jemal, A. (2019). Cancer statistics, 2019. *CA: a Cancer Journal for Clinicians, 69*(1), 7–34. https://doi.org/10.3322/caac.21551.

Singh, R. K., Kumar, S., Prasad, D. N., & Bhardwaj, T. R. (2018). Therapeutic journey of nitrogen mustard as alkylating anticancer agents: Historic to future perspectives. *European Journal of Medicinal Chemistry, 151*, 401–433. https://doi.org/10.1016/j.ejmech.2018.04.001.

Song, Y., Zhang, C., Wang, C., Zhao, L., Wang, Z., Dai, Z., et al. (2016). Ferulic acid against cyclophosphamide-induced heart toxicity in mice by inhibiting NF-B pathway.

Evidence-based Complementary and Alternative Medicine, 2016. https://doi.org/10.1155/2016/1261270.

Steinherz, L. J., Steinherz, P. G., Mangiacasale, D., O'Reilly, R., Allen, J., Sorell, M., et al. (1981). Cardiac changes with cyclophosphamide. *Medical and Pediatric Oncology, 9*(5), 417–422. https://doi.org/10.1002/mpo.2950090502.

Stein-Merlob, A. F., Rothberg, M. V., Ribas, A., & Yang, E. H. (2021). Cardiotoxicities of novel cancer immunotherapies. *Heart*. https://doi.org/10.1136/heartjnl-2020-318083.

Stewart, F. A., Seemann, I., Hoving, S., & Russell, N. S. (2013). Understanding radiation-induced cardiovascular damage and strategies for intervention. *Clinical Oncology, 25*(10), 617–624. https://doi.org/10.1016/j.clon.2013.06.012.

Swain, S. M., Whaley, F. S., & Ewer, M. S. (2003). Congestive heart failure in patients treated with doxorubicin: A retrospective analysis of three trials. *Cancer, 97*(11), 2869–2879. https://doi.org/10.1002/cncr.11407.

Swain, S. M., Whaley, F. S., Gerber, M. C., Ewer, M. S., Bianchine, J. R., & Gams, R. A. (1997). Delayed administration of dexrazoxane provides cordiaprotection for patients with advanced breast cancer treated with doxorubicin-containing therapy. *Journal of Clinical Oncology, 15*(4), 1333–1340. https://doi.org/10.1200/JCO.1997.15.4.1333.

Swain, S. M., Whaley, F. S., Gerber, M. C., Weisberg, S., York, M., Spicer, D., et al. (1997). Cardioprotection with dexrazoxane for doxorubicin-containing therapy in advanced breast cancer. *Journal of Clinical Oncology, 15*(4), 1318–1332. https://doi.org/10.1200/JCO.1997.15.4.1318.

Tadokoro, T., Ikeda, M., Ide, T., Deguchi, H., Ikeda, S., Okabe, K., et al. (2020). Mitochondria-dependent ferroptosis plays a pivotal role in doxorubicin cardiotoxicity. *JCI Insight, 5*(9). https://doi.org/10.1172/jci.insight.132747.

Tashakori Beheshti, A., Mostafavi Toroghi, H., Hosseini, G., Zarifian, A., Homaei Shandiz, F., & Fazlinezhad, A. (2016). Carvedilol administration can prevent doxorubicin-induced cardiotoxicity: A double-blind randomized trial. *Cardiology (Switzerland), 134*(1), 47–53. https://doi.org/10.1159/000442722.

Taunk, N. K., Haffty, B. G., Kostis, J. B., & Goyal, S. (2015). Radiation-induced heart disease: Pathologic abnormalities and putative mechanisms. *Frontiers in Oncology, 5*, 1–8. https://doi.org/10.3389/fonc.2015.00039.

Thomson, M. (2002). Evidence of undiscovered cell regulatory mechanisms: Phosphoproteins and protein kinases in mitochondria. *Cellular and Molecular Life Sciences, 59*(2), 213–219. https://doi.org/10.1007/s00018-002-8417-7.

Tiwari, M. (2012). Antimetabolites: Established cancer therapy. *Journal of Cancer Research and Therapeutics, 8*(4), 510–519. https://doi.org/10.4103/0973-1482.106526.

Tolcher, A. W., Appleman, L. J., Shapiro, G. I., Mita, A. C., Cihon, F., Mazzu, A., et al. (2011). A phase I open-label study evaluating the cardiovascular safety of sorafenib in patients with advanced cancer. *Cancer Chemotherapy and Pharmacology, 67*(4), 751–764. https://doi.org/10.1007/s00280-010-1372-3.

Toldo, S., & Abbate, A. (2018). The NLRP3 inflammasome in acute myocardial infarction. *Nature Reviews Cardiology, 15*(4), 203–214. https://doi.org/10.1038/nrcardio.2017.161.

Toldo, S., Mauro, A. G., Cutter, Z., & Abbate, A. (2018). Inflammasome, pyroptosis, and cytokines in myocardial ischemia-reperfusion injury. *American Journal of Physiology-Heart and Circulatory Physiology, 315*(6), H1553–H1568. https://doi.org/10.1152/ajpheart.00158.2018.

Tong, M., & Seeliger, M. A. (2015). Targeting conformational plasticity of protein kinases. *ACS Chemical Biology, 10*(1), 190–200. https://doi.org/10.1021/cb500870a.

Touyz, R. M., & Herrmann, J. (2018). Cardiotoxicity with vascular endothelial growth factor inhibitor therapy. *NPJ Precision Oncology, 2*(1). https://doi.org/10.1038/s41698-018-0056-z.

Van Der Meeren, A., Squiban, C., Gourmelon, P., Lafont, H., & Gaugler, M. H. (1999). Differential regulation by IL-4 and IL-10 of radiation-induced IL-6 and IL-8 production and ICAM-1 expression by human endothelial cells. *Cytokine*, *11*(11), 831–838. https://doi.org/10.1006/cyto.1999.0497.

Vejpongsa, P., & Yeh, E. T. H. (2014). Topoisomerase 2β: A promising molecular target for primary prevention of anthracycline-induced cardiotoxicity. *Clinical Pharmacology and Therapeutics*, *95*(1), 45–52. https://doi.org/10.1038/clpt.2013.201.

Verstovsek, S., Courby, S., Griesshammer, M., Mesa, R. A., Brachmann, C. B., Kawashima, J., et al. (2017). A phase 2 study of momelotinib, a potent JAK1 and JAK2 inhibitor, in patients with polycythemia vera or essential thrombocythemia. *Leukemia Research*, *60*, 11–17. https://doi.org/10.1016/j.leukres.2017.05.002.

Waldman, A. D., Fritz, J. M., & Lenardo, M. J. (2020). A guide to cancer immunotherapy: From T cell basic science to clinical practice. *Nature Reviews Immunology*. https://doi.org/10.1038/s41577-020-0306-5.

Weiss, R. B., Sarosy, G., Clagett-Carr, K., Russo, M., & Leyland-Jones, B. (1986). Anthracycline analogs the past, present, and future. *Cancer Chemotherapy and Pharmacology*, *18*(3), 185–197. https://doi.org/10.1007/BF00273384.

Wenningmann, N., Knapp, M., Ande, A., Vaidya, T. R., & Ait-Oudhia, S. (2019). Insights into doxorubicin-induced cardiotoxicity: Molecular mechanisms, preventive strategies, and early monitoring. *Molecular Pharmacology*, *96*(2), 219–232. https://doi.org/10.1124/mol.119.115725.

Wherry, E. J. (2011). T cell exhaustion. *Nature Immunology*, *12*(6), 492–499. https://doi.org/10.1038/ni.2035.

Wherry, E. J., & Kurachi, M. (2015). Molecular and cellular insights into T cell exhaustion. *Nature Reviews Immunology*, *15*(8), 486–499. https://doi.org/10.1038/nri3862.

Wu, P., Oren, O., Gertz, M. A., & Yang, E. H. (2020). Proteasome inhibitor-related cardiotoxicity: Mechanisms, diagnosis, and management. *Current Oncology Reports*, *22*-(7). https://doi.org/10.1007/s11912-020-00931-w.

Xiao, Y., Yin, J., Wei, J., & Shang, Z. (2014). Incidence and risk of cardiotoxicity associated with bortezomib in the treatment of cancer: A systematic review and meta-analysis. *PLoS One*, *9*(1), e87671. https://doi.org/10.1371/journal.pone.0087671.

Xu, X., Persson, H. L., & Richardson, D. R. (2005). Molecular pharmacology of the interaction of anthracyclines with iron. *Molecular Pharmacology*, *68*(2), 261–271. https://doi.org/10.1124/mol.105.013383.

Yeh, E. T. H., & Bickford, C. L. (2009). Cardiovascular complications of cancer therapy: Incidence, pathogenesis, diagnosis, and management. *Journal of the American College of Cardiology*, *53*(24), 2231–2247. https://doi.org/10.1016/J.JACC.2009.02.050.

Yellon, D. M., & Downey, J. M. (2003). Preconditioning the myocardium: From cellular physiology to clinical cardiology. *Physiological Reviews*, *83*(4), 1113–1151. https://doi.org/10.1152/physrev.00009.2003.

Zanoli, L., Lentini, P., Briet, M., Castellino, P., House, A. A., London, G. M., et al. (2019). Arterial stiffness in the heart disease of CKD. *Journal of the American Society of Nephrology*, *30*(6), 918–928. https://doi.org/10.1681/asn.2019020117.

Zhang, S., Liu, X., Bawa-Khalfe, T., Lu, L.-S. S., Lyu, Y. L., Liu, L. F., et al. (2012). Identification of the molecular basis of doxorubicin-induced cardiotoxicity. *Nature Medicine*, *18*(11), 1639–1642. https://doi.org/10.1038/nm.2919.

Zhang, L., Reynolds, K. L., Lyon, A. R., Palaskas, N., & Neilan, T. G. (2021). The evolving immunotherapy landscape and the epidemiology, diagnosis, and management of cardiotoxicity. *JACC: CardioOncology*, *3*(1), 35–47. https://doi.org/10.1016/j.jaccao.2020.11.012.

Zwick, E., Bange, J., & Ullrich, A. (2002). Receptor tyrosine kinases as targets for anticancer drugs. *Trends in Molecular Medicine*, *8*(1), 17–23. https://doi.org/10.1016/S1471-4914(01)02217-1.

CHAPTER SIX

Strategies to mitigate the toxicity of cancer therapeutics

Adriana M. Kahn[a,b,†], Kim R.M. Blenman[a,b,c,†], Steve T. Sonis[d,e,f], and Maryam B. Lustberg[a,b,*]
[a]Yale Cancer Center, New Haven, CT, United States
[b]Department of Internal Medicine, Section of Medical Oncology, Yale University, New Haven, CT, United States
[c]Department of Computer Science, Yale University, New Haven, CT, United States
[d]Department of Surgery, Divisions of Oral Medicine and Dentistry, Brigham and Women's Hospital and the Dana-Farber Cancer Institute, Boston, MA, United States
[e]Department of Oral Medicine, Infection and Immunity, Harvard School of Dental Medicine, Boston, MA, United States
[f]Biomodels LLC., Waltham, MA, United States
*Corresponding author: e-mail address: maryam.lustberg@yale.edu

Contents

Abstract

Cancer therapeutics are dynamically evolving, and include traditional chemotherapy and hormone therapy, as well as more recently developed treatment modalities, such as tyrosine kinase inhibitors, monoclonal antibodies and the revolutionary approach based on immune checkpoint inhibition. These regimens are unfortunately not free of adverse events, and patients with cancer are a susceptible population experiencing a myriad of disease and treatment toxicities combined. In this review, we present the latest overview of the management of the most common systemic cancer treatment

† Co Primary authors.

Advances in Cancer Research, Volume 155
ISSN 0065-230X
https://doi.org/10.1016/bs.acr.2022.02.006

symptoms and the science of symptom management supporting these strategies. We discuss cancer-related cognitive impairment, ocular toxicity, ototoxicity, oral mucosal toxicities, gastrointestinal toxicities, renal toxicity, aromatase inhibitor-induced musculoskeletal symptoms, chemotherapy-induced peripheral neuropathy, and immunotherapy-induced autoimmunity derived from systemic therapies for cancer. In summary, we review the future directions and ideal goals of symptom science research in order to benefit patients utilizing a comprehensive individualized approach.

1. Overview of organ specific toxicities

The remedy of symptoms and toxicities due to cancer and cancer therapy has been a long-standing goal of humans from the earliest known history. The underdiagnosis and consequently undertreatment of symptoms negatively impacts disease outcomes due treatment modifications, discontinuations, and poor quality of life (QOL). In the last several decades, advances in symptom science have paved the way for a better understanding of treatment toxicities and potential for improving management strategies.

Chemotherapy and hormone therapy, more traditionally, and newer treatment modalities, including tyrosine kinase inhibitors (TKI), monoclonal antibodies and immune checkpoint inhibitors (ICI) are often employed as part of a multimodal approach to the treatment of many tumors. These regimens can be complex, and patients with cancer are a susceptible population experiencing a myriad of symptom toxicities. In this review, we will present the latest overview of the management of the most common cancer treatment symptoms and the science of symptom management supporting these strategies. Specifically, we will discuss cancer-related cognitive impairment, ocular toxicity, ototoxicity, oral mucosal toxicities, gastrointestinal toxicities, renal toxicity, aromatase inhibitor-induced musculoskeletal symptoms, chemotherapy-induced peripheral neuropathy, and immunotherapy-induced autoimmunity, derived from systemic therapies for cancer, as briefly summarized in Fig. 1.

2. Cancer related cognitive impairment

"Chemo-brain" is a term that patients use to describe the cognitive changes that they experience with cancer and their cancer-related treatments (Mayo et al., 2021). Cancer-related cognitive impairment (CRCI), as this is formally termed, is experience in up to 75% of patients treated with chemotherapy. Of those, 35% of survivors experience CRCI months to years after

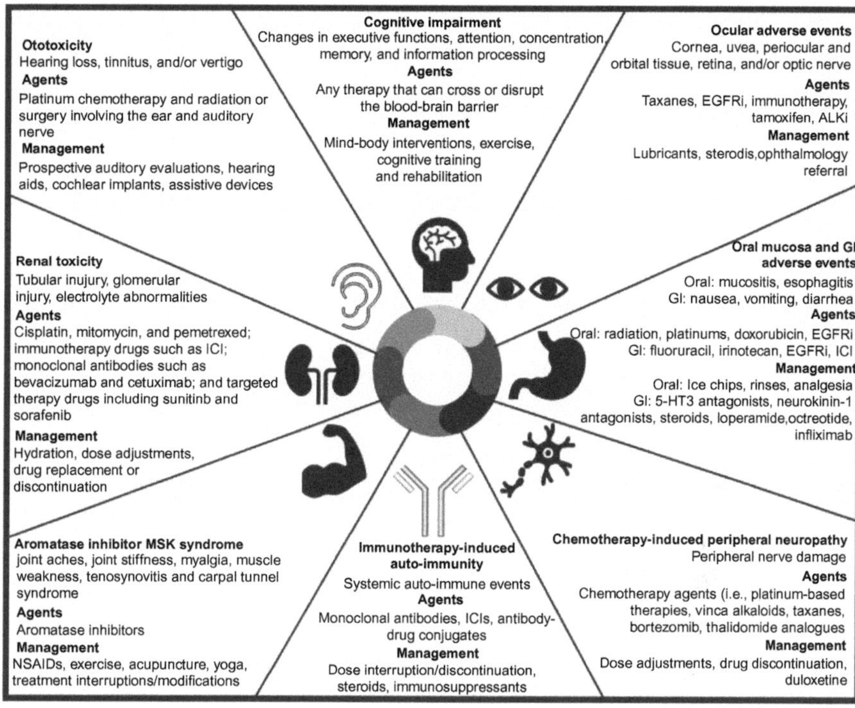

Fig. 1 Summary of organ-specific toxicities, their causal agents and management. EGFRi, epidermal growth factor receptor inhibitors; ALKi, anaplastic lymphoma kinase inhibitors; ICI, immune checkpoint inhibitors; GI, gastrointestinal; 5-HT3, 5-hydroxytriptamine; MSK, musculoskeletal; NSAIDs, non-steroidal anti-inflammatory drugs.

completing treatment (Mayo et al., 2021). Hallmarks of CRCI include changes in executive functions, duration of attention, concentration, memory, and speed of information processing. These changes impact daily living and QOL and require recognition of the problem, proper assessment, and adequate management to improve patient QOL (Mayo et al., 2021).

The underlying mechanism that leads to symptoms of CRCI is still under investigation, with preclinical and clinical data supporting the role of neuroinflammatory pathways. No definitive clinical trials have been performed to identify mechanisms of CRCI. However, there have been mechanisms theorized based on the suggested function of specific therapies (Mayo et al., 2021). For example, it is possible that any therapy (drug; radiation) that can cross or disrupt the blood–brain barrier can have direct neurotoxic effects on the brain that are associated with their intended role on cancer cells (Mayo et al., 2021).

In addition to direct treatment-related factors, several studies have investigated the role of genetic polymorphisms in CRCI. The most promising of these include apolipoprotein E (APOE) gene and the catechol-*o*-methyltransferase (COMT) gene as well as inflammatory cytokines such as TNF-alpha and IL-6. Cancer survivors with the E4 allele of the APOE gene experience worse cognitive decline with chemotherapy (Ahles et al., 2003; Mandelblatt et al., 2018). However, this association has not been shown consistently in all studies. Other findings have illustrated that single nucleotide polymorphisms in COMT gene are associated with CRCI in patients with breast cancer (Cheng et al., 2016). Prospective, larger studies are needed to confirm findings before such biomarkers are used to risk-stratify patients as part of routine care.

Patient education is the first step in management as it verifies their concerns, sets expectations, and helps them prepare coping strategies. The search for effective treatments for CRCI is still ongoing. Over 30 clinical studies have utilized strategies such as mind-body interventions, exercise, cognitive training and rehabilitation, and drug therapies (Mayo et al., 2021). However, none have been effective due to inadequate study design including small sample sizes, lack of proper controls, lack of appropriate measurement of CRCI outcomes, and the complexity of the symptoms (Mayo et al., 2021). Uncovering effective treatments may ultimately be challenging due to the temporal nature of the CRCI symptoms and the likelihood that combination and presentation of symptoms are unique to each patient. Therefore, it is possible that a more fruitful focus should be on personalized interventions.

3. Ocular toxicities

Oncological treatment-related ocular toxicities are overall rare. Most of the presenting symptoms are mild and managed symptomatically without the need of treatment discontinuation. However, some adverse events can be debilitating and irreversible. All anatomical compartments of the eye can be involved (cornea, uvea, periocular and orbital tissue, retina, and/or optic nerve) depending on the type of therapy. Therefore symptoms, signs and treatment strategies vary.

Corneal toxicity includes conjunctivitis, corneal erosion and blepharitis (Gilbard, 2005; Liu, Francis, & Abramson, 2021). Conjunctivitis can be managed with eye lubrication, omega-3 fatty acids, humidifiers, and severe cases may require topical steroid or immunosuppressive agents, as well as punctal occlusion to diminish tear outflow (Gilbard, 2005; Liu et al., 2021). Blepharitis can be managed with lid hygiene and warm compresses (Gilbard, 2005; Liu et al., 2021). These are most caused by epidermal growth factor receptor (EGFR) inhibitors, BRAF inhibitors, ICI, imatinib, tamoxifen, docetaxel, fluoropyrimidines, and anthracyclines (Ahmadi & Esmaeli, 2001; Borkar, Lacouture, & Basti, 2013; Chan et al., 2013; Choe et al., 2014; Curran & Luce, 1989; Eiseman et al., 2003; Esmaeli et al., 2002; Fraunfelder et al., 2003; Liu et al., 2021; Nayfield & Gorin, 1996; Papavasileiou et al., 2016; Saint-Jean et al., 2012; Tarafdar et al., 2012; Waikhom, Fraunfelder, & Henner, 2000).

Anterior uveitis may produce pain and redness, whereas posterior and intermediate uveitis are in general painless but may result in nonspecific visual changes, such as floaters and/or reduced visual acuity (Arepalli & Venkat, 2020; Liu et al., 2021). Upon suspicion of uveal involvement, prompt ophthalmological referral and treatment are recommended to control inflammation, pain and prevent further vision loss (Arepalli & Venkat, 2020; Liu et al., 2021). Treatment ranges from topical corticosteroid drops, periocular or intraocular steroid injections, and systemic corticosteroids (Arepalli & Venkat, 2020; Liu et al., 2021). Uveitis occurs most commonly with ipilimumab, BRAF inhibitors, EGFR inhibitor and cytarabine (Choe et al., 2014; Lim et al., 2010; Liu et al., 2021; Planer et al., 2004; Robinson et al., 2004).

Periocular toxicities include abnormalities of the eyelashes, lower lid, chalazia, excess tearing with or without nasolacrimal duct obstruction, and periorbital edema, which are in general treated symptomatically without drug discontinuation (Gilbard, 2005; Liu et al., 2021). However, if patients present with pain with eye movement, orbital congestion, proptosis of the eye, exposure keratopathy, ophthalmoplegia and diplopia, potential optic neuropathy, and blindness, orbital inflammation is suspected and is considered an ophthalmologic emergency (Liu et al., 2021). Periocular and orbital toxicities are seen with ipilimumab, EGFR inhibitors; BRAF inhibitors, imatinib, docetaxel, antimetabolites, and bortezomib (Chan et al., 2013; Choe et al., 2014; Doroshow et al., 1981; Eiseman et al., 2003; Fraunfelder & Yang, 2016; Kurata et al., 2006; Liu et al., 2021; Papavasileiou et al., 2016).

Retinal toxicities with damage to the retina and/or optic nerve may cause visual loss with or without ocular pain and warrant withholding of the suspected drug and urgent referral to an ophthalmologist (Grzybowski et al., 2015; Liu et al., 2021). Retinal toxicity is most seen with ipilimumab, trastuzumab, vemurafenib, mitogen-activated protein kinase [MEK] inhibitors, tamoxifen, platinum analogs, antimetabolites, and taxanes (Choe et al., 2014; Francis et al., 2020; Gianni et al., 1995; Johansson, 1992; Joshi & Garretson, 2007; Kaiser-Kupfer & Lippman, 1978; Liu et al., 2021; Nayfield & Gorin, 1996; Rankin & Pitts, 1993; Saleh et al., 2011; Wong, Lee, & Huang, 2012). Optic nerve toxicities are most common with anaplastic lymphoma kinase (ALK) inhibitors, imatinib, taxanes, vinca alkaloids, platinum analogs, interferon and methotrexate (MTX) (Camidge et al., 2012; Capri et al., 1994; Esmaeli et al., 2001; Fraunfelder et al., 2003; Gianni et al., 1995; Johansson, 1992; Liu et al., 2021; Rankin & Pitts, 1993; Sanderson, Kuwabara, & Cogan, 1976).

4. Ototoxicity

Ototoxicity, manifesting as hearing loss, tinnitus, and/or vertigo, is a common toxicity that manifests with a range of cancer therapies including platinum chemotherapy and radiation or surgery involving the ear and auditory nerve (Landier, 2016). Long-term effects may include permanent, severe to profound hearing loss or deafness, resulting in problems with communication and social interaction, and impaired health-related QOL. In young children, ototoxicity also negatively impacts cognitive performance and language and social skills acquisition. Risk factors for ototoxicity includes younger age (particularly <5), presence of central nervous tumors and cumulative toxicity of the drug and length of exposure. Variability of prevalence rates have been reported ranging up to 90% in some studies with platinum (Landier, 2016). Maturation of neuronal pathways and auditory structures continues during infancy and early childhood, making young children particularly vulnerable to the ototoxic effects of cancer therapies. Hence, the burden of morbidity related to ototoxic agents is substantially higher in very young children. A variety of candidate genes, including TPMT, COMT, and ACYP2, potentially involved in ototoxicity related to cancer therapy, have been explored although utility in routine clinical practice has not been established (Soliman et al., 2018).

With platinum chemotherapy, the mechanism of ototoxic injury is related to the generation of oxygen species within the cochlea, with consequent destruction of cochlear hair cells. Once the cochlear sensory hair cells are destroyed, they cannot regenerate; therefore, drug-related sensorineural hearing loss is almost always bilateral and irreversible, and also can be accompanied by tinnitus and vertigo (Kros & Steyger, 2019; Landier, 2016).

Clinical management of cancer-related ototoxicity includes surveillance for this toxicity with prospective auditory evaluations for patients receiving therapy with ototoxic potential (Brooks & Knight, 2018). For patients with a history of treatment with platinum chemotherapy, an auditory assessment is generally recommended at the baseline long-term follow-up visit and then as clinically indicated. There is currently no standardly approved prevention strategy for platinum-induced ototoxicity although a number of agents have been tested with limited success including amifostine and sodium thiosulfates (Freyer et al., 2020). Management strategies include hearing aids, cochlear implants, and assistive devices such as auditory trainers and telephone amplifiers (Landier, 2016).

5. Oral mucosal toxicities

Mucositis and esophagitis are common toxicities associated with a number of chemotherapy regimens, chemoradiation such as for patients with head and neck cancers or in bone marrow ablative regimens prior to stem cell transplantation (Sonis, 2004). Chemotherapeutic drugs across various classes have the potential to cause mucositis including cytotoxic chemotherapy (cytarabine, doxorubicin, etoposide, melphalan, MTX), and newer molecularly targeted agents like VEGF inhibitors (sunitinib, sorafenib, lenvatinib, and regorafenib), mTOR inhibitors (temsirolimus and everolimus), CDK 4/6 inhibitors (palbociclib), PARP inhibitors (niraparib), and EGFR inhibitors (cetuximab, erlotinib, dacomitinib, and mobocertinib) (Elting et al., 2013; Watters, Epstein, & Agulnik, 2011). The sequala of this toxicity is nutritional compromise due to impaired oral intake, reduced QOL and the potential for reduced drug dose intensity.

The mechanism of action relates to the cytotoxic effects of most chemotherapeutic agents leading to the upregulation and generation of messenger signals for proinflammatory cytokines. These result in tissue inflammation and, in the more severe presentations, mucosal ulceration and the loss of tissue integrity (Bowen et al., 2019; Sonis, 2004; Sonis et al., 2004). Clinically, mucositis begins with soft tissue erythema of the oral mucosa followed by development of desquamative patches that can progress to shallow ulcerations. If mucositis is erosive, it can further be complicated by secondary infection and sepsis, particularly in the setting of neutropenia. In general, mucositis is self-limiting with resolution typically within 14 days after chemotherapy administration and with neutrophil recovery (Epstein & Schubert, 2003).

For the prevention of oral mucositis, cryotherapy (sucking on ice chips) before, during, and/or after infusion of systemic therapies has been shown to be effective in patients receiving melphalan for multiple myeloma and 5-fluorouracil for solid tumors (Nawi et al., 2018; Riley et al., 2015). Palifermin, a recombinant keratinocyte growth factor (Farrell et al., 1998), has been shown to be beneficial in preventing and shortening the duration of oral mucositis in patients undergoing autologous hematopoietic cell transplantation (HCT) with total-body irradiation (Spielberger et al., 2004). Palifermin is the only mechanistically-directed biological or pharmaceutical approved by the US Food and Drug Administration (FDA), and its use is restricted to patients with hematological malignancies receiving conditioning regimens in preparation for stem cell transplants. While palifermin has apparent efficacy relative to mucositis mitigation in patients with epithelial tumors, the presence of KGF receptors in these tumors and the potential for palifermin-mediated tumor proliferation has restricted its use in this population (Rosen et al., 2006).

Photobiomodulation, an intraoral helium-neon laser (He-Ne laser) and a form of low-level laser irradiation, is another potential option for mucositis in patients undergoing conditioning therapy for HCT (Zadik et al., 2019). Photobiomodulation has been recommended in guidelines, including those by the Multinational Association of Supportive Care in Cancer (MASCC)/International Society of Oral Oncology (ISOO) and European Society for Medical Oncology (ESMO), to reduce the incidence of oral mucositis in patients receiving a conditioning regimen before HCT (Elad et al., 2020; Peterson et al., 2015). The clinical practice guidelines from the MASCC/ISOO also recommend photobiomodulation to prevent treatment-related mucositis in patients with head and neck cancer who are undergoing radiation therapy with or without chemotherapy (Elad et al., 2020). However, given the complex biological impact of photobiomodulation on pathways known to negatively impact tumor response in patients with solid tumors, the risk of the device in this patient cohort is still undefined (Sonis, 2020; Sonis et al., 2016).

The treatment of established mucositis is supportive and consists of good oral care including rinses with saline and/or sodium bicarbonate, and topical and/or systemic analgesia. A randomized study evaluated different mouthwashes consisting of salt and soda, chlorhexidine and "magic mouthwash" (i.e., lidocaine, diphenhydramine, aluminum hydroxide, and magnesium hydroxide) and found no differences in reduction of mucositis (Dodd et al., 2000). Additionally, dexamethasone mouthwash has been shown to have benefit reducing the incidence of stomatitis in patients treated with everolimus and exemestane for hormone receptor-positive, HER2-negative metastatic breast cancer (Rugo et al., 2017). However, its benefit with stomatitis associated with newer targeted therapies is not established.

Both topical and systemic analgesic approaches have been used to manage pain associated with mucositis. Topical lidocaine solutions provide pain relief; however, they require frequent administration (Carnel et al., 1990). Topical doxepin, a tricyclic antidepressant can provide clinically significant palliation for pain related to mucositis (Epstein et al., 2001). However, an updated systematic review in 2019 (Saunders et al., 2020) showed no role for topical doxepin which led to change in 2020 MASCC/ISOO guideline to "no guideline possible" (Elad et al., 2020). Opiates may be necessary for the management of severe mucositis pain, and they are an important part of the optimal care of patients with such symptoms (Coda et al., 1997). Given the significant unmet need for an effective mucositis intervention, the development of additional effective interventions is very much needed (Elad et al., 2021).

6. Gastrointestinal toxicities

Chemotherapy-induced nausea and vomiting are common complications of cancer treatment that can lead to a variety of symptoms. The most well studied mechanism involves chemotherapy-induced stimulation of the enterochromaffin cells lining the gastrointestinal tract to release of serotonin, which then stimulates vagal afferent serotonin receptors (5-HT3) located in the gastrointestinal tract, the nucleus tractus solitarius of the medulla oblongata and the chemoreceptor trigger zone outside the blood–brain barrier, ultimately sending impulses to the vomiting center (Hesketh, 2008). Another neurotransmitter with a significant role in the development of nausea and vomiting after chemotherapy exposure is substance P. This is a member of the tachykinins, including substance P, neurokinin A, and neurokinin B, which, respectively, bind to the receptors neurokinin-1, neurokinin-2, and neurokinin-3. Neurokinin-1 receptors are widely distributed throughout the central nervous system, including the area postrema and the nucleus tractus solitarius, and are also found in the gastrointestinal tract (Hesketh, 2008; Saito, Takano, & Kamiya, 2003; Severini et al., 2002). Additionally, corticosteroids have be shown to decrease the rates of nausea and vomiting, although its mechanism of action is not yet well understood (Ioannidis, Hesketh, & Lau, 2000; Italian Group for

Antiemetic Research, 1995). After elucidation of these main mechanisms, the symptomatic control of acute nausea and vomiting has dramatically improved. The most commonly combined agents are selective 5-HT3 antagonists (ondansetron, granisetron, palonosetron), neurokinin-1 antagonists (aprepitant), and corticosteroids (Bountra et al., 1993; de Wit et al., 2004; Hesketh, 2008; Hesketh et al., 2003; Ioannidis et al., 2000; Italian Group for Antiemetic Research, 1995; Poli-Bigelli et al., 2003). Less commonly used, metoclopramide, butyrophenones, phenothiazines, cannabinoids, olanzapine and benzodiazepines are additional options for refractory or anticipatory cases (Gralla, 1983; Laszlo et al., 1985; Navari et al., 2007; Tramer et al., 2001; Wampler, 1983).

Other common gastrointestinal toxicity from chemotherapy is diarrhea (Andreyev et al., 2014; Benson et al., 2004; O'Brien, Kaklamani, & Benson, 2005). Diarrhea has been described with fluoropyrimidines (particularly fluorouracil and capecitabine) and irinotecan, with severe cases happening in around 5–44% of patients (Andreyev et al., 2014; Ikuno et al., 1995). It is a known adverse events of TKIs, such as afatinib, ceritinib, erlotinib, lapatinib, sorafenib, and sunitinib; with incidence reaching around 50% of patients with some of these agents, but severe in only around 2–16% (Andreyev et al., 2014). Monoclonal antibodies directed against the EGFR may increase the incidence of diarrhea when combined with chemotherapy in around 10% of the patients (Andreyev et al., 2014; Folprecht et al., 2016; Tabernero et al., 2014). Checkpoint inhibition with the cytotoxic T lymphocyte-associated antigen 4 (CTLA4) inhibitor ipilimumab, programmed cell death protein 1 (PD-1) inhibitors nivolumab and pembrolizumab, and the anti-programmed cell death ligand 1 (PD-L1) antibodies atezolizumab and durvalumab are also a common cause of immune mediated diarrhea, with non-severe rates of up to 30% and severe cases happening in around 10% of patients (Andreyev et al., 2014; Garassino et al., 2018; Hodi et al., 2010; Massard et al., 2016).

The mechanisms of diarrhea vary depending on the causing agents. Chemotherapeutic drugs, specially fluorouracil and irinotecan, can cause damage to the intestinal mucosa, leading to loss of epithelium and secretory diarrhea, as well as decreased expression of the lactase leading to lactose intolerance and osmotic diarrhea (Andreyev et al., 2014; Ikuno et al., 1995; Parnes, Fung, & Schiffer, 1994). Irinotecan also has an acute cholinergically-mediated effect, causing acute diarrhea after infusion (Abigerges et al., 1995). It is suggested that TKIs and monoclonal antibodies potentially cause increased chloride secretion caused by dysregulation of the EGFR signaling pathway, colonic crypt damage, gut dysmotility, and alteration in gut microbiota (Andreyev et al., 2014; Bowen, 2013; Secombe et al., 2020). The diarrhea caused by ICIs differs from the previously mentioned etiologies as it is immune mediated (Andreyev et al., 2014; Kennedy & Grivas, 2020).

Patients should be asked the frequency and aspect of stool, as well as assessed for associated signs and symptoms of complication, such as abdominal cramping, nausea or vomiting, fever, hypotension, neutropenia, bleeding, dehydration (Andreyev et al., 2014; Benson et al., 2004; O'Brien et al., 2005). According to National Cancer Institute Common Terminology Criteria for Adverse Events (NCI CTCAE v5.0), diarrhea is classified as grade 1 if fewer than four stools per day or mild increase in ostomy over baseline and grade 2 when frequency increases of four to six stools per day or moderate increase in ostomy output over baseline. More worrisome, higher-grade diarrhea is categorized as grades 3 when there is an increase of seven or more stools per day or severe increase in ostomy output over baseline and grade 4 when life threatening. Patients with grade 3 or 4 diarrhea, as well as those with grade 1 or 2 complicated diarrhea, should be hospitalized for resuscitation, electrolyte replacement and clinical stability (Andreyev et al., 2014; Benson et al., 2004; O'Brien et al., 2005). Laboratory should include chemistry panel, complete blood count, electrolytes, cultures of blood and stool, fecal leukocytes and diagnostic testing for toxin-producing strains of *Clostridioides difficile* (Anand & Glatt, 1993; Andreyev et al., 2014; Benson et al., 2004). For patients with bloody diarrhea, infection should always be ruled out, with common pathogens in this setting being enterohemorrhagic *Escherichia coli* (*E. coli* O157:H7 [EHEC]), *Entamoeba histolytica*, and Shiga toxin (Andreyev et al., 2014; Benson et al., 2004; O'Brien et al., 2005). Patients with fever, peritoneal signs, or bloody diarrhea should also have a computed tomography (CT) scan of the abdomen and pelvis, and surgical consultation (Andreyev et al., 2014; Benson et al., 2004; O'Brien et al., 2005). Endoscopic assessment may be indicated if patients develop chronic diarrhea (Andreyev et al., 2014; Benson et al., 2004; O'Brien et al., 2005). Differential diagnoses include, but are not limited to small intestinal bacterial overgrowth, fat or bile acid malabsorption, lactose intolerance, inflammatory and infectious causes, most markedly cytomegalovirus, and stool impaction (Andreyev et al., 2014; Grace et al., 2013). In terms of diet, treatment is recommended with avoidance of foods that may exacerbate the diarrhea, oral rehydration, as tolerated, and intravenous hydration if necessary. Patients should be encouraged to restrict oral intake to clear liquid diet or BRAT (bananas, rice, applesauce, and toast) diet (O'Brien et al., 2005). Patients should be instructed to avoid alcohol, caffeine, and lactose-containing foods, and to stop other medications and supplements that could cause diarrhea. (O'Brien et al., 2005) Loperamide is recommended at lower doses for grades 1 and 2 uncomplicated diarrhea. For patients with complicated or grades 3 or 4 diarrhea, including patients with decreased performance status, fever, sepsis, neutropenia, frank bleeding, dehydration, or mild to moderate uncomplicated diarrhea that persists after 24 h of loperamide, higher doses of loperamide are recommended (Andreyev et al., 2014; Benson et al., 2004; O'Brien et al., 2005). Octreotide and budesonide are indicated for loperamide-refractory cases (Andreyev et al., 2014; Benson et al., 2004; O'Brien et al., 2005). Antibiotics should be considered for patients with fever, neutropenia, hypotension, peritoneal signs, or bloody diarrhea (Andreyev et al., 2014; Benson et al., 2004; O'Brien et al., 2005). In case of ICI-related diarrhea, loperamide is usually recommended for grade 1 diarrhea. For grade 2, corticosteroids with re-initiation of therapy once symptoms improve can be considered (Brahmer et al., 2018). For grades 3 and 4, patients should be treated with intravenous corticosteroids, and, if steroid refractory, with a corticosteroid sparing agent, such as infliximab, with permanent discontinuation of the immunotherapy agent (Brahmer et al., 2018). Ongoing studies assessing nausea and vomiting prevention and treatment involve dronabinol (NCT03984214), capsaicin (NCT04918069), acupuncture (NCT04116697), aromatherapy (NCT04116697, NCT04478630), pregabalin (NCT04181346) and intermittent fasting (NCT04141514). For diarrhea, examples of agents under investigation are crofemeler (NCT02910219, NCT04538625) and probiotics (NCT02819960).

7. Renal toxicity

Several classes of chemotherapy agents have been associated with various renal toxicities. These included traditional cytotoxic drugs such as cisplatin, mitomycin, and pemetrexed; immunotherapy drugs such as ICI; monoclonal antibodies such as bevacizumab and cetuximab; and targeted therapy drugs including sunitinb and sorafenib (Rosner & Perazella, 2017). Renal toxicities can be broadly categorized as those with tubular injury (acute tubular necrosis; acute interstitial nephritis), glomerular injury (thrombotic microangiopathy), and electrolyte abnormalities (hypomagnesemia; hypophosphatemia) (Santos et al., 2020). The prevalence of these toxicities is variable and ranges from 5% to 40% depending on the type of renal injury.

The underlying mechanism of injury varies depending on the type of renal insult encountered. In platinum induced tubular toxicity, the increase in reactive oxygen species can lead to mitochondrial dysfunction, impaired solute transport and decrease glomerular filtration (Manohar & Leung, 2018). Consequently, salt and water excretion are increased, leading to polyuria and magnesium wasting (Yao et al., 2007). Renal function generally improves, although a subset of patients will develop permanent renal function impairment and or continued magnesium wasting, which can persist years after initial drug exposure. Clinical management strategies include use of intravenous hydration. Mannitol and loop diuretics have not been shown to be beneficial in randomized studies although used clinically (Manohar & Leung, 2018). Amifostine is the only US FDA approved drug for prevention of nephrotoxicity from cisplatin therapy through the neutralization of reactive oxygen species induced by cisplatin. However, this agent is not widely used due to cost and concern for potentially reducing anti-tumor efficacy (Hensley et al., 2009). In certain cancer regimens, substitution of cisplatin by carboplatin, which has less nephrotoxic effects, is considered on individual cases depending on data of carboplatin efficacy per tumor type.

Although cisplatin is the most commonly recognized agent causing chemotherapy induced tubular toxicity, several other agents are implicated, including ifosfamide, MTX, and ipilimumab. The alkylating agent ifosfamide commonly reduces glomerular filtration rates in pediatric and adult patients (Akilesh et al., 2014). Rarely, this toxicity can progress to end stage renal disease requiring hemodialysis. Fanconi syndrome, which is associated with proximal tubular dysfunction, is also seen at times with ifosfamide (Berns et al., 1995). Risk of injury is increased with platinum exposure associated with pelvis radiation and/or other co-existing kidney injuries.

High-dose MTX is another agent associated with tubular toxicity. Urinary alkalinization, which increases MTX solubility, and aggressive hydration are important for the prevention of MTX precipitation in the tubules. Hemodialysis and hemoperfusion have been used in attempt to remove MTX from circulation (Widemann & Adamson, 2006). Glucarpidase, a recombinant bacterial enzyme that rapidly metabolizes MTX to inactive compounds, can decrease MTX plasma level within 15 min after administration and is effective as a single dose (Widemann et al., 2010).

Increased use of ICIs has led to increased recognition of immune mediated toxicities that may involve the kidneys. The anti-CTLA4 antibody ipilimumab can cause renal injury in the form of acute interstitial nephritis or acute tubular nephritis, and the gold standard for diagnosis is a kidney biopsy, although not all patients have a confirmatory diagnosis (Fadel, El Karoui, & Knebelmann, 2009). Most patients do improve with prompt discontinuation of ipilimumab and initiation of steroid therapy. Other classes of ICIs have been also associated with kidney injury (Thompson et al., 2020).

Several antineoplastic agents lead to glomerular toxicity, including gemcitabine, bevacizumab, and mitomycin. The primary manifestation is of a thrombotic microangiopathy with manifestation of clinical syndromes such as acute kidney injury, hypertension, proteinuria and microangiopathic hemolytic anemia (Richmond, Gilbar, & Abro, 2013). Early recognition and discontinuation of drug are the principal management strategies, although there has been limited evidence for plasmapheresis, particularly for gemcitabine. Eculizumab, a monoclonal antibody targeting the complement protein C5 approved for treatment of atypical hemolytic uremic syndrome, has been used to treat gemcitabine-induced microangiopathy. (Schulte-Kemna et al., 2020) As with other toxicities discussed here, the onset of such complications negatively impacts cancer therapy itself, as it leads to discontinuation of a therapy that may be otherwise effective.

8. Aromatase inhibitor-associated musculoskeletal syndrome

Aromatase inhibitors (AI) are associated with the development of musculoskeletal (MSK) adverse events, including joint aches, joint stiffness, myalgia, muscle weakness, tenosynovitis and carpal tunnel syndrome, which have been described as the AI-associated musculoskeletal syndrome (AIMSS). (Crew et al., 2007; Din et al., 2010; Gupta, Henry, & Loprinzi, 2020; Henry et al., 2008; Mao et al., 2009; Morales et al., 2008; Sestak et al., 2008; Spagnolo et al., 2016; Wagner et al., 2018) These symptoms occur in approximately 40% of the patients, may be severe in almost 30% of patients and can be responsible for treatment discontinuation in 10–20% of patients (Crew et al., 2007; Din et al., 2010; Gupta et al., 2020; Presant et al., 2007). In most patients, the onset of symptoms is within 3 months of starting therapy and treatment discontinued or replacement occurs approximately 6 months after starting AIs (Henry et al., 2008; Mao et al., 2009; Wagner et al., 2018).

The etiology of AIMSS is suggested to be associated with the decrease in estrogen levels. The assessment of QOL in patients on AI prompted clinical and laboratory evaluation of patients by rheumatology in one study, with evaluation suggesting either non-inflammatory musculoskeletal symptoms or inflammation localized to tenosynovial structures (Henry et al., 2008). Inflammatory findings were seen in a study where women receiving an AI experienced significant reduction in grip strength and tenosynovial changes on magnetic resonance imaging (MRI) (Morales et al., 2008). Protective factors so far identified in association with decreased level of AIMSS were prior tamoxifen therapy, BMI $\geq$25, no previous chemotherapy, and postmenopausal status >5 years, suggesting that a slower and progressive decline in estrogen levels over time may decrease AIMSS development and intensity (Crew et al., 2007; Mao et al., 2009; Sestak et al., 2008).

Many strategies for managing AIMSS have been studied and applied to patients, with varying levels of evidence. The use of nonsteroidal anti-inflammatory drugs and acetaminophen may be useful for symptom control in the short-term setting (Gupta et al., 2020). Local treatments for carpal tunnel syndrome and trigger finger with night-time splints, release surgery or steroid injections can be considered (Gupta et al., 2020; Sestak et al., 2008). Omega-3 fatty acids may be helpful in treating AIMSS in individuals who are obese (Shen et al., 2018). Exercise, acupuncture, and yoga were shown to improve overall QOL and reduce AIMSS symptoms (Arem et al., 2016; Galantino et al., 2012; Hershman et al., 2018). Duloxetine significantly decreases joint pain scores in patients on AIs, although drug-related adverse events may be limiting (Henry et al., 2018). If these strategies are not efficacious or feasible, temporary discontinuation or change of treatment may be necessary. Short breaks from AI therapy for 2–4 weeks and re-initiation with either a different or the same AI appears to be a beneficial approach (Briot et al., 2010; Henry et al., 2012). In terms of treatment efficacy, intermittent pauses on AI therapy does not seem to affect patient outcomes. In the Study of Letrozole Extension (SOLE) trial (NCT00553410), postmenopausal women with breast cancer in remission after 4–6 years of adjuvant endocrine therapy were randomly assigned to receive either continuous letrozole (for 5 years) or intermittent letrozole (9 months on followed by 3 months off in years 1–4 and treating throughout year 5), with similar 5-year disease-free survival. (Colleoni et al., 2018) Switch to tamoxifen for those who are unable or unwilling to continue treatment with an AI is also an option (Early Breast Cancer Trialists' Collaborative Group, 2011). New treatment strategies for AIMSS are currently being tested. Examples are cannabidiol (NCT04754399), Tai Chi

(NCT04716920), targeted physical therapy (NCT04560699), curcumin (NCT03865992), and auriculotherapy (NCT03096041).

9. Chemotherapy induced peripheral neuropathy (CIPN)

Chemotherapy-induced peripheral neuropathy (CIPN) is a common disabling neurological adverse event induced by standard of care chemotherapy agents (i.e., platinum-based therapies, vinca alkaloids, taxanes, bortezomib, thalidomide analogs) (Chan et al., 2019). CIPN is estimated to occur in 68% of patients with 30% having symptoms for at least 6 months (Chan et al., 2019). There are no US FDA or European EMA therapies approved to treat this condition.

Dose, older age, obesity, and diabetes may be clinical factors that increase risk for CIPN. Many molecular targets and pathways have been identified that are not specific to CIPN but could potentially impact neuron function, including inflammation, repair, signaling, and degeneration, and induce nerve damage (Chan et al., 2019). However, their role in CIPN pathogenesis has not been confirmed.

There are no therapeutic agents recommended for prophylaxis of CIPN (Jordan et al., 2020; Loprinzi et al., 2020). Chronic neurotoxicity is managed by changes in dose, administration, or regimen. Duloxetine is recommended for symptom management of painful CIPN (Jordan et al., 2020; Loprinzi et al., 2020). Over 40 randomized controlled clinical trials for prophylactic and therapeutic treatment of CIPN have been completed but showed no large benefit (Hu et al., 2019). Potentially, the key to producing appreciable benefit is identifying molecular targets that are specific to CIPN and/or entities that can directly protect the neuronal system from the toxic effects of therapy. It would also be helpful to develop assays that can predict risk of CIPN before, during, and after treatment. There are new and emerging clinical trials for prophylaxis (nilotinib; dasatinib; calmangafodipir; APX3330; fingolimod; GM-1; duloxetine) and treatment (fingolimod; duloxetine) of CIPN (Hu et al., 2019).

10. Immunotherapy-induced autoimmunity

The most common toxicities of monoclonal antibody type therapies including ICI and antibody-drug conjugates (ADCs) are autoimmune diseases induced by non-specific activation of the patient's immune system (Carlino, Larkin, & Long, 2021; Tarantino et al., 2021). These immune related adverse events (irAEs) are most frequently found in colon, liver, skin, pituitary gland, thyroid, and lungs. They generally occur within the first few weeks to months of treatment (Weber, Kähler, & Hauschild, 2012). The severity varies and in rare cases can be fulminant and fatal (Wang et al., 2018). No prospective trials have been conducted to inform on the management of these irAEs, however, practice guidelines have been suggested through the Society for Immunotherapy of Cancer (SITC), ESMO, and American Society of Clinical Oncology (ASCO) (Brahmer et al., 2018; Haanen et al., 2017; Puzanov et al., 2017). In general, in most cases, immunosuppression with corticosteroids, cytokine inhibitors, and immunomodulators are found to be effective. Due to the wide range of severity, early recognition and management of irAEs are critical to reduce overall morbidity and rare cases of mortality associated with these events (Brahmer et al., 2018; Haanen et al., 2017; Puzanov et al., 2017; Wang et al., 2018).

Immunotherapy infusion reactions, mostly mild, occur within the first 2h. Infusion reactions can present as pruritis, fever, changes in heart rate/blood pressure, gastrointestinal manifestations, skin rashes, and in rare cases life-threatening anaphylaxis. They are a result of the immune system's recognition of the monoclonal antibody as a foreign entity and are responding to components that make up the antibody (e.g., specific carbohydrates) that the patient may have pre-existing antibodies against before introduction of treatment (Chung et al., 2008). Management includes temporarily withdrawal of the therapy, using slower infusion rate, or addition of concomitant therapy (antipyretics; antihistamines) for grade 1 and 2 infusion reactions. For severe (grade 3) or life-threating (grade 4) infusion reactions, it is recommended that treatment be permanently discontinued. Prophylactic treatment (antihistamines; acetaminophen; corticosteroids) have been recommended for drugs with high incidence of infusion reactions to potentially reduce the severity of infusion reactions.

Skin irAEs (e.g., rashes) are the most common toxicities associated with immunotherapies. The highest number of T cells, B cells, or their secreted products, are located in the skin and associated mucosal areas as these large surface areas are the human body's first layer of defense against pathogens and toxins. It is therefore plausible that rashes and other dermatological manifestations could be an evolutionary mechanism to alert the body to systemic toxicity (Sharma et al., 2019). Clinical presentation of skin irAEs include pruritis, inflammatory dermatoses (eruptions), immunobullous diseases, vitiligo, mucosal symptoms, and hair/nail symptoms. Guidelines have been published for management of these events, which include antihistamines and topical corticosteroids for low grade symptoms and systemic treatments (corticosteroids; antibiotics; other immunosuppressants), interruption of immunotherapy, or permanent discontinuation of immunotherapy for high disruptions to patient QOL or severe manifestations (grade 2 and above) (Brahmer et al., 2018; Choi et al., 2020; Haanen et al., 2017; Puzanov et al., 2017; Thompson et al., 2019; Thompson et al., 2020).

Immunotherapy-induced colitis is clinically defined by radiological findings or symptoms of inflammation (Marin-Acevedo, Harris, & Burton, 2018). Symptoms include abdominal pain, fever, nausea/vomiting, and bloody stool with rare severe cases at risk for bowel perforations (Marin-Acevedo et al., 2018). Research has shown that there are key inflammatory (CD8; Foxp3) similarities in the immune microenvironment of patients with immunotherapy-induced colitis and classic autoimmune inflammatory bowel disease that could be exploited as hallmarks and novel targets for therapy (Hone Lopez et al., 2021; Lo et al., 2021; Sasson et al., 2021). Management for grade 1 toxicity include standard supportive care (e.g., loperamide; atropine sulfate; diphenoxylate hydrochloride) and corticosteroids. For grade 2 and above, corticosteroids with or without infliximab/vedolizumab are recommended dependent on severity and number of days with lack of improvement (Gong & Wang, 2020; Marin-Acevedo et al., 2018).

Immunotherapy hepatotoxicity presents as elevated serum alanine aminotransferase (ALT) and aspartate aminotransferase (AST) with rare elevations in total bilirubin associated with long periods of elevated hepatic enzymes. It has been proposed that immune activation against hepatocytes stemming from shared antigens with tumor and subsequent autoantibody development is potentially responsible for hepatitis and hepatocyte cell death that is presented as increased ALT and AST levels (Remash et al., 2021). Management is primarily corticosteroids and/or other additional immunosuppressive agents with temporary discontinuation (grade 2) or permanent discontinuation (grade 3 and above) of immunotherapy treatment (Brahmer et al., 2018; Haanen et al., 2017; Puzanov et al., 2017).

The most common immunotherapy associated endocrinopathies are hypophysitis, hypothyroidism, and hyperthyroidism. Type 1 diabetes mellitus and adrenal insufficiency are seen at significantly lower rates. The signs and symptoms of endocrinopathies in the context of cancer are nonspecific and include weight loss, vision changes, nausea, headache, and fatigue. Therefore, routine monitoring is recommended. Unique autoimmune mechanisms are likely responsible for each of the diseases.

Management of hypophysitis and hypothyroidism is typically replacement of the affected hormone. Hypothyroidism is typically treated by thyroid hormone supplementation (Brahmer et al., 2018; Haanen et al., 2017; Puzanov et al., 2017). Hyperthyroidism is treated with β-blockers for symptomatic relief and ultimately methods to decrease thyroid hormone production (thionamide; radioiodine ablation; surgery) with hospitalization recommended for severe cases (Brahmer et al., 2018; Haanen et al., 2017; Puzanov et al., 2017; Ross et al., 2016). These endocrinopathies require lifelong treatment and management.

ICIs and ADCs are associated with pneumonitis and interstitial lung disease. Pneumonitis is the most common ICI-induced lung toxicity associated with mortality. ADCs such as tratuzumab emtansine (T-DM1), trastuzumab deruxtecan (T-Dxd; DS-8201), and trastuzumab duocarmazine (SYD985) have been linked to interstitial lung disease mortality (Hackshaw et al., 2020). Common symptoms are dry cough, dyspnea, chest pain, fever, fatigue, and skin rash. The mechanisms responsible for immunotherapy-induced lung toxicity is unclear since chemotherapy and non-immunotherapy targeted therapies also present with pneumonitis and interstitial lung disease. General mechanisms suggested for ADC induced interstitial lung disease include (1) target-dependent and -independent uptake of ADC in healthy cells and (2) non-specific cell death from free (unconjugated) ADC payload. However, a recent study in cynomolgus monkeys showed that T-Dxd and not its unconjugated payload produced interstitial lung disease (Kumagai et al., 2020). There are no prospective clinical trials for treatment that have led to optimal approaches. However, guidelines have been suggested. For asymptomatic cases (grade 1), immunotherapy should be stopped until there is evidence of improvement (for ICIs) or resolution (for ADCs) (Brahmer et al., 2018; Puzanov et al., 2017; Tarantino et al., 2021). However, it is recommended that treatment be permanently discontinued for T-DM1 (Tarantino et al., 2021). For symptomatic cases (grade 2) in patients with ICIs, it is recommended to withhold immunotherapy treatment, start the patient on prednisolone, and consider hospitalization (Brahmer et al., 2018; Puzanov et al., 2017; Tarantino et al., 2021). It is also suggested that patients may be restarted on ICIs if symptoms and imaging abnormalities resolve (Brahmer et al., 2018; Puzanov et al., 2017). For patients on ADCs with symptomatic lung toxicity (grade 2), it is recommended that the ADC be permanently discontinued, and the lung toxicity treated with oral corticosteroids initially and intravenous administration for no improvement or worsening conditions (Tarantino et al., 2021). For patients with grade 3 and above lung toxicity, both ICI and ADC treatments should permanently discontinued and it is recommended that the patient be hospitalized (Brahmer et al., 2018; Puzanov et al., 2017; Tarantino et al., 2021).

11. Putting it all together. What is the future of symptom science?

With the growing number of cancer survivors of all ages, there is growing recognition of the importance of personalizing both cancer treatments as well cancer-related toxicity management. The various toxicities of cancer and cancer treatments range from acute to late effects long after completion of treatment or remain persistent in those with incurable diseases or ongoing treatment. The field has advanced to begin to understand the foundation of many of the cancer treatment side effects. However, as we have described, a complete biological understanding of the mechanisms of injury, predictors of toxicity and personalized management recommendations still elude us and are the basis of ongoing multidisciplinary studies. Further, as new treatments enter the horizon and standards of care shift, the field of symptom science remains a dynamic area for growth and progress.

The concept of reductionism espoused by Descartes in 1637 has for years played a critical role in medical science. Based on the premise that the best way to understand the whole is to study its parts, the continuous quest to identify and characterize increasingly smaller and smaller units has become an ongoing trend leading to specialization and subspecialization of both basic and clinical science. While we owe much to this approach, it is inherently problematic since it minimizes the dynamic interrelationships that characterize the development and course of disease and toxicity manifestations.

Clinically, we know that it is highly unusual for patients undergoing cancer treatment (whether cytotoxic, targeted, or immunomodulatory therapies) to develop a single standalone toxicity. Rather, while one might dominate, toxicities tend to appear in clusters and patients experience multiple co-occurring symptoms that are related to each other (i.e., symptom clusters) (Miaskowski et al., 2017). Compared with a single symptom, the occurrence of symptom clusters is associate with worse patient outcomes and symptom burden. For example, the symptom cluster of pain, fatigue, sleep disturbance, and mood disturbance can all negative impact QOL (Illi et al., 2012).

The observation of symptom clusters has multiple implications. First, it is likely that toxicities share common pathobiological underpinnings. A spike in pro-inflammatory cytokines or activation of the innate immune response have multiple implications across systems with phenotypic consequences. Second, we have all seen patients who coast through treatment and those who are "train wrecks." What characterizes these differences? Why does the toxicity threshold differ from one patient to the next when clinically they seem identical? Third, does our current symptom-focused approach ultimately make sense or would a broader systems medicine strategy in which therapies target common toxicity nodes lead us toward precision intervention and less polypharmacy? And if that is the case, how do we get there?

Findings over the past few decades have changed how we think about the pathoetiology of regimen-related toxicities, what determines differential patient risk, and how we view the dynamics and inter-relationships of toxicity incidence, severity, and trajectory. Over the past three decades, it has become increasingly clear that toxicity pathogenesis is biologically complex. The old concept that tissue-centric regimen-related damage was mostly attributable to non-specific cytotoxic effects on normal tissues has been shown to impact only a minority of the replenishing cells associated with oral and gastrointestinal mucositis, dermatitis, and pneumonitis for example. Likewise, the recognition of the gut-brain axis provides a natural platform for shared toxicity mechanisms and provides insight in the multiplex nature of toxicity risk and behavior. The application of "omics" to address issues of toxicity risk and tumor response has been enthusiastically embraced. And yet, even as the number of elements contributing to toxicity phenotypes has proliferated, the course of discovery still trends toward more siloed granularity.

We now have the opportunity and challenge of putting the pieces of the puzzle together, assigning "weights" to the contribution of each element and understanding that toxicity development and even risk are not static. Specifically, in multiplex analyses we evaluate the overlap of symptom clusters, biological pathways, and the contribution of genomics (both germline and somatic), microbiomics, proteomics, and metabolomics interactions as risk and response predictors (Fig. 2). We also include comorbidities, concomitant medications, and behaviors to evolve hierarchical definitions as we seek to understand the multifactorial nature of regimen-related toxicity risk. Mapping regimen-related toxicities in such a way not only provides a chance to predict risk, but also to identify best druggable targets. Assessing these variables at a point in time fails to recognize the dynamics associated with the continuing changes in cancer patients' physiology, "omic" status, and microbiome. To think that a patient is the same on day 1 of cancer therapy vs at the completion of a chemotherapy cycle is naïve.

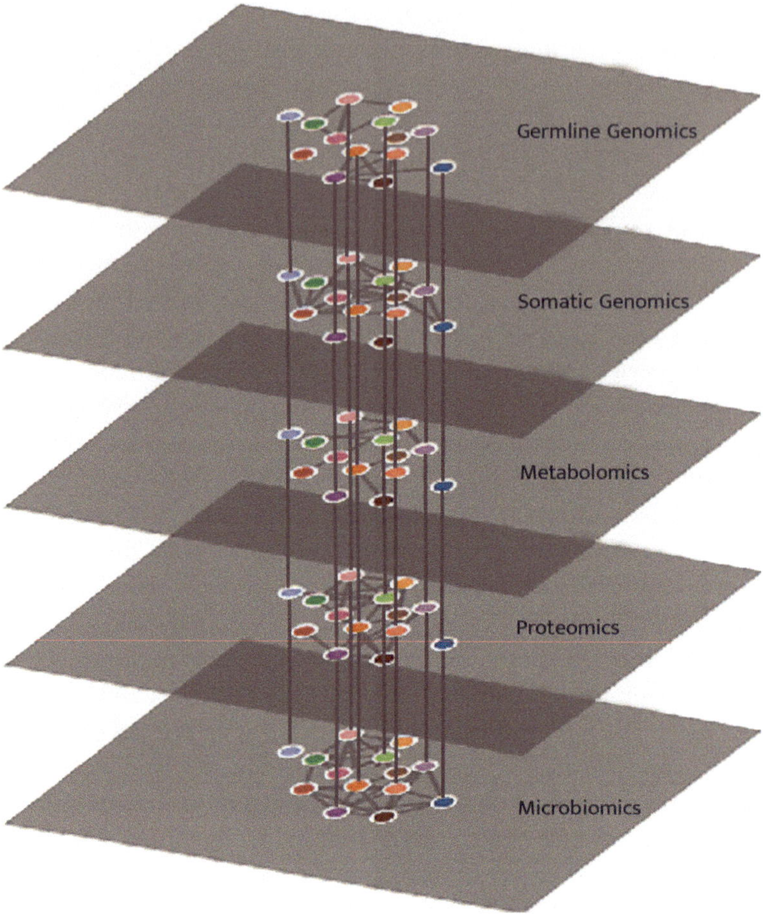

Fig. 2 Proposal of interactions for symptom science. Multiplex networks recognize the complexities of diseases and provide a conduit to comprehensively assess the interaction of multiple layers of disease features and risk contributors. In this example, five networks which contribute to regimen-related toxicity risk are assessed, not only independently, but in the aggregate so that overlapping features can be identified. The more "horizontal networks" are defined (specific cancer regimen, comorbidities, etc.), the greater the opportunity for integrating component elements vertically to assess toxicity risks and behaviors and to evaluate drivers of clustering.

Thus, the future of symptom science is largely dependent on our ability to dynamically define and integrate systemic biological changes, clinical phenotypes, and patient-centric factors (omics) over the course of treatment and after treatment for survivorship. Such an objective is no small undertaking and requires an interdisciplinary approach built on "big data," uniform reporting standards, and robust biological sampling.

References

Abigerges, D., et al. (1995). Phase I and pharmacologic studies of the camptothecin analog irinotecan administered every 3 weeks in cancer patients. *Journal of Clinical Oncology, 13*(1), 210–221.

Ahles, T. A., et al. (2003). The relationship of APOE genotype to neuropsychological performance in long-term cancer survivors treated with standard dose chemotherapy. *Psychooncology, 12*(6), 612–619.

Ahmadi, M. A., & Esmaeli, B. (2001). Surgical treatment of canalicular stenosis in patients receiving docetaxel weekly. *Archives of Ophthalmology, 119*(12), 1802–1804.

Akilesh, S., et al. (2014). Chronic Ifosfamide toxicity: Kidney pathology and pathophysiology. *American Journal of Kidney Diseases, 63*(5), 843–850.

Anand, A., & Glatt, A. E. (1993). Clostridium difficile infection associated with antineoplastic chemotherapy: A review. *Clinical Infectious Diseases, 17*(1), 109–113.

Andreyev, J., et al. (2014). Guidance on the management of diarrhoea during cancer chemotherapy. *The Lancet Oncology, 15*(10), e447–e460.

Arem, H., et al. (2016). Exercise adherence in a randomized trial of exercise on aromatase inhibitor arthralgias in breast cancer survivors: The hormones and physical exercise (HOPE) study. *Journal of Cancer Survivorship, 10*(4), 654–662.

Arepalli, S., & Venkat, A. G. (2020). Uveitis secondary to cancer therapeutics. *Annals of Eye Science, 5*, 19.

Benson, A. B., 3rd, et al. (2004). Recommended guidelines for the treatment of cancer treatment-induced diarrhea. *Journal of Clinical Oncology, 22*(14), 2918–2926.

Berns, J. S., et al. (1995). Severe, irreversible renal failure after ifosfamide treatment. A clinicopathologic report of two patients. *Cancer, 76*(3), 497–500.

Borkar, D. S., Lacouture, M. E., & Basti, S. (2013). Spectrum of ocular toxicities from epidermal growth factor receptor inhibitors and their intermediate-term follow-up: A five-year review. *Support Care Cancer, 21*(4), 1167–1174.

Bountra, C., et al. (1993). Anti-emetic profile of a non-peptide neurokinin NK1 receptor antagonist, CP-99,994, in ferrets. *European Journal of Pharmacology, 249*(1), R3–R4.

Bowen, J. M. (2013). Mechanisms of TKI-induced diarrhea in cancer patients. *Current Opinion in Supportive and Palliative Care,* 7(2), 162–167.

Bowen, J., et al. (2019). The pathogenesis of mucositis: Updated perspectives and emerging targets. *Support Care Cancer, 27*(10), 4023–4033.

Brahmer, J. R., et al. (2018). Management of Immune-Related Adverse Events in patients treated with immune checkpoint inhibitor therapy: American Society of Clinical Oncology clinical practice guideline. *Journal of Clinical Oncology, 36*(17), 1714–1768.

Briot, K., et al. (2010). Effect of a switch of aromatase inhibitors on musculoskeletal symptoms in postmenopausal women with hormone-receptor-positive breast cancer: The ATOLL (articular tolerance of letrozole) study. *Breast Cancer Research and Treatment, 120*(1), 127–134.

Brooks, B., & Knight, K. (2018). Ototoxicity monitoring in children treated with platinum chemotherapy. *International Journal of Audiology, 57*(sup4), S34–S40.

Camidge, D. R., et al. (2012). Activity and safety of crizotinib in patients with ALK-positive non-small-cell lung cancer: Updated results from a phase 1 study. *The Lancet Oncology, 13*(10), 1011–1019.

Capri, G., et al. (1994). Optic nerve disturbances: A new form of paclitaxel neurotoxicity. *Journal of the National Cancer Institute, 86*(14), 1099–1101.

Carlino, M. S., Larkin, J., & Long, G. V. (2021). Immune checkpoint inhibitors in melanoma. *Lancet, 398*(10304), 1002–1014.

Carnel, S. B., et al. (1990). Treatment of radiation-and chemotherapy-induced stomatitis. *Otolaryngology and Head and Neck Surgery, 102*(4), 326–330.

Chan, A., et al. (2013). Prevalence of excessive tearing in women with early breast cancer receiving adjuvant docetaxel-based chemotherapy. *Journal of Clinical Oncology, 31*(17), 2123–2127.

Chan, A., et al. (2019). Biological predictors of chemotherapy-induced peripheral neuropathy (CIPN): MASCC neurological complications working group overview. *Support Care Cancer, 27*(10), 3729–3737.

Cheng, H., et al. (2016). The COMT (rs165599) gene polymorphism contributes to chemotherapy-induced cognitive impairment in breast cancer patients. *American Journal of Translational Research, 8*(11), 5087–5097.

Choe, C. H., et al. (2014). Ocular toxicity in BRAF mutant cutaneous melanoma patients treated with vemurafenib. *American Journal of Ophthalmology, 158*(4), 831–837 (e2).

Choi, J., et al. (2020). Multinational Association of Supportive Care in cancer (MASCC) 2020 clinical practice recommendations for the management of severe dermatological toxicities from checkpoint inhibitors. *Support Care Cancer, 28*(12), 6119–6128.

Chung, C. H., et al. (2008). Cetuximab-induced anaphylaxis and IgE specific for galactose-alpha-1,3-galactose. *The New England Journal of Medicine, 358*(11), 1109–1117.

Coda, B. A., et al. (1997). Comparative efficacy of patient-controlled administration of morphine, hydromorphone, or sufentanil for the treatment of oral mucositis pain following bone marrow transplantation. *Pain, 72*(3), 333–346.

Colleoni, M., et al. (2018). Extended adjuvant intermittent letrozole versus continuous letrozole in postmenopausal women with breast cancer (SOLE): A multicentre, open-label, randomised, phase 3 trial. *The Lancet Oncology, 19*(1), 127–138.

Crew, K. D., et al. (2007). Prevalence of joint symptoms in postmenopausal women taking aromatase inhibitors for early-stage breast cancer. *Journal of Clinical Oncology, 25*(25), 3877–3883.

Curran, C. F., & Luce, J. K. (1989). Ocular adverse reactions associated with adriamycin (doxorubicin). *American Journal of Ophthalmology, 108*(6), 709–711.

de Wit, R., et al. (2004). The oral NK(1) antagonist, aprepitant, given with standard antiemetics provides protection against nausea and vomiting over multiple cycles of cisplatin-based chemotherapy: A combined analysis of two randomised, placebo-controlled phase III clinical trials. *European Journal of Cancer, 40*(3), 403–410.

Din, O. S., et al. (2010). Aromatase inhibitor-induced arthralgia in early breast cancer: What do we know and how can we find out more? *Breast Cancer Research and Treatment, 120*(3), 525–538.

Dodd, M. J., et al. (2000). Randomized clinical trial of the effectiveness of 3 commonly used mouthwashes to treat chemotherapy-induced mucositis. *Oral Surgery, Oral Medicine, Oral Pathology, Oral Radiology, and Endodontology, 90*(1), 39–47.

Doroshow, J. H., et al. (1981). Ocular irritation from high-dose methotrexate therapy: Pharmacokinetics of drug in the tear film. *Cancer, 48*(10), 2158–2162.

Early Breast Cancer Trialists' Collaborative Group, et al. (2011). Relevance of breast cancer hormone receptors and other factors to the efficacy of adjuvant tamoxifen: Patient-level meta-analysis of randomised trials. *Lancet, 378*(9793), 771–784.

Eiseman, A. S., et al. (2003). Ocular surface, ocular adnexal, and lacrimal complications associated with the use of systemic 5-fluorouracil. *Ophthalmic Plastic & Reconstructive Surgery, 19*(3), 216–224.

Elad, S., et al. (2020). MASCC/ISOO clinical practice guidelines for the management of mucositis secondary to cancer therapy. *Cancer, 126*(19), 4423–4431.

Elad, S., et al. (2021). The broadening scope of oral mucositis and oral ulcerative mucosal toxicities of anticancer therapies. *CA: a Cancer Journal for Clinicians*.
Elting, L. S., et al. (2013). Risk of oral and gastrointestinal mucosal injury among patients receiving selected targeted agents: A meta-analysis. *Support Care Cancer, 21*(11), 3243–3254.
Epstein, J. B., & Schubert, M. M. (2003). Oropharyngeal mucositis in cancer therapy. Review of pathogenesis, diagnosis, and management. *Oncology (Williston Park), 17*(12), 1767–1779 (discussion 1779-82, 1791-2).
Epstein, J. B., et al. (2001). Oral topical doxepin rinse: Analgesic effect in patients with oral mucosal pain due to cancer or cancer therapy. *Oral Oncology, 37*(8), 632–637.
Esmaeli, B., et al. (2001). Interferon-induced retinopathy in asymptomatic cancer patients. *Ophthalmology, 108*(5), 858–860.
Esmaeli, B., et al. (2002). Docetaxel secretion in tears: Association with lacrimal drainage obstruction. *Archives of Ophthalmology, 120*(9), 1180–1182.
Fadel, F., El Karoui, K., & Knebelmann, B. (2009). Anti-CTLA4 antibody-induced lupus nephritis. *The New England Journal of Medicine, 361*(2), 211–212.
Farrell, C. L., et al. (1998). Keratinocyte growth factor protects mice from chemotherapy and radiation-induced gastrointestinal injury and mortality. *Cancer Research, 58*(5), 933–939.
Folprecht, G., et al. (2016). Oxaliplatin and 5-FU/folinic acid (modified FOLFOX6) with or without aflibercept in first-line treatment of patients with metastatic colorectal cancer: The AFFIRM study. *Annals of Oncology, 27*(7), 1273–1279.
Francis, J. H., et al. (2020). MEK inhibitor-associated central retinal vein occlusion associated with Hyperhomocysteinemia and MTHFR variants. *Ocular Oncology and Pathology, 6*(3), 159–163.
Fraunfelder, F. W., & Yang, H. K. (2016). Association between bortezomib therapy and eyelid Chalazia. *JAMA Ophthalmology, 134*(1), 88–90.
Fraunfelder, F. W., et al. (2003). Ocular side-effects associated with imatinib mesylate (Gleevec). *Journal of Ocular Pharmacology and Therapeutics, 19*(4), 371–375.
Freyer, D. R., et al. (2020). Prevention of cisplatin-induced ototoxicity in children and adolescents with cancer: A clinical practice guideline. *The Lancet Child & Adolescent Health, 4*(2), 141–150.
Galantino, M. L., et al. (2012). Impact of yoga on functional outcomes in breast cancer survivors with aromatase inhibitor-associated arthralgias. *Integrative Cancer Therapies, 11*(4), 313–320.
Garassino, M. C., et al. (2018). Durvalumab as third-line or later treatment for advanced non-small-cell lung cancer (ATLANTIC): An open-label, single-arm, phase 2 study. *The Lancet Oncology, 19*(4), 521–536.
Gianni, L., et al. (1995). Paclitaxel in metastatic breast cancer: A trial of two doses by a 3-hour infusion in patients with disease recurrence after prior therapy with anthracyclines. *Journal of the National Cancer Institute, 87*(15), 1169–1175.
Gilbard, J. P. (2005). The diagnosis and management of dry eyes. *Otolaryngologic Clinics of North America, 38*(5), 871–885.
Gong, Z., & Wang, Y. (2020). Immune checkpoint inhibitor-mediated diarrhea and colitis: A clinical review. *The Journal of Oncology Practice, 16*(8), 453–461.
Grace, E., et al. (2013). Review article: Small intestinal bacterial overgrowth- -prevalence, clinical features, current and developing diagnostic tests, and treatment. *Alimentary Pharmacology & Therapeutics, 38*(7), 674–688.
Gralla, R. J. (1983). Metoclopramide. A review of antiemetic trials. *Drugs, 25*(Suppl 1), 63–73.
Grzybowski, A., et al. (2015). Toxic optic neuropathies: An updated review. *Acta Ophthalmologica, 93*(5), 402–410.

Gupta, A., Henry, N. L., & Loprinzi, C. L. (2020). Management of Aromatase Inhibitor-Induced Musculoskeletal Symptoms. *The Journal of Oncology Practice, 16*(11), 733–739.

Haanen, J., et al. (2017). Management of toxicities from immunotherapy: ESMO clinical practice guidelines for diagnosis, treatment and follow-up. *Annals of Oncology, 28*(suppl_4), iv119–iv142.

Hackshaw, M. D., et al. (2020). Incidence of pneumonitis/interstitial lung disease induced by HER2-targeting therapy for HER2-positive metastatic breast cancer. *Breast Cancer Research and Treatment, 183*(1), 23–39.

Henry, N. L., et al. (2008). Prospective characterization of musculoskeletal symptoms in early stage breast cancer patients treated with aromatase inhibitors. *Breast Cancer Research and Treatment, 111*(2), 365–372.

Henry, N. L., et al. (2012). Predictors of aromatase inhibitor discontinuation as a result of treatment-emergent symptoms in early-stage breast cancer. *Journal of Clinical Oncology, 30*(9), 936–942.

Henry, N. L., et al. (2018). Randomized, multicenter, placebo-controlled clinical trial of duloxetine versus placebo for aromatase inhibitor-associated arthralgias in early-stage breast cancer: SWOG S1202. *Journal of Clinical Oncology, 36*(4), 326–332.

Hensley, M. L., et al. (2009). American Society of Clinical Oncology 2008 clinical practice guideline update: Use of chemotherapy and radiation therapy protectants. *Journal of Clinical Oncology, 27*(1), 127–145.

Hershman, D. L., et al. (2018). Effect of acupuncture vs sham acupuncture or waitlist control on joint pain related to aromatase inhibitors among women with early-stage breast cancer: A randomized clinical trial. *JAMA, 320*(2), 167–176.

Hesketh, P. J. (2008). Chemotherapy-induced nausea and vomiting. *The New England Journal of Medicine, 358*(23), 2482–2494.

Hesketh, P. J., et al. (2003). The oral neurokinin-1 antagonist aprepitant for the prevention of chemotherapy-induced nausea and vomiting: A multinational, randomized, double-blind, placebo-controlled trial in patients receiving high-dose cisplatin- -the Aprepitant protocol 052 study group. *Journal of Clinical Oncology, 21*(22), 4112–4119.

Hodi, F. S., et al. (2010). Improved survival with ipilimumab in patients with metastatic melanoma. *The New England Journal of Medicine, 363*(8), 711–723.

Hone Lopez, S., et al. (2021). Immune checkpoint inhibitor treatment induces colitis with heavy infiltration of CD8 + T cells and an infiltration pattern that resembles ulcerative colitis. *Virchows Archiv*.

Hu, S., et al. (2019). Recent developments of novel pharmacologic therapeutics for prevention of chemotherapy-induced peripheral neuropathy. *Clinical Cancer Research, 25*(21), 6295–6301.

Ikuno, N., et al. (1995). Irinotecan (CPT-11) and characteristic mucosal changes in the mouse ileum and cecum. *Journal of the National Cancer Institute, 87*(24), 1876–1883.

Illi, J., et al. (2012). Association between pro- and anti-inflammatory cytokine genes and a symptom cluster of pain, fatigue, sleep disturbance, and depression. *Cytokine, 58*(3), 437–447.

Ioannidis, J. P., Hesketh, P. J., & Lau, J. (2000). Contribution of dexamethasone to control of chemotherapy-induced nausea and vomiting: A meta-analysis of randomized evidence. *Journal of Clinical Oncology, 18*(19), 3409–3422.

Italian Group for Antiemetic Research. (1995). Dexamethasone, granisetron, or both for the prevention of nausea and vomiting during chemotherapy for cancer. *The New England Journal of Medicine, 332*(1), 1–5.

Johansson, B. A. (1992). Visual field defects during low-dose methotrexate therapy. *Documenta Ophthalmologica, 79*(1), 91–94.

Jordan, B., et al. (2020). Systemic anticancer therapy-induced peripheral and central neurotoxicity: ESMO-EONS-EANO clinical practice guidelines for diagnosis, prevention, treatment and follow-up. *Annals of Oncology, 31*(10), 1306–1319.

Joshi, M. M., & Garretson, B. R. (2007). Paclitaxel maculopathy. *Archives of Ophthalmology, 125*(5), 709–710.

Kaiser-Kupfer, M. I., & Lippman, M. E. (1978). Tamoxifen retinopathy. *Cancer Treatment Reports, 62*(3), 315–320.

Kennedy, L. C., & Grivas, P. (2020). Immunotherapy-related colitis: An emerging challenge and a quest for prospective data. *The Journal of Oncology Practice, 16*(8), 464–465.

Kros, C. J., & Steyger, P. S. (2019). Aminoglycoside- and cisplatin-induced ototoxicity: Mechanisms and otoprotective strategies. *Cold Spring Harbor Perspectives in Medicine, 9*(11).

Kumagai, K., et al. (2020). Interstitial pneumonitis related to trastuzumab deruxtecan, a human epidermal growth factor receptor 2-targeting ab-drug conjugate, in monkeys. *Cancer Science, 111*(12), 4636–4645.

Kurata, T., et al. (2006). Pemetrexed-induced edema of the eyelid. *Lung Cancer, 54*(2), 241–242.

Landier, W. (2016). Ototoxicity and cancer therapy. *Cancer, 122*(11), 1647–1658.

Laszlo, J., et al. (1985). Lorazepam in cancer patients treated with cisplatin: A drug having antiemetic, amnesic, and anxiolytic effects. *Journal of Clinical Oncology, 3*(6), 864–869.

Lim, L. T., et al. (2010). Bilateral anterior uveitis secondary to erlotinib. *European Journal of Clinical Pharmacology, 66*(12), 1277–1278.

Liu, C.Y., J.H. Francis, and D.H. Abramson, Ocular side effects of systemically administered chemotherapy, in UpToDate, Post TW (Ed). 2021.

Lo, Y. C., et al. (2021). Checkpoint inhibitor colitis shows drug-specific differences in immune cell reaction that overlap with inflammatory bowel disease and predict response to colitis therapy. *American Journal of Clinical Pathology, 156*(2), 214–228.

Loprinzi, C. L., et al. (2020). Prevention and Management of Chemotherapy-Induced Peripheral Neuropathy in survivors of adult cancers: ASCO guideline update. *Journal of Clinical Oncology, 38*(28), 3325–3348.

Mandelblatt, J. S., et al. (2018). Cancer-related cognitive outcomes among older breast cancer survivors in the thinking and living with cancer study. *Journal of Clinical Oncology*, JCO1800140.

Manohar, S., & Leung, N. (2018). Cisplatin nephrotoxicity: A review of the literature. *Journal of Nephrology, 31*(1), 15–25.

Mao, J. J., et al. (2009). Patterns and risk factors associated with aromatase inhibitor-related arthralgia among breast cancer survivors. *Cancer, 115*(16), 3631–3639.

Marin-Acevedo, J. A., Harris, D. M., & Burton, M. C. (2018). Immunotherapy-induced colitis: An emerging problem for the hospitalist. *Journal of Hospital Medicine, 13*(6), 413–418.

Massard, C., et al. (2016). Safety and efficacy of durvalumab (MEDI4736), an anti-programmed cell death Ligand-1 immune checkpoint inhibitor, in patients with advanced urothelial bladder cancer. *Journal of Clinical Oncology, 34*(26), 3119–3125.

Mayo, S. J., et al. (2021). Cancer-related cognitive impairment in patients with non-central nervous system malignancies: An overview for oncology providers from the MASCC neurological complications study group. *Support Care Cancer, 29*(6), 2821–2840.

Miaskowski, C., et al. (2017). Advancing symptom science through symptom cluster research: Expert panel proceedings and recommendations. *Journal of the National Cancer Institute*, 109(4).

Morales, L., et al. (2008). Prospective study to assess short-term intra-articular and tenosynovial changes in the aromatase inhibitor-associated arthralgia syndrome. *Journal of Clinical Oncology, 26*(19), 3147–3152.

Navari, R. M., et al. (2007). A phase II trial of olanzapine, dexamethasone, and palonosetron for the prevention of chemotherapy-induced nausea and vomiting: A Hoosier oncology group study. *Support Care Cancer, 15*(11), 1285.
Nawi, R. I. M., et al. (2018). Oral cryotherapy: Prevention of oral mucositis and pain among patients with colorectal cancer undergoing chemotherapy. *Clinical Journal of Oncology Nursing, 22*(5).
Nayfield, S. G., & Gorin, M. B. (1996). Tamoxifen-associated eye disease. A review. *Journal of Clinical Oncology, 14*(3), 1018–1026.
O'Brien, B. E., Kaklamani, V. G., & Benson, A. B., 3rd. (2005). The assessment and management of cancer treatment-related diarrhea. *Clinical Colorectal Cancer, 4*(6), 375–381 (discussion 382-3).
Papavasileiou, E., et al. (2016). Ipilimumab-induced ocular and orbital inflammation- -a case series and review of the literature. *Ocular Immunology and Inflammation, 24*(2), 140–146.
Parnes, H. L., Fung, E., & Schiffer, C. A. (1994). Chemotherapy-induced lactose intolerance in adults. *Cancer, 74*(5), 1629–1633.
Peterson, D. E., et al. (2015). Management of oral and gastrointestinal mucosal injury: ESMO clinical practice guidelines for diagnosis, treatment, and follow-up. *Annals of Oncology, 26*(Suppl 5), v139–v151.
Planer, D., et al. (2004). Anterior uveitis as a complication of treatment with high dose cytosine-arabinoside. *American Journal of Hematology, 76*(3), 304–306.
Poli-Bigelli, S., et al. (2003). Addition of the neurokinin 1 receptor antagonist aprepitant to standard antiemetic therapy improves control of chemotherapy-induced nausea and vomiting. Results from a randomized, double-blind, placebo-controlled trial in Latin America. *Cancer, 97*(12), 3090–3098.
Presant, C. A., et al. (2007). Aromatase inhibitor-associated arthralgia and/ or bone pain: Frequency and characterization in non-clinical trial patients. *Clinical Breast Cancer, 7*(10), 775–778.
Puzanov, I., et al. (2017). Managing toxicities associated with immune checkpoint inhibitors: Consensus recommendations from the Society for Immunotherapy of cancer (SITC) toxicity management working group. *Journal for Immunotherapy of Cancer, 5*(1), 95.
Rankin, E. M., & Pitts, J. F. (1993). Ophthalmic toxicity during carboplatin therapy. *Annals of Oncology, 4*(4), 337–338.
Remash, D., et al. (2021). Immune checkpoint inhibitor-related hepatotoxicity: A review. *World Journal of Gastroenterology, 27*(32), 5376–5391.
Richmond, J., Gilbar, P., & Abro, E. (2013). Gemcitabine-induced thrombotic microangiopathy. *Internal Medicine Journal, 43*(11), 1240–1242.
Riley, P., et al. (2015). Interventions for preventing oral mucositis in patients with cancer receiving treatment: Oral cryotherapy. *Cochrane Database of Systematic Reviews, 12*.
Robinson, M. R., et al. (2004). Cytotoxic T lymphocyte-associated antigen 4 blockade in patients with metastatic melanoma: A new cause of uveitis. *Journal of Immunotherapy, 27*(6), 478–479.
Rosen, L. S., et al. (2006). Palifermin reduces the incidence of oral mucositis in patients with metastatic colorectal cancer treated with fluorouracil-based chemotherapy. *Journal of Clinical Oncology, 24*(33), 5194–5200.
Rosner, M. H., & Perazella, M. A. (2017). Acute kidney injury in patients with cancer. *The New England Journal of Medicine, 377*(5), 500–501.
Ross, D. S., et al. (2016). 2016 American Thyroid Association guidelines for diagnosis and Management of Hyperthyroidism and Other Causes of thyrotoxicosis. *Thyroid, 26*(10), 1343–1421.
Rugo, H. S., et al. (2017). Prevention of everolimus-related stomatitis in women with hormone receptor-positive, HER2-negative metastatic breast cancer using dexamethasone mouthwash (SWISH): A single-arm, phase 2 trial. *The Lancet Oncology, 18*(5), 654–662.

Saint-Jean, A., et al. (2012). Ocular adverse events of systemic inhibitors of the epidermal growth factor receptor: Report of 5 cases. *Ophthalmology, 119*(9), 1798–1802.
Saito, R., Takano, Y., & Kamiya, H. O. (2003). Roles of substance P and NK(1) receptor in the brainstem in the development of emesis. *Journal of Pharmacological Sciences, 91*(2), 87–94.
Saleh, M., et al. (2011). Bilateral macular ischemia and severe visual loss following trastuzumab therapy. *Acta Oncologica, 50*(3), 477–478.
Sanderson, P. A., Kuwabara, T., & Cogan, D. G. (1976). Optic neuropathy presumably caused by vincristine therapy. *American Journal of Ophthalmology, 81*(2), 146–150.
Santos, M. L. C., et al. (2020). Nephrotoxicity in cancer treatment: An overview. *World Journal of Clinical Oncology, 11*(4), 190–204.
Sasson, S. C., et al. (2021). Interferon-gamma-producing CD8(+) tissue resident memory T cells are a targetable Hallmark of immune checkpoint inhibitor-colitis. *Gastroenterology, 161*(4), 1229–1244 (e9).
Saunders, D. P., et al. (2020). Systematic review of antimicrobials, mucosal coating agents, anesthetics, and analgesics for the management of oral mucositis in cancer patients and clinical practice guidelines. *Support Care Cancer, 28*(5), 2473–2484.
Schulte-Kemna, L., et al. (2020). Eculizumab in chemotherapy-induced thrombotic microangiopathy. *Clinical Nephrology – Case Studies, 8*, 25–32.
Secombe, K. R., et al. (2020). Diarrhea induced by small molecule tyrosine kinase inhibitors compared with chemotherapy: Potential role of the microbiome. *Integrative Cancer Therapies, 19* (1534735420928493).
Sestak, I., et al. (2008). Risk factors for joint symptoms in patients enrolled in the ATAC trial: A retrospective, exploratory analysis. *The Lancet Oncology, 9*(9), 866–872.
Severini, C., et al. (2002). The tachykinin peptide family. *Pharmacological Reviews, 54*(2), 285–322.
Sharma, A., et al. (2019). The skin as a metabolic and immune-competent organ: Implications for drug-induced skin rash. *Journal of Immunotoxicology, 16*(1), 1–12.
Shen, S., et al. (2018). Omega-3 fatty acid use for obese breast cancer patients with aromatase inhibitor-related arthralgia (SWOG S0927). *Breast Cancer Research and Treatment, 172*(3), 603–610.
Soliman, S. E., et al. (2018). Clinical and genetic associations for carboplatin-related ototoxicity in children treated for retinoblastoma: A retrospective noncomparative single-institute experience. *Pediatric Blood & Cancer, 65*(5), e26931.
Sonis, S. T. (2004). The pathobiology of mucositis. *Nature Reviews. Cancer, 4*(4), 277–284.
Sonis, S. (2020). Could the impact of photobiomodulation on tumor response to radiation be effected by tumor heterogeneity? *Support Care Cancer, 28*(2), 423–424.
Sonis, S. T., et al. (2004). Perspectives on cancer therapy-induced mucosal injury: Pathogenesis, measurement, epidemiology, and consequences for patients. *Cancer, 100*(9 Suppl), 1995–2025.
Sonis, S. T., et al. (2016). Could the biological robustness of low level laser therapy (Photobiomodulation) impact its use in the management of mucositis in head and neck cancer patients. *Oral Oncology, 54*, 7–14.
Spagnolo, F., et al. (2016). Anastrozole-induced carpal tunnel syndrome: Results from the international breast cancer intervention study II prevention trial. *Journal of Clinical Oncology, 34*(2), 139–143.
Spielberger, R., et al. (2004). Palifermin for oral mucositis after intensive therapy for hematologic cancers. *The New England Journal of Medicine, 351*(25), 2590–2598.
Tabernero, J., et al. (2014). Aflibercept versus placebo in combination with fluorouracil, leucovorin and irinotecan in the treatment of previously treated metastatic colorectal cancer: Prespecified subgroup analyses from the VELOUR trial. *European Journal of Cancer, 50*(2), 320–331.

Tarafdar, S., et al. (2012). Tamoxifen keratopathy as seen with in-vivo confocal microscopy. *Seminars in Ophthalmology*, *27*(1–2), 27–28.

Tarantino, P., et al. (2021). Interstitial lung disease induced by anti-ERBB2 antibody-drug conjugates: A review. *JAMA Oncology*, 7(12), 1873–1881.

Thompson, J. A., et al. (2019). Management of Immunotherapy-Related Toxicities, version 1.2019. *Journal of the National Comprehensive Cancer Network*, *17*(3), 255–289.

Thompson, J. A., et al. (2020). NCCN guidelines insights: Management of Immunotherapy-Related Toxicities, version 1.2020. *Journal of the National Comprehensive Cancer Network*, *18*(3), 230–241.

Tramer, M. R., et al. (2001). Cannabinoids for control of chemotherapy induced nausea and vomiting: Quantitative systematic review. *BMJ*, *323*(7303), 16–21.

Wagner, L. I., et al. (2018). Patient-reported predictors of early treatment discontinuation: Treatment-related symptoms and health-related quality of life among postmenopausal women with primary breast cancer randomized to anastrozole or exemestane on NCIC clinical trials group (CCTG) MA.27 (E1Z03). *Breast Cancer Research and Treatment*, *169*(3), 537–548.

Waikhom, B., Fraunfelder, F. T., & Henner, W. D. (2000). Severe ocular irritation and corneal deposits associated with capecitabine use. *The New England Journal of Medicine*, *343*(10), 740–741.

Wampler, G. (1983). The pharmacology and clinical effectiveness of phenothiazines and related drugs for managing chemotherapy-induced emesis. *Drugs*, *25*(Suppl 1), 35–51.

Wang, D. Y., et al. (2018). Fatal toxic effects associated with immune checkpoint inhibitors: A systematic review and meta-analysis. *JAMA Oncology*, *4*(12), 1721–1728.

Watters, A. L., Epstein, J. B., & Agulnik, M. (2011). Oral complications of targeted cancer therapies: A narrative literature review. *Oral Oncology*, *47*(6), 441–448.

Weber, J. S., Kähler, K. C., & Hauschild, A. (2012). Management of immune-related adverse events and kinetics of response with ipilimumab. *Journal of Clinical Oncology*, *30*(21), 2691–2697.

Widemann, B. C., & Adamson, P. C. (2006). Understanding and managing methotrexate nephrotoxicity. *The Oncologist*, *11*(6), 694–703.

Widemann, B. C., et al. (2010). Glucarpidase, leucovorin, and thymidine for high-dose methotrexate-induced renal dysfunction: Clinical and pharmacologic factors affecting outcome. *Journal of Clinical Oncology*, *28*(25), 3979–3986.

Wong, R. K., Lee, J. K., & Huang, J. J. (2012). Bilateral drug (ipilimumab)-induced vitritis, choroiditis, and serous retinal detachments suggestive of vogt-koyanagi-harada syndrome. *Retinal Cases & Brief Reports*, *6*(4), 423–426.

Yao, X., et al. (2007). Cisplatin nephrotoxicity: A review. *The American Journal of the Medical Sciences*, *334*(2), 115–124.

Zadik, Y., et al. (2019). Systematic review of photobiomodulation for the management of oral mucositis in cancer patients and clinical practice guidelines. *Support Care Cancer*, *27*(10), 3969–3983.

FSC
www.fsc.org
MIX
Papier | Fördert
gute Waldnutzung
FSC® C083411

Kolibri 360 GmbH
Ferdinand-Jühlke-Straße 7
99095 Erfurt, Deutschland
produktsicherheit@kolibri360.de